AF732612

LABORATORY GUIDE FOR CONDUCTING SOIL TESTS AND PLANT ANALYSIS

LABORATORY GUIDE FOR CONDUCTING SOIL TESTS AND PLANT ANALYSIS

J. Benton Jones, Jr.

CRC

CRC Press
Boca Raton London New York Washington, D.C.

Library of Congress Cataloging-in-Publication Data

Jones, J. Benton, 1930-
Laboratory guide for conducting soil tests and plant analysis / J. Benton Jones, Jr.
p. cm.
Includes bibliographical references and index.
ISBN 0-8493-0206-4 (alk. paper)
1. Soils--Analysis--Laboratory manuals. 2. Plants--Analysis--Laboratory manuals. I. Title.

S593 .J615 2001
631.4′072--dc21

2001025446
CIP

Direct all inquiries to CRC Press LLC, 2000 N.W. Corporate Blvd., Boca Raton, Florida 33431.

International Standard Book Number 0-8493-0206-4
Library of Congress Card Number 2001025446
Printed in the United States of America 1 2 3 4 5 6 7 8 9 0
Printed on acid-free paper

Preface

Soil analysis (frequently referred to as soil testing) and plant (leaf) analysis (including tissue testing) play major roles in crop production decision making, providing the means for assessing the nutrient element status of the soil/crop environment, and establishing the basis for making lime and fertilizer recommendations. These analyses are also used for diagnosing nutrient element–caused stress by identifying the element(s) involved, forming the basis for supplemental applications of elements needed to correct uncovered or confirmed insufficiencies. More recently, soil analysis is becoming a major technique for measuring the impact soil characteristics and amendments will have on environmental water quality issues. Soil fertility and plant nutrition research requires the use of standard methods of analysis to generate reliable analytical data that can be universally interpreted by the scientific community.

This laboratory guide provides some historical background for the assay methods more commonly in use today, describing the basis and range of application, plus the requirements for conducting the test. Although not an all-inclusive text on the subject, the techniques for sampling, sample preparation, and laboratory analysis of soil and plant tissue, including some of the more commonly used instrumental methods of analysis, analytical procedures for determining the physical and chemical composition of soils and the elemental content of plant tissues, are described in some detail. Related interpretative data and basic concepts of soil and plant nutrition are also given.

This laboratory guide is designed (1) for instruction in soil and plant analysis procedures, (2) for use by growers, crop consultants, county agents, etc., who rely on soil and plant analysis data for managing the nutrient element status of soils and crops, and (3) for use by the scientific community that requires and relies on soil/plant analysis data in research.

The Author

J. Benton Jones, Jr., Ph.D., is Professor Emeritus at the University of Georgia (UGA), retiring from the university in 1989 after completing 21 years of service plus 10 years as Professor of Agronomy at the Ohio Agricultural Research and Development Center (OARDC). While at the OARDC, he established the Ohio Plant Analysis Laboratory, the first of its kind providing analytical and interpretative services dealing primarily with agronomic crops. In September 1968, Dr. Jones accepted a position with the UGA, supervising the construction of the Georgia Soil Testing and Plant Analysis Laboratory in Athens, serving as its first director until 1974, when he became Division Chairman and Head of the Division and Department of Horticulture. During that time, he assisted the UGA Institute of Ecology in its establishment of an analytical laboratory, the first to employ a new instrumental procedure that is widely used today.

Dr. Jones has written extensively on analytical methods and has developed a number of analytical procedures for the assay of soil and plant tissue, as well as techniques for the interpretation of soil and plant analyses for their application in crop production decision making.

Dr. Jones was the first president and then served until 1998 as secretary-treasurer of the Soil and Plant Analysis Council, a scientific society that was founded in 1969. He is an author of more than 200 scientific articles and 15 book chapters, and has written five books. He established two international journals, *Communications in Soil Science and Plant Analysis,* serving as its editor for 24 years, and the *Journal of Plant Nutrition,* serving as its editor for 19 years.

Dr. Jones received his B.S. degree in agricultural science from the University of Illinois in 1952 and his M.S. and Ph.D. degrees in agronomy from Pennsylvania State University in 1956 and 1959, respectively. He has traveled widely with consultancies in the Soviet Union, China, Taiwan, South Korea, Saudi Arabia, Egypt, Costa Rica, Cape Verde, India, Hungary, Kuwait, and Indonesia.

Dr. Jones has received many awards and recognitions for his service to the science of soil testing and plant analysis. He is a Fellow of the American Association for the Advancement of Science, the American Society of Agronomy, and the Soil Science Society of America. An award in his honor, the "J. Benton Jones, Jr. Award," was established in 1989 by the Soil and Plant Analysis Council. Dr. Jones received an Honorary Doctorate from the University of Horticulture, Budapest, Hungary. He is a member of three honorary societies, Sigma Xi, Gamma Sigma Delta, and Phi Kappa Phi, and he is listed in *Who's Who in America* as well as in a number of other similar biographical listings.

Dr. Jones currently resides in Anderson, South Carolina, is still writing and advising growers, and is experimenting with various hydroponic growing systems for use in the field and greenhouse.

Contents

Chapter 1. Introduction 1
A. Reference Methods 2
B. Reagents, Standards, and Water 3
1. Reagents 3
2. Standards 3
3. Water 4
C. Elemental and Compound Designation 5
D. Other Considerations 6
E. Interpretation of a Soil Test/Plant Analysis Result 6
F. Units 7
G. Disclaimer 7
References 7

Chapter 2. Soil Analysis (Testing) 11
A. History and Purpose 11
B. Sequence of Procedures 15
C. Sampling 16
D. Transport to the Laboratory 19
E. Preparation of the Laboratory Sample 20
1. Drying 20
2. Crushing/Grinding/Sieving 21
F. Sample Aliquot Determination 22
1. Weighing vs. Scooping 22
2. Estimated Weight Scoops 23
3. NCR-13 Scoops 24
4. Procedure for Using a Soil Scoop 24
G. Laboratory Factors 25
1. Extraction Reagents 25
2. Extraction Procedure 26
3. Reagents and Standards 26
H. Long-Term Storage 27

I. Soil pH 27
1. Introduction 27
2. Using a pH Meter 28
a. Calibration of the pH Meter 28
b. Electrode Positioning 30
c. Preparation of pH Buffers 30
d. pH Determination in Water 31
e. pH Determination in 0.01 *M* Calcium Chloride ($CaCl_2$) 33
f. pH Determination in 1 *N* Potassium Chloride (KCl) 34
3. pH Determination Using Indicators 35
4. Interpretation 36
5. Soil Storage 41
J. Soil Buffer pH (Lime Requirement) 41
1. Introduction 41
2. SMP Buffer 42
3. Adams–Evans Buffer 46
4. Mehlich Buffer 48
5. Titratable and Exchangeable Acidity 53
a. Determination of Exchangeable Acidity Using Barium Chloride–TEA Buffer 53
b. Determination of Exchangeable Acidity and Exchangeable Aluminum Using 1 *N* Potassium Chloride 54
6. Interpretation 56
7. Definition of Liming Materials 57
8. Acid-Neutralizing Values for AgLime Materials 60
9. Adjusting the AgLime Rate for Different Depths of Incorporation 61
10. Effect of Fineness on Availability of AgLime 62
K. Extractable Phosphorus 62
1. Introduction 62
2. Extraction Reagents and Procedures 66
a. Morgan 66
b. Bray P1 67
c. Bray P2 68
d. Mehlich No. 1 (North Carolina Double Acid) 70
e. Olsen's Sodium Bicarbonate 71
f. Ammonium Bicarbonate–DTPA 72
g. Mehlich No. 3 73
h. 0.01 *M* Calcium Chloride 75
3. Methods of Phosphorus Determination 76
a. UV-VIS Spectrophotometry 76
b. Plasma Emission Spectrometry 77
4. Conversion Factors 77
5. Interpretation 78
6. Fertilizer Recommendations 78
7. Effect of Fertilizer Phosphorus on Phosphorus Soil Test Level 79

L. Major Cations (K, Ca, Mg, and Na) ... 79
1. Introduction ... 79
2. Extraction Reagents and Procedures ... 81
a. Neutral Normal Ammonium Acetate (K, Ca, Mg, and Na) ... 81
b. Mehlich No. 1 (North Carolina Double Acid) (K, Ca, Mg, and Na) ... 82
c. Mehlich No. 3 (K, Ca, Mg, and Na) ... 83
d. Morgan (K, Ca, and Mg) ... 84
e. Ammonium Bicarbonate–DTPA (K) ... 86
f. Water (K, Ca, Mg, and Na) ... 87
g. 0.01 *M* Calcium Chloride (K, Mg, and Na) ... 87
3. Methods of Cation Determination ... 88
4. Methods of Expression ... 89
5. Interpretation ... 89
6. Effect of Fertilizer Potassium on the Potassium Soil Test Level ... 91
7. Calculation of Cation Exchange Capacity (CEC) ... 92
8. Determination of Percent Base Saturation and Use ... 93
M. Micronutrients (B, Cl, Cu, Fe, Mn, and Zn) ... 93
1. Introduction ... 93
2. Extraction Reagents and Procedures ... 95
a. Hot Water (B) ... 95
b. Mehlich No. 1 (Zn) ... 97
c. 0.1 *N* Hydrochloric Acid (Zn) ... 98
d. Mehlich No. 3 (B, Cu, Mn, and Zn) ... 99
e. Ammonium Bicarbonate–DTPA (Cu, Fe, Mn, and Zn) ... 101
f. DTPA (Cu, Fe, Mn, and Zn) ... 103
g. 0.01 *M* Calcium Chloride (B, Cu, Fe, Mn, and Zn) ... 104
h. Morgan (B, Cu, Fe, Mn, and Zn) ... 104
i. 0.01 *M* Calcium Nitrate (Cl) ... 105
j. 0.5 *M* Potassium Sulfate (Cl) ... 106
k. Saturated Calcium Hydroxide (Cl) ... 107
3. Methods of Micronutrient Determination in Soil Extracts ... 108
4. Methods of Expression ... 109
5. Cleaning Laboratory Ware ... 109
6. Interpretation ... 110
N. Trace Heavy Metals ... 115
1. Introduction ... 115
2. Extraction Reagents and Procedures ... 115
a. Ammonium Bicarbonate–DTPA (AB–DTPA) ... 115
b. DTPA ... 117
c. 0.01 *M* Calcium Chloride ... 118
3. Method of Heavy Metal Determination ... 118
4. Interpretation ... 119
O. Extractable Nitrate–Nitrogen (NO_3–N) ... 121
1. Introduction ... 121

2. Extraction Reagents and Procedures 122
a. Ammonium Bicarbonate–DTPA 122
b. 2 *M* Potassium Chloride 123
c. 0.01 *M* Calcium Sulfate 123
d. 0.04 *M* Ammonium Sulfate 124
e. 0.01 *M* Calcium Chloride 124
3. Nitrate Standards 125
4. Methods of Nitrate Determination 126
a. UV-VIS Spectrophotometric Determination Procedure 126
b. Specific-Ion Electrode Determination 127
c. Kjeldhal Distillation 128
d. Ion Chromatography 129
5. Effects of Storage 129
6. Interpretation 129
P. Extractable Sulfate–Sulfur (SO_4–S) 129
1. Introduction 129
2. Extraction Reagent and Procedures 130
a. Monocalcium Phosphate 130
b. 0.5 *M* Ammonium Acetate–0.25 *M* Acetic Acid 130
c. 0.01 *M* Calcium Chloride 131
3. Sulfate–Sulfur (SO_4–S) Standards 132
4. Determination Procedures 132
a. Turbidity 132
b. Inductively Coupled Plasma Emission Spectrometry (ICP-AES) 133
c. Determination by Ion Chromatography 135
5. Interpretation 135
Q. Testing Organic Soils and Soilless Media 135
1. Introduction 136
2. Extraction with Water 137
3. Extraction with 0.005 *M* DTPA to Improve Extraction of Micronutrients 137
4. Nutrient Element Assay Procedures 138
5. Interpretation 138
R. Organic Matter and Humic Matter Content Determinations 139
1. Introduction 139
2. Methods of Organic Matter Determination 140
a. Wet Digestion 140
b. Loss-on-Ignition (LOI) 143
c. Humic Matter by 0.2 *N* Sodium Hydroxide Extraction 144
d. Determination of Soluble Organic Carbon 148
S. Soluble Salt Determination 150
1. Introduction 150
2. The Conductivity Meter 152
3. Standard Calibration Solution 152
4. Procedures 152

5. Interpretation 153
6. Electrical Conductivity Units and Conversions 154
7. Salinity (NaCl) and Alkalinity ($NaHCO_3$) 155
T. Soil Texture (Mechanical Analysis) 155
1. Separate Characteristics 155
2. Textural Classifications 156
3. Method of Determination 156
4. Hydrometer Procedure 159
a. Soil Preparation 159
b. Hydrometer Readings 159
c. Blank Hydrometer Determination 160
d. Calculation for % Sand, Silt, and Clay 160
References 160

Chapter 3. Plant Analysis 191
A. Purpose 191
B. Sampling 192
1. Techniques 192
2. Number of Plants to Sample 197
3. Lack of Homogeneity 198
4. Petioles 199
5. Compound Leaves 199
6. Comparative Sampling 199
7. Inappropriate Plant Tissue 200
C. Sample Preparation 200
1. Introduction 200
2. Initial Handling 200
3. Decontamination (Washing) 201
4. Moisture Removal (Oven-Drying) 202
5. Particle Size Reduction (Grinding) 202
6. Organic Matter Destruction 203
a. High-Temperature Thermal Oxidation (Dry Ashing) 204
b. Wet-Acid Digestion (Wet Ashing) 205
D. Elemental Analysis for the Mineral Elements 207
E. Total Nitrogen (N) Determination 209
1. Introduction 209
2. Kjeldahl Methods 211
3. Determination of Ammonium in Kjeldahl Digest 212
4. Non-Kjeldahl Methods 212
F. Total Sulfur (S) Determination 213
1. Introduction 213
2. Interpretation 215
G. Methods for Expressing Elemental Content 215
H. Interpretation of Results 216
I. Extractable Elements 221
1. Introduction 221

2. Extraction Procedures for Nitrate (NO_3) in Plant Tissue 222
3. Extraction of Phosphorus (PO_4) in Plant Tissue Using 2% Acetic Acid 223
4. Interpretation for Extractable Nitrogen (NO_3) and Phosphorus (PO_4) 223
5. Extractable Ammonium (NH_4) in Plant Tissue 226
6. Extractable Sulfate (SO_4) in Plant Tissue 226
7. Extractable Chloride (Cl) in Plant Tissue 227
8. Extraction of Chloride (Cl), Nitrate (NO_3), Orthophosphate (PO_4), Potassium (K), and Sulfate (SO_4) in Plant Tissue Using 2% Acetic Acid 228
9. Extractable Iron (Fe) in Plant Tissue 229

References 229

Chapter 4. Tissue Testing 247

A. Introduction 247
B. Testing Kits 248
C. Preparation of Reagents for Conducting Tissue Tests Using Filter Paper 249
D. Sampling Techniques 250
E. Testing Procedures 251
   1. General Test Procedures for Paper and Vial-Type Kits 251
      a. Nitrate–Nitrogen Test 251
      b. Phosphorus Test 251
      c. Potassium Test 252
      d. Nitrate–Nitrogen Stalk Test 252
      e. Phosphorus Stalk Test 252
      f. Other Tests 252
F. Methods of Interpretation 253
   1. When and How to Use Tissue Tests 253
G. Use of a Specific-Ion Nitrate Meter 255
H. Sources for Kits and Instruments 256

References 256

Chapter 5. Principles of Instrumental Analysis 259

A. Introduction 259
B. UV-VIS Spectrophotometry (Colorimetry) 262
C. Emission Spectrophotometry 264
   1. Flame Emission Spectrophotometry 265
   2. Atomic Absorption Spectrophotometry 266
   3. Inductively Coupled Plasma Emission Spectrometry 268
      a. Introduction 268
      b. Spectrometer Designs 268
   4. Operating Characteristics of an ICP-AES 269
      a. Advantageous Characteristics 269
      b. Disadvantages 270

5. Standard Preparation 270
6. Calibration Techniques 271
7. Common Operating Problems 272
8. Important General Points 273
9. Spectrophotometry Terms 274
D. pH and Specific-Ion Electrodes 275
E. Ion Chromatography 277
References 278

Chapter 6. Quality Assurance/Quality Control in the Laboratory 285
A. Introduction 285
B. Accuracy and Precision 287
C. Standards 287
D. Instrumentation 289
E. Laboratory Procedures 289
F. Participation in Proficiency Testing Programs 290
References 290

Appendix A. List of Reagents, Standards, pH Buffers, Acids, and Indicators, and Preparation of Standard Acids, Bases, and Buffers Required in the Instruction Guide 293
A. Reagents 293
B. Reagents for Preparation of Standards 296
C. Reagents for Preparation of pH Buffers 297
D. Concentrated Acids 297
E. Indicators 298
F. Standard Acids, Bases, and Buffers 298

Appendix B. Standards and Standard Preparation 301
A. Purpose 301
B. Primary Standards 301
C. Working Standards 306
D. Matrix Effects 307
E. Matrix Modifiers 307
F. Blanks 308
G. Commercial Sources for Primary and Working Standards 308
H. Soil and Plant Tissue Standards 309
References 309

Appendix C. Extraction Reagents and Procedures 311

Appendix D. The North American Proficiency Testing Program for Soil, Plant, and Water Analysis Laboratories (NAPT) 323

Appendix E. ASI Extraction Reagent Method for Soil Analysis 327

Appendix F. Reference Texts 333

Appendix G. Definitions 339

Appendix H. Conversion Factors 351

Index 357

Chapter 1

Introduction

This laboratory guide instructs the reader on procedures to collect, prepare, and analyze soil and plant tissue for the determination of their physical properties and chemical (elemental) composition. The soil analysis (testing) procedures described in this laboratory guide are the more commonly used procedures applicable to most soil types.

Since applied analytical chemistry and methods of instrumental analysis are essential in soil and plant analysis procedures, the principles of operation for the more important instrumental analysis techniques are given in some detail in Chapter 5.

This laboratory guide contains a wide coverage of the pertinent literature on laboratory procedures for the analysis of soils and plant tissues so that the reader can investigate in greater detail the bases for the test methodologies described. Several basic texts are frequently referenced, including selected references from the current literature. Many of the soil analysis procedures have been taken from the recently revised edition of the *Soil Analysis Handbook of Reference Methods* (Anonymous, 1999) and those for plant analysis and tissue testing from the *Handbook of Reference Methods for Plant Analysis* (Kalra, 1998).

Houba et al. (1994) have discussed the future role of soil and plant analyses focusing on the increasing demand for reliable and timely analytical data. Environmental concerns have spurred interest in nutrient management plans whose foundations are based on soil and plant assays (Campbell, 1994; Varallyay, 1994; Häni, 1996; Sparrow et al., 2000). The Council for Agricultural Science and Technology (CAST) has recently published an issue paper on the "Relevance of Soil Testing to Agriculture and the Environment," focusing on the value of soil tests to identify the potential for an environmental impact, pointing out the need for improving soil tests for both agricultural and environmental purposes (CAST, 2000).

To assist the user of this laboratory guide, interpretative data are given for most test procedures as are soil fertility and plant nutrition concepts that relate to the interpretation and application of soil and plant analysis data.

A. Reference Methods

The need to standardize soil and plant analysis procedures and methods is more apparent today, although there is little unanimity on the subject. In the United States, much of the evaluation of analysis methods and the setting of parameters for each laboratory procedure are performed by Land-Grant College and University regional research committees on soil and plant analysis. Manuals and guides that have been published by these committees may be found in the list of reference texts in Appendix F.

In addition, a number of other scientific and industrial societies have been engaged in developing and publishing reference methods of analysis. The Association of Official Analytical Chemists (AOAC), organized in 1884, is the oldest of these societies in the United States. The 8th edition of the *Methods of Analysis of the AOAC* (Horwitz, 1955) was the last edition that included methods of soil analysis. Procedures for plant analysis have been and still are given in the AOAC manual (Horwitz, 2000). The American Society for Testing and Materials (ASTM), the American Public Health Association (Anonymous, 1989), and, more recently, the Intersociety Committee (Houba et al., 1996) have been engaged in researching and publishing reference methods of analysis for a wide variety of substances, including soils. The American Society of Agronomy (ASA) and the Soil Science Society of America (SSSA) have published a number of books on methods of soil analysis and interpretation (see Appendix F). In 1990, SSSA and AOAC established a joint committee "to conduct validation studies for methods of soil analysis" (Kalra, 1996). The first validation was for soil pH (Kalra, 1995), and future additions of the AOAC manual will include most of the commonly used soil analysis procedures.

The Soil and Plant Analysis Council (initially the Council on Soil Testing and Plant Analysis, which is currently headquartered at 621 Rose St., Lincoln, NE 68502-2040) was organized in 1970, and one of its primary goals is to research and publish reference methods for soil analysis (testing) and plant analysis, resulting in its most recent publications: the *Soil Analysis Handbook of Reference Methods* (Anonymous, 1999) and the *Handbook of Reference Methods for Plant Analysis* (Kalra, 1998). Its quarterly newsletter *The Soil-Plant Analyst* frequently includes newly gathered information on soil and plant analysis techniques.

The potential environmental role, particularly for soil analysis, demands reference methods. The growing interest in the environment (CAST, 2000), the concern about overdosing soils with fertilizer (Sims, 1998) and/or using soil for waste product disposition (Hue, 1995), and the need for care in using fertilizer materials economically demand more uniformly applied analysis methods. Standardization of methodology is indeed necessary if soil analysis is to be used as a valid monitoring tool. In the near future, regulating agencies may dictate the methods of soil analysis, as various governmental agencies have required and still require the use of AOAC methods for the analysis of fertilizers, lime, and other substances (Horwitz, 2000). If soil analysis is to have a strong scientific base, standardization of procedures is essential.

Much of the scientific research findings published on soil fertility and crop production frequently contain soil and plant analysis data that could be of doubtful value because of several factors, i.e., either because the use of test procedures was not applicable or because the test was not sufficiently identified for proper interpretation by the reader. Frequently, these articles neither include references to the particular test procedures used nor provide a detailed description of the method(s) used. An article may merely refer to a particular test procedure by name, such as "Bray P1" for the determination of soil P, or may describe a method as "modified" without indicating what aspect of the test procedure was modified. Although this guide does not solve this problem, it does assist those using a soil and/or plant analysis method, the results of which may eventually be used in published findings, by describing the essential requirements of the test procedure, the range of its use, and the generally accepted interpretation values.

B. Reagents, Standards, and Water

For all the analytical procedures described in this laboratory guide, reagents, standards, and water used must be of the highest quality and have characteristics that will not interfere with the analytical procedure.

1. Reagents

A list of all the reagents required to conduct the analytical procedures described in this laboratory guide is given in Appendix A. Reagents should be of *reagent* or *analytical* grade. The storage requirements for many of the reagents are frequently specified to ensure reliable performance. Commercially prepared reagents are sometimes available, particularly extraction reagents and standards; however, users are advised to test the quality of these reagents and standards before use.

2. Standards

The source, preparation, and testing of standards are described in all the procedures given in this laboratory guide. For many, the use of commercially prepared standards, whose reliability is high, is convenient and saves time both in the preparation and verification testing required for user-prepared standards. Therefore, whenever possible, the use of commercially prepared standards is highly recommended. However, the source and labeling of standards are important considerations, ensuring freedom from analytes in a standard that may be included in a multielement assay, as well as ensuring that the characteristics of the matrix (mix of cations and anions) and the acid content, whether nitric (HNO_3) or hydrochloric (HCl), or both, will not affect or interfere with the analytical procedure being used. The preparation and use of standards are discussed in some detail in Appendix B.

Reference soils for use in verifying an analytical result for most of the analytical methods given in this guide can be obtained from the Soil and Plant Analysis Council, and standard reference plant tissue can be obtained from the Standard Reference Testing Program, National Institute of Standards and Technology (NIST), Room 204, Bldg. 202, Gaithersburg, MD 20899. Ihnat (1993) has published a list of reference soil materials and sources; Quevauviller (1996) a list of the trace elements in soil materials; and Ihnat (1998) an extensive list of plant materials that can be used for verification of plant analysis analytical procedures.

3. Water

The quality of water used in the preparation of reagents and standards is critical to ensure reliability of the analytical procedure conducted. When the word *water* is used in this text, it refers to pure water, water free from any dissolved ions or other substances. Such water may be obtained commercially or by means of distillation (single or double), ion exchange, and/or reverse osmosis (Anonymous, 1997). The water used in a procedure should be tested, especially when the analytical procedure is one where the presence of a low ion concentration can significantly affect the analytical result. An example is the determination of P by the molybdenum blue spectrophotometric procedure; in this case, low levels of either the arsenate (AsO_4^{2-}) and/or silicate (SiO_4^{4-}) anion can generate the same blue color as that of the orthophosphate (PO_4^{3-}) anion. Glassware washing procedures have been presented by Kammin et al. (1995) and Tucker (1992). The quality of water for the final wash of glassware is equally important and should be of the highest purity. Rinsing glassware using several repeated small aliquots of pure water gives better results than one or two rinses with large aliquots of water. Allowing the rinse water to drain completely from the rinsed item, rather than immediately wiping dry or oven-drying, is the recommended procedure.

C. Elemental and Compound Designation

In this text, all elements are designated by their symbols, whereas reagents and compounds are named and their symbol compositions shown when first mentioned in that portion of the text. The symbols for those elements and compound elements found in this text are as follows:

Element	Symbol	Element	Symbol
Aluminum	Al	Manganese	Mn
Antimony	Sb	Magnesium	Mg
Arsenic	As	Molybdenum	Mo
Boron	B	Nitrogen	N
Bromine	Br	Nickel	Ni
Cadmium	Cd	Phosphorus	P
Chlorine	Cl	Potassium	K
Chromium	Cr	Selenium	Se
Cobalt	Co	Sodium	Na
Copper	Cu	Sulfur	S
Fluoride	F	Titanium	Ti
Indium	In	Uranium	U
Iron	Fe	Vanadium	V
Iodine	I	Yttrium	Y
Lead	Pb	Zinc	Zn
Lithium	Li		

Compounds	Symbol
Acetate	$C_2H_3O_2^-$
Ammonium	NH_4^+
Arsenate	AsO_4^{2-}
Bicarbonate	HCO_3^-
Borate	BO_3^{3-}
Carbonate	CO_3^{2-}
Cyanide	CN^-
Nitrate	NO_3^-
Nitrite	NO_2^-
Phosphate (ortho)	PO_4^{3-}
Silicate	SiO_4^{4-}
Sulfate	SO_4^{2-}
Thiocyanate	CNS^-

D. Other Considerations

The ruggedness of an analytical procedure, that is, the exactness required for each parameter, its tolerance for variance, is important and the parameters given with the procedure should be strictly followed to ensure reliable performance of the method. Factors, such as the condition of the assayed sample, pH and composition of reagents, time, temperature, physical parameters in terms of shaking speeds, characteristics of storage and extraction vessels, weight and volume measurements of samples, reagents, and standards, instrument settings, and methods of instrument calibration and operation, are normally specified and should be exactly followed. What might be perceived as an acceptable variance by an analyst may invalidate the analytical result obtained.

Verification of the analytical result requires application of the principles of quality assurance (QA) and quality control (QC), frequently referred to as QA/QC laboratory procedures, a topic discussed in some detail in Chapter 6. Laboratory accreditation has been been one of the recommended devices for ensuring reliable laboratory performance, and the Soil and Plant Analysis Council has developed an accreditation program for soil/plant analysis laboratories (Jones and White, 1994). One means of ensuring reliable laboratory performance is participation in a proficiency testing program, such as the North American Proficiency Testing Program, described in Appendix D. Miller et al. (1996), Wolf et al. (1996), and Wolf and Miller (1998) have described details of the North American Proficiency Testing Program. Proficiency testing programs exist in many countries (Rayment et al., 2000); the WAPL (Wageningen Evaluating Programmes for Analytical Laboratories) is the only international program (Houba et al., 1996; van Dijk and Houba, 2000).

For those looking for analytical assistance, the recently published *Registry of Soil and Plant Analysis Laboratories in the United States and Canada* provides a listing of laboratories, giving information on analytical services provided, contact person, etc.

E. Interpretation of a Soil Test/Plant Analysis Result

The interpretation data given in this laboratory guide have been gathered from a number of sources and are provided for general use only. Sources of interpretative data are given in each section so that the user can turn to these references for verification. Even the terms that classify an assay result as belonging in a particular category have varying meanings; therefore, the user must use caution when applying suggested interpretative data given in this laboratory guide. Excellent review articles on soil test interpretation have been written by Graham (1959) and Conyers (1999), for plant analysis by Smith and Loneragan (1997) and Mills and Jones (1996), and for correlating soil and plant analyses to fertilizer strategies by Black (1993), Jones (1985; 1993; 1998), and van Erp and van Beusichem (1998).

The concept of intensity and balance as a means of evaluating a soil test result has been proposed by Geraldson (1970), and Baker (1973; 1977; 1990) has expanded this concept by considering ionic balance as an important factor. An alternative to extraction is the use of either resins or electro-ultrafiltration (EUF), procedures that have been compared with traditional extraction procedures; a summary is provided by van Raij (1998). However, these various alternative soil analysis procedures have not been widely accepted or used.

Computerization and data processing of soil and plant analysis results are the common means of reporting soil test and plant analysis results to farmers and growers, as well as of evaluating data for research purposes. Donohue and Gettier (1990) have reviewed commonly used procedures for data processing of soil tests and plant analyses.

F. Units

Units of length, area, volume, mass, and yield are given in either English and/or SI units, normally using those units initially given with the method. For temperature, both Centigrade (C) and Fahrenheit (F) values are given. Conversion factors are found in Appendix H.

G. Disclaimer

The naming and identification of products given in this laboratory guide do not constitute endorsement. Most of the analysis procedures described have been taken from current publications found in the public domain.

References

Anonymous 1989. *Standard Methods for the Examination of Water and Waste Water*, 17th ed., American Public Health Association, Washington, D.C.

Anonymous 1997. Water: what's in it and how to get it out, *Today's Chem.*, 6(1):16–19.

Anonymous 1999. *Soil Analysis Handbook of Reference Methods*, CRC Press, Boca Raton, FL.

Baker, D.E. 1973. A new approach to soil testing: II. Ionic equilibrium involving H, K, Ca, Mg, Mn, Fe, Cu, Zn, Na, P, and S, *Soil Sci. Soc. Am. Proc.,* 37:537–541.

Baker, D.E. 1977. Ion activities and ratios in relation to corrective treatments of soil, in T.R. Peck et al., Eds., *Soil Testing: Correlating and Interpreting the Analytical Results*, ASA Special Publication 29, American Society of Agronomy, Madison, WI, 55–74.

Baker, D.E. 1990. Baker soil test theory and applications, *Commun. Soil Sci. Plant Anal.,* 21:981–1008.

Black, C.A. 1993. *Soil Fertility Evaluation and Control*, Lewis Publishers, Boca Raton, FL.

Campbell, L.C. 1994. Beneficial impact of precision nutrient management on the environment and future needs, *Commun. Soil Sci. Plant Anal.,* 25:889–908.

CAST. 2000. *Relevance of Soil Testing to Agriculture and the Environment*, Issue Paper Number 15, June 2000, Council for Agricultural Science and Technology, Ames, IA.

Conyers, M.K. 1999. Factors affecting soil test interpretation, in K.I. Peverill, L.A. Sparrow, and D.J. Reuter, Eds., *Soil Analysis: An Interpretation Manual*, CSIRO Publishing, Collingwood, Australia, 23–34.

Donohue, S.J. and S.W. Gettier. 1990. Data processing in soil testing and plant analysis, in R.L. Westerman, Ed., *Soil Testing and Plant Analysis,* 3rd ed., SSSA Book Series No. 3, Soil Science Society of America, Madison, WI, 741–755.

Geraldson, C.E. 1970. Intensity and balance concept as an approach to optimum vegetable production, *Commun. Soil Sci. Plant Anal.,* 1:187–196.

Graham, E.R. 1959. An Explanation of Theory and Methods of Soil Testing, Missouri Agricultural Experiment Station Bulletin 734, Missouri Agricultural Experiment Station, Columbia.

Häni, H. 1996. Soil analysis as a tool to predict effects on the environment, *Commun. Soil Sci. Plant Anal.,* 27:289–306.

Horwitz, W. Ed. 1955. *Methods of Analysis of the AOAC,* 8th ed., Association of Official Analytical Chemists, Washington, D.C.

Horwitz, W. Ed. 2000. *Official Methods of Analysis of the AOAC International,* 17th ed., Association of Official Analytical Chemists, Arlington, VA.

Houba, V.J.G., I. Novozamsky, and J.J. van der Lee. 1994. Status and future of soil and plant analysis, *Commun. Soil Sci. Plant Anal.,* 25:753–765.

Houba, V.J.G., J. Uittenbogaard, and P. Pellen. 1996. Wageningen evaluating programmes for analytical laboratories (WEPAL), organization and purpose, *Commun. Soil Sci. Plant Anal.*, 27:421–431.

Hue, N.V. 1995. Sewage sludge, in J.E. Rechcigl, Ed., *Soil Amendments and Environmental Quality*, CRC Press, Boca Raton, FL, 199–147.

Ihnat, M. 1993. Reference materials for data quality, in M.R. Carter, Ed., *Soil Sampling and Methods of Analysis*, Lewis Publishers, Boca Raton, FL, 247–262.

Ihnat, M. 1998. Plant and related reference materials for quality control of element content, in Y.P. Kalra, Ed., *Handbook of Reference Methods for Plant Analysis*, CRC Press, Boca Raton, FL, 235–284.

Jones, J.B., Jr. 1985. Soil testing and plant analysis: guides to the fertilization of horticultural crops, *Hortic. Rev.*, 7:1–68.

Jones, J.B., Jr. 1993. Modern interpretation systems for soil and plant analysis in the United States, *Aust. J. Exp. Agric.*, 33:1039–1043.

Jones, J.B., Jr. 1998. *Plant Nutrition Manual*, CRC Press, Boca Raton, FL.

Jones, J.B., Jr. and W.C. White. 1994. An accreditation program for soil and plant analysis laboratories, *Commun. Soil Sci. Plant Anal.*, 25:843–857.

Kalra, Y.P. 1995. Determination of pH of soils by different methods: collaborative study, *J. Assoc. Off. Anal. Chem.*, 78:310–324.

Kalra, Y.P. 1996. Soil pH: first soil analysis methods validated by the AOAC International, *J. For. Res.*, 1:61–64.

Kalra, Y.P., Ed. 1998. *Handbook of Reference Methods for Plant Analysis*, CRC Press, Boca Raton, FL.

Kammin, W.R., S. Cull, R. Knox, J. Ross, M. McIntosh, and D. Thomson. 1995. Labware cleaning protocols for the determination of low-level metals by ICP-MS, *Am. Environ. Lab.*, 11/95–12:95:1–3.

Miller, R.O., J. Kotuby-Amacher, and N.B. Dellevalle. 1996. A proficiency testing program for the agricultural laboratory industry: results for the 1994 program, *Commun. Soil Sci. Plant Anal.*, 27:451–461.

Mills, H.A. and J.B. Jones, Jr. 1996. *Plant Analysis Handbook II*, MicroMacro Publishing, Athens, GA.

Quevauviller, P. 1996. Certified reference materials for the quality control of total and extractable trace element determination in soils and sludges, *Commun. Soil Sci. Plant Anal.*, 27:403–418.

Rayment, G.E., R.O. Miller, and E. Sulaeman. 2000. Proficiency testing and other interactive measures to enhance analytical quality in soil and plant laboratories, *Commun. Soil Sci. Plant Anal.*, 31:1513–1530.

Sims, J.T. 1998. Phosphorus soil testing: innovations for water quality protection, *Commun. Soil Sci. Plant Anal.*, 29:1471–1489.

Smith, F.W. and J.F. Loneragan. 1997. Interpretation of plant analysis: concepts and principles, in D.J. Reuter and J.B. Robinson, Eds., *Plant Analysis: An Interpretation Manual,* CSIRO Publishing, Collingwood, Australia, 1–33.

Sparrow, L.A., A.N. Sharpley, and D.J. Reuter. 2000. Safeguarding soil and water quality, *Commun. Soil Sci. Plant Anal.*, 31:1717–1742.

Tucker, M.R. 1992. Determination of zinc, manganese, and copper by Mehlich 3 extraction, in S.J. Donohue, Ed., *Reference Soil and Media Diagnostic Procedures for the Southern Region of the United States*, Southern Cooperative Series Bulletin Number 374, Virginia Agricultural Experiment Station, Blacksburg, 19–22.

van Dijk, D. and V.J.G. Houba. 2000. Homogeneity and stability of materials distributed within the Wageningen Evaluating Programs for Analytical Laboratories, *Commun. Soil Sci. Plant Anal.,* 31:1745–1756.

van Erp, P.J. and M.L. van Beusichem. 1998. Soil and plant testing program as a tool for optimizing fertilizer strategies, in Z. Rengel, Ed., *Nutrient Use in Crop Production*, Haworth Press, Binghamton, NY, 53–80.

van Raij, B. 1998. Bioavailable tests: alternatives to standard soil extractions, *Commun. Soil Sci. Plant Anal.,* 29:1553–1570.

Varallyay, G. 1994. Precision nutrient management — impact on the environment and needs for the future, *Commun. Soil Sci. Plant Anal.,* 25:909–930.

Wolf, A.M. and R.O. Miller. 1998. Development of a North American Proficiency Testing Program for Soil and Plant Analysis, *Commun. Soil Sci. Plant Anal.,* 29:1685–1690.

Wolf, A.M., J.B. Jones, Jr., and T. Hood. 1996. Proficiency testing for improving analytical performance in soil testing proficiency testing programs. *Commun. Soil Sci. Plant Anal.,* 27:1611–1622.

Chapter 2

Soil Analysis (Testing)

A. History and Purpose

> *There is good evidence that the competent use of soil tests can make a valuable contribution to the more intelligent management of the soil.*

This statement by the National Soil Test Workgroup in its 1951 report (Nelson et al., 1951) is still applicable today. The objectives of soil testing have changed little since they were first presented almost 50 years ago by two North Carolina researchers, Fitts and Nelson (1956):

1. To group soil into classes for the purpose of suggesting fertilizer and lime practices.
2. To predict the probability of getting a profitable response to the application of plant nutrient elements.
3. To help evaluate soil productivity.
4. To determine specific soil conditions that may be improved by addition of soil amendments or cultural practices.

Soil testing as it is practiced today would best fit Objectives 1 and 2, farmers and growers testing soil to determine lime and fertilizer needs. Although acceptance of the first two objectives is nearly unanimous, there is still considerable disparity of opinion about the practical application of Objective 2, soil test interpretation measured in terms of the recommended application rates of fertilizer (Liebhardt, 1981/1982; Black, 1993a; van Erp and van Beusichem, 1998; Voss, 1998; Helyar and Price, 1999). Adjustments may be made on the basis of crop requirement, anticipated yield, management skill of the farmer, and economic goals, each factor affecting the recommendation even with a similar soil test result.

Objectives 3 and 4 describe the soil test in diagnostic terms showing how test results can describe the more general condition of the soil. From a long-term standpoint, these objectives have far more importance than is generally recognized. Further discussion of this topic is beyond the scope of this laboratory guide and users should consult other sources for details on soil test interpretation and application (Davidescu and Davidescu, 1972; Peck et al., 1977; Cottenie, 1980; Jones, 1985; Halliday and Trenkel, 1992; Black, 1993b; Barber, 1995; Rengel, 1998).

For the extractable elements, the extraction procedure should meet the following criteria:

1. *The procedure should extract the element from the same labile nutrient element pool in the soil that plants do.* Some would maintain that the closer the amount of element extracted approaches that absorbed by the plant, the better the soil test procedure; however, this is an unrealistic objective as many of the elements exist in the soil in various forms, frequently in complex equilibria with shifts in form influenced by pH, temperature, water status, biological activity, past fertilization and cropping practices, and an extraction reagent may only tap a portion of a particular form(s) that constitutes the nutrient pool available for plant use. McLean (1982a) and Corey (1990) have discussed how recent soil testing procedures were devised to estimate the size and intensity of these nutrient pools (Barber, 1995; Tan, 1998).
2. *A good soil test should be cheap, reproducible in different laboratories, and easily adapted to routine laboratory procedures.* Developments in analytical chemistry, synthesis of synthetic chelating agents, and an ever-increasing understanding of the chemistry of the essential plant nutrient elements have resulted in the development of good soil testing procedures, while atomic absorption spectrometry (Wright and Stuczynski, 1996), the use of an AutoAnalyzer® (Flannery and Markus, 1972; 1980; Tel and Heseltine, 1990a; b) or flow injection analyzer (Ruzicka and Hansen, 1988), and plasma emission spectrometry (Soltanpour et al., 1996; 1998) have mechanized and increased the ease as well as the speed and sensitivity with which the elemental content of extracting solutions can be determined; soil tests for the micronutrients, difficult to determine just a few years ago, are now routine procedures in today's soil testing laboratories.

Melsted and Peck (1973) and Peck and Soltanpour (1990) have discussed the basic principles of soil testing, which has been practiced with some degree of success for almost 50 years, their reviews covering the topics from sampling to making fertilizer recommendations. Mehlich (1974) looked at the uniformity of soil test results as influenced by extraction reagents and soil properties. One result of such evaluations has led to the effort to standardize analysis procedures and emphasize the importance of the relationship between extraction reagent selection and soil properties. Peck (1990) suggested that the history of soil testing in the United States has been interwoven with the growth and development of soil science and, therefore, is dependent on the availability and quality of research data on soil chemistry and the interpretation of soil test values and their correlation to crop response. Unfortunately, in the past several decades, much of this needed research has slowed as a result of changing priorities at land-grant colleges and universities which in the past have conducted much of this research. Therefore, future developments in soil testing procedures and interpretation will come from other agencies.

Soil tests can be grouped into several categories based on objective:

Soil Test	Objective
Water, salt, and buffer pHs	Soil reaction and lime requirement
Extractable elements	
Major elements (P, K, Ca, Mg, NO_3, SO_4)	Nutrient element status
Micronutrients (B, Cl, Cu, Fe, Mo, Mn, Zn)	Nutrient element status
Other elements (Al, Na)	Toxicity
Trace elements and heavy metals (As, Cd, Co, Cr, Cu, Mn, Pb, Ni)	Toxicity
Organic matter content	Physical and chemical characteristics
Mechanical analysis	Soil texture classification
Soluble salts	Total salts in the soil solution

All these determinations can be performed via a number of laboratory procedures; the method selected is determined, in part, by the physical and chemical characteristics of the soil. Therefore, there is no such thing as a soil test, that is, a single method of laboratory analysis applicable to all soils. However, there are some general criteria that have guided the development of soil testing procedures, particularly the extraction procedures that are used to evaluate the nutrient element status of the soil.

The goals of those engaged in soil testing research are twofold:

1. To improve the correlation of a soil test result to crop response over the entire response range from deficiency to excess (or toxicity).
2. To develop soil testing procedures that have wider adaptation in terms of range of soil properties and elements included, making current or new testing procedures more universal.

A discussion of the first goal which is a topic that is a subject in and of itself is beyond the scope of this laboratory guide, but its importance is not to be ignored (Davidescu and Davidescu, 1972; Peck, 1977). The second goal has significance for the objectives and procedures described in this laboratory guide. Good examples are widening the ratio of soil to extraction reagent, as is done in Mehlich No. 2 (Mehlich, 1978), and combining extraction reagents, as is done by adding together ammonium bicarbonate (NH_4HCO_3) and DTPA (diethylenetriaminepentaacetic acid) (Soltanpour and Schwab, 1977), Morgan extraction reagent (Morgan, 1932; 1941; Lunt et al., 1950) and DTPA (Wolf, 1982), and the new "combination" Mehlich No. 3 extraction reagent (Mehlich, 1984a). The goal in every case is to have one extraction reagent for as many elements as possible, with applicability to a wide range of soil types. There is renewed interest in using water as an extracting reagent for P (Luscombe et al., 1979). The use of an equilibrium solution has interesting implications and promise as a universal soil testing procedure (Baker, 1971; 1973; 1990; Baker and Amacher, 1981; Houba et al., 1990).

These examples demonstrate that, although soil testing has had a long development and application history, there is still need to improve test performance. Today, analytical capabilities are advancing faster than test methodology. However, most of the soil testing procedures in use today are sufficient to evaluate the fertility status of the soil. Change is occurring in several different directions, toward universal single-extraction reagent methods and the use of repeated extractions and equilibrium solutions (Baker, 1973; 1990; van Erp et al., 1998; Houba et al., 2000).

Soil testing is the only means of specifying lime and fertilizer needs and is the technique required to describe the nutrient element fertility status of the soil correctly (Melsted and Peck, 1973; Jones, 1983; Peck and Soltanpour, 1990; Campbell, 1994; 1998; Voss, 1998). Without a soil test result and/or without following the recommendation given by a soil test, lime and fertilizer use would be indiscriminate and particularly hazardous to a successful crop yield free from nutrient element stress. Unfortunately, such stresses are commonplace on many cropland soils. On a worldwide basis, about one quarter of the world's land surface is affected by some type of naturally occurring elemental stress (Dudal, 1976; Gardner, 1996; Brown, 1997). With intensive cropping of even the best natively fertile soils, stress eventually occurs if the proper procedures are not followed (1) to replace crop-removed

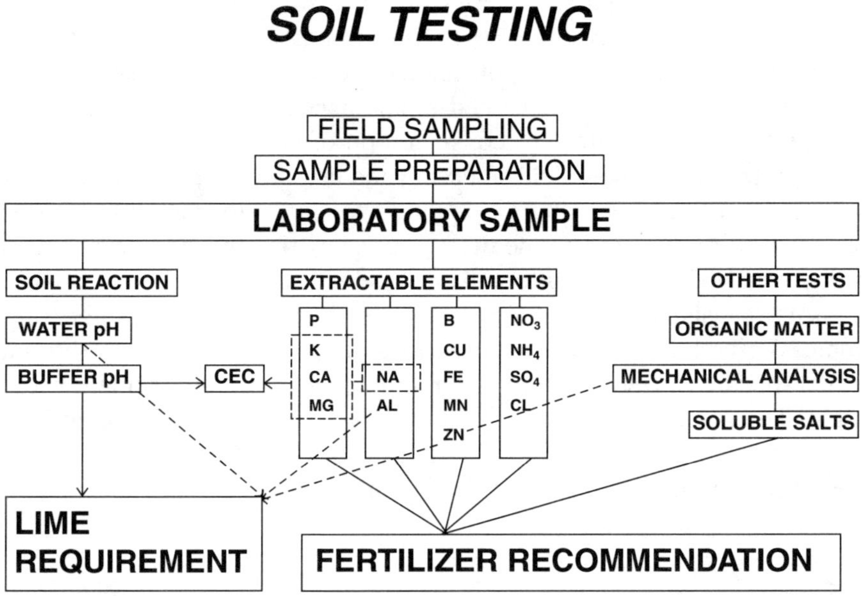

Figure 2.1
Sequence of procedures for conducting a soil test.

nutrient elements, (2) to counter acidification, and (3) to maintain the proper nutrient element balance for optimum plant growth.

B. Sequence of Procedures

The value of a soil analysis result is no better than the quality of the sample assayed, determined by:

1. How sample was taken from the field (James and Wells, 1990; Crépin and Johnson, 1993; Peterson and Calvin, 1996; Peck and Beck, 1998; Schnug et al., 1998; Wright, 1998; Brown, 1999; Radojevic and Baskin, 1999).
2. What conditions existed in transport to the laboratory.
3. The type of preparation techniques used to prepare the laboratory sample (Bates, 1993; Anonymous, 1994a; Hoskins and Ross, 1995; Gelderman and Mallarino, 1998; Brown, 1999; Radojevic and Baskin, 1999).
4. Sample aliquot measurement (Mehlich, 1972; 1973; van Lierop, 1981; 1989; Bates, 1993; Peck, 1998).
5. Laboratory factors (Eliason, 1998).
6. Sample storage (Bates, 1993; Houba and Novozamsky, 1998; Brown, 1999).

Soil testing encompasses a series of steps from field sampling to laboratory analysis and, eventually, to interpretation, as is illustrated in Figure 2.1. None of the steps is independent of the others; the care taken at one point affects the result obtained at another. Therefore, a discussion of one aspect necessarily involves consideration of all aspects.

A soil test begins with field sampling and preparation prior to laboratory analysis. Once the soil has been prepared for laboratory analysis, various tests are performed to determine those characteristics needed to evaluate the fertility status of the soil and to make a lime and fertilizer recommendation.

In most instances, a dual system of weighed and/or volume-measured samples is presented. This rationale is necessary in cases in which the original method specified a weight of sample or volume of known or assumed specific weight. The reader may refer to Mehlich (1972; 1973) for additional information on volume-weight considerations and to Tucker (1984) and Peck (1998) for more details on scoop design and use.

C. Sampling

Soils are naturally variable horizontally as well as vertically, which requires careful consideration in terms of sampling technique. Topography and soil type are common factors for determining where, within sampling boundaries, to collect a single soil composite. There are three commonly used sampling strategies:

1. Simple random sampling
2. Stratified random sampling — selecting individual soil cores in a random pattern within a designated area
3. Systematic or grid sampling

There are statistical concepts in soil sampling that will determine which method of sampling best defines the area under test evaluation. Since any detailed discussion is beyond the scope of this laboratory guide, readers are referred to the review articles on this topic by Peck and Melsted (1973), Sabbe and Marx (1987), James and Wells (1990), Crépin and Johnson (1993), Peterson and Calvin (1996), Peck and Beck (1998), Radojevic and Baskin (1999), and Brown (1999) for general sampling considerations and by Schnug et al. (1998), Nowak (1998),Wright (1998), and Crépin and Johnson (1993) and Anonymous (1999a) for systematic or grid sampling procedures. A new publication edited by Westervelt and Reetz (2000) describes geographic information systems (GIS) applicable to site-specific agriculture. Soil sampling procedures adapted for soils in the southern region of the

United States — but equally applicable to other regions — may be found in a bulletin edited by Thom and Sabbe (1994).

The depth of sampling is determined by one of several factors (Brown, 1999): horizonal characteristics (limiting depth to but one soil horizon), depth of soil mixing for land preparation, and rooting depth of the crop growing or to be grown.

Since most field soils are not homogeneous, naturally or due to past and/or current cultural practices (Woodruff, 1994; Kovar, 1994), the challenge for the sampler is to obtain a sample that is representative of the field under test. The common procedure is to take a number of individual cores to form a composite; the number of cores required to make one composite sample ranges from as few as 4 to as many as 16.

Several studies have shown that the variance for a determined soil test parameter is not substantially reduced by increasing the number of cores composited over 8. It would be more desirable to composite fewer cores and to submit more than one composite to the laboratory for analysis. Therefore, the mean analysis result for several composite samples becomes the soil test value accompanied by a variance or range. Although this practice would increase the time and cost of the soil test, the variance in the soil test value can be determined for the field under test, adding a valuable factor to the obtained test result.

The area represented by one composite soil sample is also an important consideration. Here again, there is considerable variance of opinion regarding the best procedure to follow, some recommending at least one composite per 5 acres (2 ha), others one composite per 100 acres (40 ha). The decision becomes one of management choice with or without past experience or knowledge of the homogeneity or lack of it for the field under test. Until the soil test level of a field has become firmly established, it would be best to divide the field into equal-sized sections, with each section no more than 10 acres (4 ha), and to gather a composite from each section. The soil test level is then determined by averaging the sum (with outliers discarded) of the test values of all the composites collected.

Coring should be at random, avoiding areas in the field that are markedly different in elevation and soil type. Coring should not be done near roads, fence rows, buildings, or tree lines. In fields being treated as a single unit but with soil type differences, cores from these differing soil types should not be mixed, but composites made from each major soil type for separate laboratory analysis. Some have suggested that, instead of dividing the field into equal-sized blocks as stated above, it be first divided based on differences in soil type, and then further subdivided into equal-sized blocks for soil sampling and compositing. Such a procedure would be repeated when next sampled until a pattern of homogeneity is established and previously separate sections can be combined for establishing new boundaries for compositing.

The recommended procedure is to core to the plow (mixing) depth or to that depth of soil occupied by the majority of plant roots in unplowed soils (Kovar, 1994). Surface and subsurface (below the plow depth) layers should not be mixed, keeping these two horizons separate for individual analysis and interpretation. Normally, subsurface soils are not collected for analysis unless for specific purposes, such as for deep-rooted crops or when past nutrient element crop stress suggests a possible significant subsoil infertility problem. Deep soil profile soil samples are required for tests such as profile NO_3–N (Camberato and Deaton, 1994; Griffin et al., 1995), a test procedure that will be discussed in more detail later.

Normally, sampling instructions do not specify a particular "best time" to collect soil samples, although there are seasonal cycles in some soil test parameters (Lockman and Molloy, 1984). The best time, when seasonal effects are minimal, is in midsummer to early fall. Some recommend sampling when plant tissue samples are being collected for analysis, a time normally during the mid- or late-summer months. However, the time of the year best for taking soil samples is probably of less importance than that the time be the same each year so that a track of test results can be maintained (Jones, 1983).

The following soil sampling procedures for field sampling are recommended:

Location	Procedure
Plowed fields	Core to the plow depth; in fields planted or to be planted in row crops but not plowed, core to the depth where at least 75% of the plant roots will be found
Turf	Core to 4 in. (10 cm) into the soil (the surface of the soil would begin just below the root mat)
Orchards and vineyards	Core to 18 to 20 in. (46 to 51 cm), staying within the plant canopy

By using varying rate applicators, lime and fertilizer application rates can be based on prepared grid maps that outline the areas of similar soil pH and levels of extractable elements. A range of sampling techniques can be used to base the grid patterns (Crépin and Johnson, 1993; Schnug et al., 1998; Wright, 1998; Cook and Bramley, 2000), and lime and fertilizer application rates may be adjusted to either maximize probable crop response and/or effect a reduction in soil pH and level of element variability (Haneklaus and Schnug, 2000).

Various devices can be used to collect soil cores; the more commonly used is some type of Hoffer Soil Sampling Tube or soil auger. The following are sources for obtaining hand-driven or mechanically driven soil tubes or auger:

Soil Sampling Devices Suppliers
Clements Associates, 1992 Hunter Ave., Newton, IA 50208-8652 (800-247-6630; fax: 515-792-1361)
Concord Environmental Equipment, RR1, Box 78, Hawley, MN 56449-9739 (218-937-5100; fax: 218-937-5101)
Geophyta, 2685 County Road 254, Vickery, OH 43464-9775 (419-547-8538)
Linco Equipment, Inc., I-39 and U.S. 24W, El Paso, IL 61738 (309-527-6455; fax: 309-527-660)
Oakfield Apparatus, P.O. Box 65, Oakfield, WI 53065-0065 (414-583-4114; fax: 414-583-4166)
Western Ag Innovations, 217 Badger Ct., Saskatoon, SK, Canada S7N 2X2 (306-249-3237; fax: 306-249-3237)

The collected cores are put into a clean bucket, thoroughly mixed, and transferred to the soil sample bag for transport to the laboratory. If only a portion of the collected cores are being saved for analysis, the sample cores must be thoroughly mixed. A small hand trowel is a helpful tool for accomplishing this required mixing. It is best to use a clean plastic bucket to receive the collected cores.

To avoid possible contamination, clean tools should be used when collecting soil samples in the field. Metal devices should be made of tool or stainless steel. Galvanized or brass devices will contaminate soil samples with Zn and/or Cu.

A collected soil sample should be placed in a clean paper bag and kept in a cool place until delivered to the laboratory. The glue used in some makes of paper bags may contain substantial quantities of B, which can contaminate soil samples when they are unusually wet during temporary storage in the bag. A quick test of the bag can determine whether B is present in the glue. Wet soil samples should not be placed in plastic or waterproof bags unless the time period for storage is short (48 h) and the samples are kept cool (10°C; 50°F).

D. Transport to the Laboratory

If the period of time between field sample collection and arrival at the laboratory will be more than several days, field-moist soil when placed in an air-tight container can undergo significant biological changes at room and/or elevated temperatures. Organic matter decomposition can release elements (ions), such as PO_4^{3-}, SO_4^{2-}, BO_3^{3-}, and NH_4^+, into the soil solution, while anaerobic conditions can result in organic matter decomposition and loss of N from the soil. For long-term transport, the collected soil should be kept in a cool environment (5 to 10°C; 40 to 50°F) and excess water removed by partial drying, keeping the soil just moist.

Freezing a soil sample will maintain soil biological integrity, but it may significantly alter the physiochemical properties, as freezing has the same effect on soil as high temperature (>32°C; >90°F) drying.

E. Preparation of the Laboratory Sample

1. Drying

The conventional procedure is to air-dry field soil samples at ambient laboratory temperature (21 to 27°C; 70 to 80°F) prior to crushing and sieving (Anonymous, 1994a). The drying process should be done as promptly and rapidly as possible to minimize microbial activity (mineralization). The time required to bring a soil sample to an air-dried condition is determined by its moisture, organic matter content, and texture. Soils high in clay and/or organic matter content require a considerably longer time to bring to an air-dried condition than do sandy-textured soils.

Drying can be facilitated by exposing as much surface of the soil to circulating air as possible and by elevating the drying temperature, but not to exceed 38°C (100°F), because significant changes in the physiochemical properties of the soil can occur at elevated drying temperatures. Field soils should not be oven-dried at elevated temperatures or if frozen.

The drying of some types of soils will result in a significant release or fixation of K (Goulding, 1987; Sparks, 1987); therefore, for some determinations, the arriving soil sample may be assayed as received without removing field moisture (Goulding, 1987; Bates, 1993). In addition, the determination of the micronutrients Cu, Fe, Mn, and Zn can be affected by the drying process (Kahn and Soltanpour, 1978; Shuman, 1980). Since significant changes do occur when soil is dried (Hanway et al., 1962; Murphy et al., 1983), there was a time when some soil testing laboratories took field soils as received for analysis, using a slurry method of sample preparation. However, the method proved cumbersome and time-consuming for processing large numbers of samples.

The moisture content of an air-dried soil is determined by the physiochemical properties of the soil and the relative humidity of air surrounding

Figure 2.2
Soil grinding and sieving device. (Courtesy of Custom Laboratory Equipment, Orange City, FL.)

the sample. This variability has little effect on most soil analysis procedures, the minimal effect occurring when the soil aliquot is measured by volume rather than by weight.

2. Crushing/Grinding/Sieving

Following drying, the soil sample is crushed, either by hand or by using a mechanical device (Figure 2.2), and then passed through a 10-mesh (2-mm) screen (Anonymous, 1994a). The grinding process can have an effect on AB–DTPA-extractable Fe, Zn, Mn, Cu, P, and K (Soltanpour et al., 1979).

Sieving through a 10-mesh (2-mm) screen removes stones and other extraneous substances, yielding a uniform sample that can be easily handled in the laboratory and stored indefinitely. This preparation procedure can contaminate a soil sample, either from the composition of the contacting surfaces or from deposition of dust and/or previous sample residue. The crushing and sieving devices must be free of elements that might be determined in the analysis. For example, brass sieves should not be used if Cu and Zn are elements to be determined.

Although crushing and sieving can also be a mixing process, sample size reduction may be necessary and care must be exercised to ensure that the sample is thoroughly mixed before dividing.

Particle size reduction can have an effect on some elemental determinations, as discussed by Kahn (1979) for the determination of Cu, Fe, and Zn and by Houba et al. (1993) for equilibrium extraction reagent procedures.

In general, once the soil sample has been air-dried, crushed, and screened, it can be stored indefinitely in a dry environment without significant changes in soil test values (Bates, 1993; Houba and Novozamsky, 1998; Houba et al., 2000).

F. Sample Aliquot Determination

1. Weighing vs. Scooping

In most soil testing laboratories, analyte sample aliquots are obtained by scooping rather than by weighing, primarily because of the time required to weigh samples. Normally, scoops are designed to deliver an estimated weight rather than a specific volume of sample. Scoop size will vary depending on the estimated volume-weight (bulk density) for the soil being scooped.

Assumed volume-weights range from a low of 1.18 to a high of 1.33 (a 1-cm^3 volume of soil would weigh from 1.18 to 1.33 g). The volume-weight is determined in part by texture and organic matter content; sandy, low-organic-matter content soils have a higher volume-weight than soils high in clay and organic matter content. Peck (1980), in a study of volume-weight determinations for soils from the north-central region of the United States, defines a "typical" soil as a medial silt loam texture with 2.5% organic matter content crushed to pass a 10-mesh screen. The volume-weight (bulk density) was found to be 1.18 for this "typical" soil as compared with a volume–weight of 1.32 for "undisturbed" soil. This compares with the estimated volume-weight of 1.25 for the sandy soils found mostly in the southeastern coastal plain area of the United States.

The design of the scoop itself is an important factor that can affect the ability of the scoop to deliver the same "estimated" weight of sample each time. In general, a scoop whose radius is equal to its height is more consistent in its delivery than a scoop whose height is greater than its radius. Peck (1980) describes the best scoop design for use with prepared (dried and passed through a 10-mesh screen) soils that have an approximate volume-weight of 1.18 as those whose height and radius are approximately equal.

Soil aliquot transfer to a saturation or extracting vessel is commonly done by weighing. The use of volume as the measurement for aliquot amount has been recommended by Mehlich (1972; 1973). Bates (1993) has discussed weight vs. volume measurement considerations and van Lierop (1989) has compared weight vs. volume measurement of soil aliquots on accuracy of the assay result.

In this laboratory guide, both volume and estimated-weight scoops are used to obtain the soil aliquot for many determinations as well as determinations based on weighed samples. In most instances, the method most commonly associated with that procedure is specified.

2. Estimated Weight Scoops

Scoop size is based on an assumed "average" volume-weight of prepared sample, air-dried, 10-mesh-sieved (2-mm) soil. The typical soil prepared for analysis, as described in this instruction guide, has an assumed weight-to-volume ratio of 1.18 for silt loam and clay-textured soils, and 1.25 for sandy soils. Therefore, those soil test procedures adapted to a soil with a particular texture will designate scoop volumes that match the assumed weight-to-volume ratio:

Silt loam and clay-textured soils		Sandy soils	
Weight, g	Scoop size, cm^3	Weight, g	Scoop size, cm^3
2.5	1.70	5.0	4.0
5.0	4.25		
10.0	8.50		

Scoops are of a fixed volume and do not necessarily yield an estimated or assumed weight. However, when the volume-weight of a soil sample is known, a specific volume of that soil can be scooped to give an estimated weight.

In most instances, a dual system of weighed and/or volume-measured samples is presented. This rationale is necessary in cases in which the original method specified a weight of sample or a volume of known or assumed specific weight. The reader may refer to Mehlich (1972; 1973) and van Lierop (1981; 1989) for additional information on volume-weight considerations and to Peck (1998) for more details on scoop design and use.

Another scoop is designed with a rounded or "cup-shaped" bottom to avoid the possibility of unfilled cavities in the base of the scoop. Tucker (1984) describes a technique for making scoops with 1-, 2.5-, 5.0-, and 10.0-cm capacities, as well as a technique for calibrating prepared scoops.

Some have recommended the use of a round surface, such as a glass rod, as the leveling tool, which allows the soil particles at the edge of the leveling tool to roll under the moving edge, thus reducing the possibility of creating small cavities in the planed surface after leveling.

To the purist, scooping is anathema, introducing error into the analysis as a result of variations in sample densities (Glenn, 1983). However, experience has shown that scooping, if properly done, can be an adequate substitute for weighing, producing equivalent analytical results. The major sources for error are in the design of the scoop and its improper use.

Mehlich (1973) has proposed a system of soil testing based entirely on scooped samples, a volume method of analysis and interpretation that will be discussed in greater detail later. Similarly, Wolf (1982) has a soil testing methodology based entirely on a scooped sample for laboratory analysis. In addition, the Adams–Evans Lime Buffer Test (Adams and Evans, 1962) is performed with a volume (scooped) sample.

Although scooping does have some unique advantages, convention has dictated that the laboratory aliquot be measured by weight unless the test itself or operational conditions dictate otherwise.

3. NCR-13 Scoops

The design specifications of the NCR-13 scoops commonly used by soil testing laboratories in the north-central region of the United States, described by Peck (1998), are as follows:

NCR-13 Standard Soil Scoop Specifications (manufactured from stainless steel)

Scoop size[a], g	Scoop capacity, cc	Outside diameter, in.	Inside diameter, in.	Inside diameter, in.
1	0.85	5/8	1/2	17/64
2	1.70	3/4	5/8	22/64
5	4.25	1	7/8	28/64
10	8.50	1 1/4	1 1/8	34/64

[a] Grams of soil in terms of the "typical" soil (defined as a medial silt loam texture with 1.25% organic matter crushed to pass a 10-mesh screen, bulk density of crushed "typical" soil approximates 1.18 compared with 1.32 for "undisturbed" soil) weighing 2,000,000 lb/acre in the top 6 2/3-in. layer.

The NCR-13 standard soil scoop is shown in Figure 2.3.

4. Procedure for Using a Soil Scoop

- Stir the crushed and screened sample with a spatula to loosen soil prior to measuring.
- Dip into the center of the soil sample with the soil scoop, filling it heaping full without pushing against the side of the soil container.

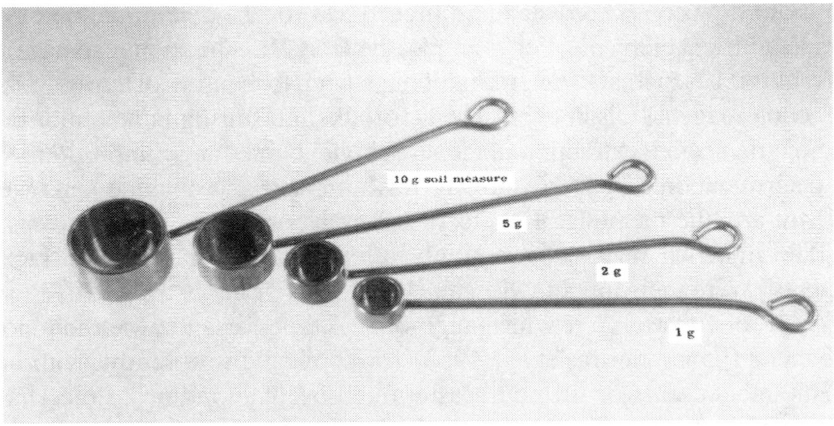

Figure 2.3
NCR-13 standard soil scoops. (Courtesy of T. Peck, University of Illinois, Urbana.)

- Hold the scoop firmly and tap the handle three times with a spatula from a distance of 2 or 3 in. from the soil-filled scoop.
- Hold the spatula blade perpendicular to the top of the scoop and strike off excess soil. A flat spatula blade may be replaced by a round rod, which protects against scarring the leveled surface.
- Empty the scoop into an appropriate extraction vessel.

Since an accurate measure for a scooped sample is essential, scoop design is a very important factor. The diameter of the scoop should be twice its height to ensure the most efficient packing density in the scoop.

Variance among repeated scoopings of a soil sample will be within 2 to 3% of the same volume or estimated weight. In general, scooping of soil samples has been found to yield results comparable to weighed samples in repeated analyses of the same soil sample.

G. Laboratory Factors

1. Extraction Reagents

Many of the extraction reagents currently in use today (Jones, 1990; 1998a; Anonymous, 1999b) reflect the history of their development and use; extraction reagents that were developed for specific applications in the 1940s and

1950s are now considered standard procedures for the determination of one or more nutrient elements. For example, the Bray P1 extraction reagent (Bray and Kurtz, 1945) for P determination in acid to neutral pH soils, Olsen extraction reagent (Olsen et al., 1954) for alkaline soils, and neutral normal ammonium acetate extraction reagent (Schollenberger and Simon, 1954) for the determination of K, Ca, and Mg for both acid and alkaline soils were, and still are, the methods of choice in many laboratories.

The first two reagents commonly referred to as "universal extraction reagents" were the Morgan (Morgan, 1932; 1941; Lunt et al., 1950) extraction reagent for use on a wide range of soil types and the Mehlich No. 1 (Mehlich, 1953a; Nelson et al., 1953) for application on sandy, acid, low-organic-matter soils of the southeastern coastal plain region of the United States. With the introduction of multielement analyzers, such as various forms of autoanalyzers (Watson and Isaac, 1990) and inductively coupled plasma emission spectrometers (Watson and Isaac, 1990; Soltanpour et al., 1996; 1998), one extraction reagent for the determination of many elements, including the major elements (P, K, Ca, and Mg) as well as the micronutrients (B, Cu, Fe, Mn, and Zn), resulted in the development of Mehlich No. 3 extraction reagent (Mehlich, 1984a) for many different types of soils and the AB–DTPA extraction reagent for alkaline soils (Soltanpour, 1991). Jones (1990; 1998a) and van Raij (1994) have written reviews on the development and use of the universal extraction reagents.

Recently, the adaptation of the 0.01 *M* $CaCl_2$ extraction reagent procedure for multielement determination has been proposed by Houba and colleagues (1990; 2000). A list of these extraction reagents and their procedure for use are given in Appendix C.

2. Extraction Procedure

Extraction procedure parameters, such as the shape and size of the extraction vessel (Wheaton bottle vs. Erlenmeyer flask), shaking speed, and temperature, can have a significant effect on the extraction of P and K from a soil by most of the commonly used extraction procedures (Eliason, 1998). Therefore, control of these factors is essential if the assay result is to be reliable.

3. Reagents and Standards

Careful preparation, storage, and use of reagents and standards are critical to successfully carry out the procedures described in this guide. One frequently overlooked factor that can affect the analytical result is the pH of the extraction reagent. For those assay procedures given in this guide, the user should be aware of these effects and carefully follow the procedures as given without modification to ensure that reliable assay results are obtained.

A list of reagents to carry out the methods given in this laboratory guide is given in Appendix A, preparation procedures for standards are given in Appendix B, and the procedures for use of extractant reagents are given in Appendix C.

H. Long-Term Storage

Since there have been no specific studies that define the required storage conditions and identify those changes that will occur from long-term storage of laboratory-prepared soil samples, it has been suggested that soils are best able to maintain their original integrity when stored in an air-dried condition at low humidity and just above the freezing temperature; however, some result data indicate that some soil parameters may change during such storage conditions (Bates, 1993; Houba and Novozamsky, 1998; Brown, 1999).

I. Soil pH

1. Introduction

Soil pH is a measure of the hydronium ion (H_3O^+ or, more commonly, the H^+) activity in the soil solution (Peech, 1965; Bates, 1973; Thomas and Hargrove, 1984; Thomas, 1996; Tan 1998), and pH is defined as the negative logarithm (base 10) of the H^+ activity (moles per liter) in the soil solution, expressed as follows:

pH is the negative $\log_{10}$ of the hydrogen (H^+) ion concentration:

$$pH = 1/\log_{10} (H^+)$$

The soil is either acidic, having ionized (or free) H^+ ions, or basic, having ionized (or free) OH^- ions. Therefore, pH is a measure of the soil acidity or basicity measured on a scale from 0 to 14, with a pH of 7.0 the neutral point that is neither acidic nor basic. Because pH is a log scale of the H^+ ion concentration, a change of one unit of pH is a tenfold change in H^+ ion concentration.

Making an accurate and consistent measurement of soil pH is not easily done, as there are a number of factors that can significantly affect the determination. The use of a salt solution — 0.01 *M* calcium chloride ($CaCl_2{\cdot}2H_2O$) or

1 *N* potassium chloride (KCl) — is one means of overcoming the "salt effect" on pH determination, particularly when determining the pH for sandy soils or those soils with relatively low (<10 meq/100 g) cation exchange capacities (CEC).

Soil pH can be determined using chemical dyes (Jackson, 1958; Woodruff, 1961; Hesse, 1971). However, a pH meter equipped with a glass electrode and calomel reference cell is the normal procedure for making this measurement. These electrodes may be physically separate or combined into one electrode body. Experience has shown that the combination electrode (one-body electrode) is more prone to difficulties; therefore, the two separate electrode combination is preferred (Peck, 1983). Excellent discussions on the measurement of soil pH are provided by McLean (1973), Fisher (1984), van Lierop (1990), Gascho et al. (1996), and Sumner (1994).

pH can be difficult to measure accurately depending on soil characteristics and the technique used to make the measurement (Schofield and Taylor, 1955; McLean, 1973; Conyers and Darey, 1988; van Lierop, 1990; Slattery et al., 1999). When using a pH meter, the pH is determined in a soil/water slurry, which can be of various ratios, but normally 1:1 or 1:2 (Anonymous, 1994b; Watson and Brown, 1998), or in a soil slurry of either 0.01 *M* $CaCl_2{\cdot}2H_2O$ (van Lierop, 1990; Plank, 1992a; Houba et al., 2000) or 1 *N* KCl (Moore and Loeppert, 1987; Plank, 1992b). If one of these salt solutions is used, the determined pH value will be different. Normally the pH determined in 0.01 *M* $CaCl_2$ is from 0.3 to 0.5 units less than that determined in water, and 0.7 to 1.0 units less when determined in 1 *N* KCl. Therefore, an interpretation of a pH value must be identified based on the solution in which it was determined, water or otherwise.

The determination of the soil pH is a three-step procedure:

1. Prepare the soil–water, 0.01 *M* $CaCl_2{\cdot}2H_2O$, or 1 *N* KCl slurry.
2. Calibrate the pH meter.
3. Place the electrodes into the prepared slurry and read the pH.

2. Using a pH Meter

a. Calibration of the pH meter

The calibration procedure given in the operating manual for the pH meter should be followed. In general, calibration requires the use of two buffer standards of known pH, bracketing the expected range of pH for the soils to be measured. For most acid soils, a buffer of pH 7.0 and another of pH 4.0 or 5.0 will be adequate to calibrate the pH meter. For alkaline soils, pH 10.0 and pH 6.0 buffers are required.

When calibrating, the procedure calls for adjusting the meter reading with the higher pH buffer first, followed by meter adjustment with the lower pH buffer. Before placing the electrodes into the buffer solutions, the lower surface is thoroughly rinsed with water. When the electrodes are placed into the buffer solution, the solution is swirled to ensure complete contact between the buffer solution and the surface of the electrodes. The meter pointer should be steady shortly after initial contact if in proper operating order. The electrode surfaces are flushed with flowing water between determinations.

If the pH meter cannot be adjusted using the buffers, then the electrodes (usually the glass electrode) may need replacement. If the meter is slow to reach the proper meter reading, this may indicate that the small opening at the base of the calomel reference electrode is clogged. Gentle polishing of the tip of this electrode with emery paper may correct the problem. A recent study of electrode difficulties associated with soil–water pH determinations revealed that the flow rate of filling solution from the calomel reference electrode is frequently the culprit, a high flow rate resulting in low pH readings, a low flow rate resulting in high pH readings (Peck, 1983). A recent innovation that ensures an adequate flow rate of filling solution from the calomel reference electrode employs a microplunger that is turned slightly each time the electrode is used, placing a slight positive pressure on the filling solution, which activates its flow from the electrode into the soil–water slurry.

The glass electrode should never be wiped with a dry tissue; this will remove the water film from the glass membrane, and that water layer is essential for proper functioning of the electrode. A new glass electrode should be placed in water at least 12 h before use and kept in water between uses to establish and maintain proper hydration.

Even though the pH meter may appear to be operating correctly after calibration, this does not ensure that it will accurately determine soil–water pH. Therefore, suitable soil standards of known water pH should be employed to verify the accuracy of the pH meter readings.

It is wise to set the position of the two electrodes so that the calomel cell electrode is slightly lower than the glass electrode. If contact is made with the bottom of the soil–water slurry container, the fragile glass electrode will be protected by the more rugged calomel cell electrode making the contact.

Once the pH meter has been calibrated, it is ready for use to determine the pH of prepared soil–water slurries. Placement of the electrodes into the soil–water slurry is such that the glass electrode is just at the soil–water interface of the standing slurry. Moving the vessel containing the soil–water slurry in a horizontal rotating motion or mechanically stirring to put the soil–water slurry in motion and activating the pH meter will cause the needle on the dial to begin to move.

Once the needle stops its movement or begins to hover around a certain dial reading, record the dial reading to the nearest 0.1 unit. This is the pH. The needle may never settle or stop on a particular spot on the dial; therefore, an average estimated reading is made within the range of oscillations. It may require several seconds before needle movement begins to oscillate or hover over a particular place on the dial scale.

Movement of the soil–water slurry is essential to remove the water film on the glass electrode membrane, replacing it with soil–water suspension and increasing contact with H^+ ions in the soil solution held in close proximity to the soil colloids. The pH measured in a standing soil–water slurry will generally be higher than in a moving slurry.

b. Electrode positioning

The positioning of the pH meter electrodes as well as the stirring of the slurry can significantly influence the pH determination. An excellent reference on the measurement of a soil pH and the maintenance of pH electrodes is given by McLean (1982b) and Sumner (1994). The basic chemistry of soil acidity has been described by Peech (1965), Thomas and Hargrove (1984), Hendershot et al. (1993), and Thomas (1996). General instructions on the measurement of pH and factors influencing such measurements have been described by Meier et al. (1989), Sartoretto (1991), and Sims and Eckert (1995).

c. Preparation of pH buffers

Buffers can either be purchased from a chemical supplier or made by the analyst as follows:

pH 10.0 Buffer Solution

Weigh 1.756 g sodium hydroxide (NaOH) and 3.092 g boric acid (H_3BO_3) into a 1000-mL volumetric flask and bring to volume with water.

pH 9.22 Buffer Solution

Weigh 3.80 g disodium tetraborate decahydrate ($Na_2B_4O_7 \cdot 10H_2O$) into a 1000-mL volumetric flask and bring to volume with water.

Comments: *(1) Disodium tetraborate may lose crystal water if stored for a long time and (2) the buffer solution is stable for 1 month if stored in a polyethylene bottle.*

pH 7.0 Buffer Solution

Weigh 3.3910 g citric acid ($C_6H_8O_7$) and 23.3844 g disodium phosphate ($Na_2HPO_4 \cdot 12H_2O$) into a 1000-mL volumetric flask and bring to volume with water.

pH 7.0 Buffer Solution

Weigh 3.800 g potassium dihydrogen phosphate (KH_2PO_4) and 3.415 g disodium hydrogen phosphate (Na_2HPO_4) into a 1000-mL volumetric flask and bring to volume with water.

Note: *Disodium hydrogen phosphate and potassium dihydrogen phosphate may be dried first for 2 h at 110 to 120°C (230 to 248°F).*

pH 4.0 Buffer Solution

Weigh 11.8060 g citric acid ($C_6H_8O_7$) and 10.9468 g disodium phosphate ($Na_2HPO_4 \cdot 12H_2O$) into a 1000-mL volumetric flask and bring to volume with water.

pH 4.0 Buffer Solution

Weigh 10.21 g potassium hydrogen phthalate ($C_8H_5KO_4$) into a 1000-mL volumetric flask and bring to volume with water.

Note: *Potassium hydrogen phthalate may be dried first for 2 h at 110 to 120°C (230 to 248°F).*

d. pH determination in water

Soil water pH is an important test parameter, determined by placing a soil sample into a specified volume of water and measuring the pH of the resultant slurry. Normally, the ratio of soil to water is 1:1. The procedure sounds simple enough, but the actual determination is as difficult to make as any other soil testing procedure.

If buffer pH is also to be determined, the sample size and the amount of water added will be determined by the buffer pH method to be used. The soil–water slurry is saved for the determination of the buffer pH if the soil–water pH is less than 6.0 (Section J of this chapter).

Soil–water and buffer pH methods used in the laboratories located in the northeastern United States are given in the Northeast Regional Bulletin 453 (Anonymous, 1995).

Determination for Following Adams–Evans Buffer

Scoop 10 cm^3 air-dried <10-mesh-sieved (2-mm) soil into a cup.
Pipette 10 mL water into the cup and stir for 5 s.
Let stand for 10 min.
Calibrate the pH meter according to the instructions supplied with the specific meter.
Stir the soil and water slurry.
Lower the electrodes into the soil–water slurry so that the electrode tips are at the soil–water interface.
While stirring the soil–water slurry, read the pH and record to the nearest tenth of a pH unit.

Note: *Erratic movement of the pH meter dial or number may be due to faulty operating electrodes or to lack of sufficient junction potential (Sumner, 1994).*

Determination for Following SMP Buffer

Weigh 5 g or scoop 4.25 cm^3 air-dried <10-mesh-sieved (2-mm) soil into a cup.
Pipette 5 mL water into the cup and stir for 5 s.
Let stand for 10 min.
Calibrate the pH meter according to the instructions supplied with the specific meter.
Stir the soil and water slurry.
Lower the electrodes into the soil–water slurry so that the electrode tips are at the soil–water interface.
While stirring the soil–water slurry, read the pH and record to the nearest tenth of a pH unit.

Note: *Erratic movement of the pH meter dial or number may be due to faulty operating electrodes or to lack of sufficient junction potential (Sumner, 1994).*

In an agreement between the Association of Official Analytical Chemists (AOAC) and the Soil Science Society of America (SSSA) described by Kalra (1996), a standardized water pH procedure based on a collaborative study is as follows (Kalra, 1995).

Standardized Water pH Determination

Weigh 10 g air-dried <10-mesh-sieved (2-mm) soil into a cup.
Pipette 10 mL water into the cup and mix thoroughly for 5 s with a glass rod or mechanical stirrer.

Let the soil–water suspension stand for 30 min. Perform pH measurement at 20 to 25°C (68 to 77°F). Before analysis stir the soil–water suspension with glass rod or mechanical stirrer.

Insert electrodes of calibrated pH meter into the cup and swirl soil–water suspension slightly.

Note: *The position of the reference electrode with respect to the glass electrode and the flow rate from the reference electrode may affect pH determination; follow manufacturer's instructions for electrode(s).*

Read pH immediately (after 30 to 60 s) to the nearest 0.1 pH unit.

e. pH determination in 0.01 *M* calcium chloride ($CaCl_2$)

To mask the variability in salt content of soils, maintain the soil in a flocculated condition, and decrease the junction potential effect, the soil pH measurement is made in a 0.01 *M* $CaCl_2$ solution (Schofield and Taylor, 1955; van Lierop, 1990; Plank, 1992a; Houba et al., 2000). The same procedure is followed as that used for the measurement of pH in water, except 0.01 *M* $CaCl_2$ is substituted for water in the preparation of the soil slurry. The pH values obtained will be lower (usually from 0.3 to 0.5 of a pH unit) than that measured in water. The pH meter needle will move less with agitation of the soil slurry and quickly settle on the pH reading. A different set of pH interpretative values must be used than those previously given when the pH is measured in 0.01 *M* $CaCl_2$ (Fotyma et al., 1998).

Reagents

0.01 M Calcium Chloride

Weigh 1.47 g calcium chloride dihydrate ($CaCl_2{\cdot}2H_2O$) into a 1000-mL volumetric flask and bring to volume with water.

1.0 M Calcium Chloride

Weigh 147 g calcium chloride dihydrate ($CaCl_2{\cdot}2H_2O$) into a 1000-mL volumetric flask and bring to volume with water.

pH Determination

Weigh 5 g or scoop 4.25 cm^3 air-dried <10-mesh-sieved (2-mm) soil into a 50-mL cup.

Pipette 5 mL 0.01 *M* $CaCl_2$ solution into the cup and stir for 30 min on a mechanical stirrer or shaker (or stir periodically with a glass rod for a period of 30 min).
Calibrate the pH meter according to the instructions supplied with the specific meter.
Stir the soil and 0.01 *M* $CaCl_2$ slurry.
Lower the electrodes into the soil–0.01 *M* $CaCl_2$ slurry so that the electrode tips are at the soil–water interface.
While stirring the soil–water slurry, read the pH and record to the nearest tenth of a pH unit.

Note: *For laboratories desiring both soil pH in water and 0.01 M* $CaCl_2$*, 5 mL water can be substituted for the 5 mL 0.01 M* $CaCl_2$*. After the water pH is determined, add one drop of 1 M* $CaCl_2$ *to the soil–water suspension, stir or shake for 30 min, and then read the pH of the suspension and designate as* pH_{CaCl_2}.

f. pH determination in 1 *N* potassium chloride (KCl)

To establish a significant salt content in the soil slurry for measuring pH, the soil pH measurement is made in a 1 *N* KCl solution (Moore and Loeppert, 1987; Plank, 1992b). This procedure is commonly used in Europe and many other sections of the world for soil pH determination. The same procedure is followed as that used for the measurement of pH in water, except 1 *N* KCl is substituted for water in the preparation of the soil slurry. The pH values obtained will be lower (usually from 0.5 to 1.0 of a pH unit) than that measured in water. The pH meter needle will move less with agitation of the soil slurry and quickly settle on the pH reading. A different set of pH interpretative values must be used when the pH is measured in 1 *N* KCl.

Reagent

1 N Potassium Chloride

Weigh 74.56 g potassium chloride (KCl) into a 1000-mL volumetric flask and bring to volume with water.

pH Determination

Weigh 5 g or scoop 4.25 cm^3 air-dried <10-mesh-sieved (2-mm) soil into a cup.
Pipette 5 mL 1 *N* KCl into the cup and stir for 5 s.
Let stand for 10 min.
Calibrate the pH meter according to the instructions supplied with the specific meter.

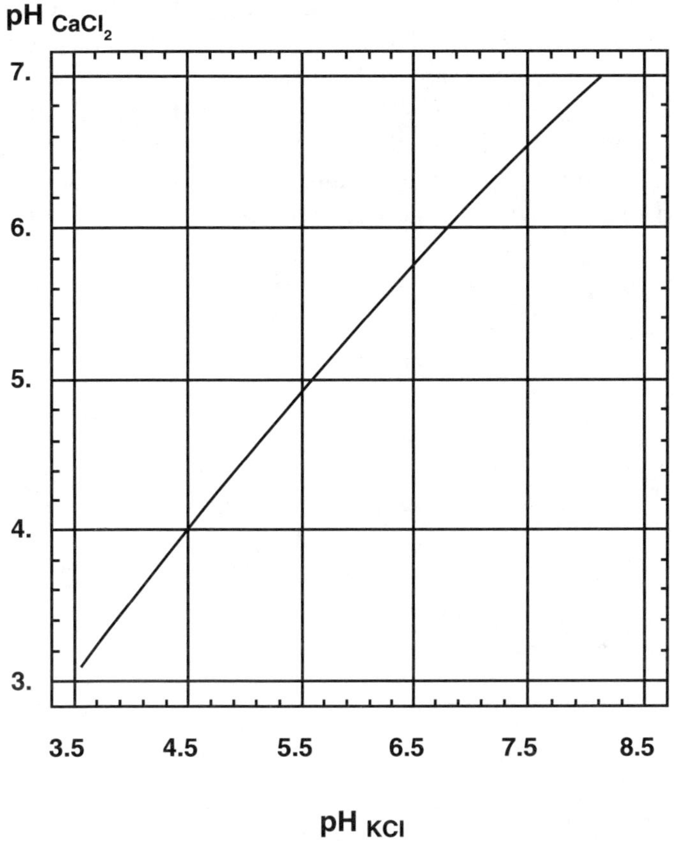

Figure 2.4
Regression line for the relationship between pH-$CaCl_2$ and pH-KCl.

Stir the soil and 1 *N* KCl slurry.
Lower the electrodes into the soil–1 *N* KCl slurry so that the electrode tips are at the soil–water interface.
While stirring the soil–1 *N* KCl slurry, read the pH and record to the nearest tenth of a pH unit.

A comparison of the pH measurement in 0.01 *M* $CaCl_2 \cdot 2H_2O$ and 1 *N* KCl is shown in Figure 2.4 (Fotyma et al., 1998).

3. pH Determination Using Indicators

The use of litmus paper and pH indicators (dyes) was the common procedure for determining the soil pH before the glass electrode pH meter was developed. Indicators are still useful for field application if a portable pH meter is not available. Lists of useful indicators are given by Jackson (1958), Woodruff (1961), and Hesse (1971).

One of the more commonly used indicators is bromocresol purple whose color range is from yellow (pH < 5.2) to purple (pH > 6.5), a range in color change suitable for determing the pH of most acid soils. A wide-range indicator can be prepared from a mixture of two parts of bromothymol blue and one part of methyl red. The ranges of color change in relation to the soil pH are:

Color	pH
Brilliant red	<3.0
Red	3.1–4.0
Red-orange	4.1–4.7
Orange	4.8–5.2
Orange-yellow	5.3–5.7
Yellow	5.8–6.1
Greenish-yellow	6.2–6.4
Yellowish-green	6.5–6.7
Green	6.8–7.3
Greenish-blue	7.4–7.8
Blue	>7.9

pH Determination

Place a small aliquot of soil into a depression on a white spot plate and add a solution of the selected indicator dropwise until the soil is thoroughly saturated and then stir to mix.

Allow to stand for a few moments until the soil settles and the color of the indicator is seen around the edge.

Using a color comparator chart, determine the pH.

Note: *Using a series of indicators, the full range of pH can be covered, giving a result not too dissimilar from that obtained using a glass electrode-equipped pH meter. However, more than one indicator and good color perception by the user are required to use the indicator technique successfully.*

4. Interpretation

A 1997 survey report by the Potash & Phosphate Institute (PPI) (Anonymous, 1998a) reported on the percent of soils in the United States that has water

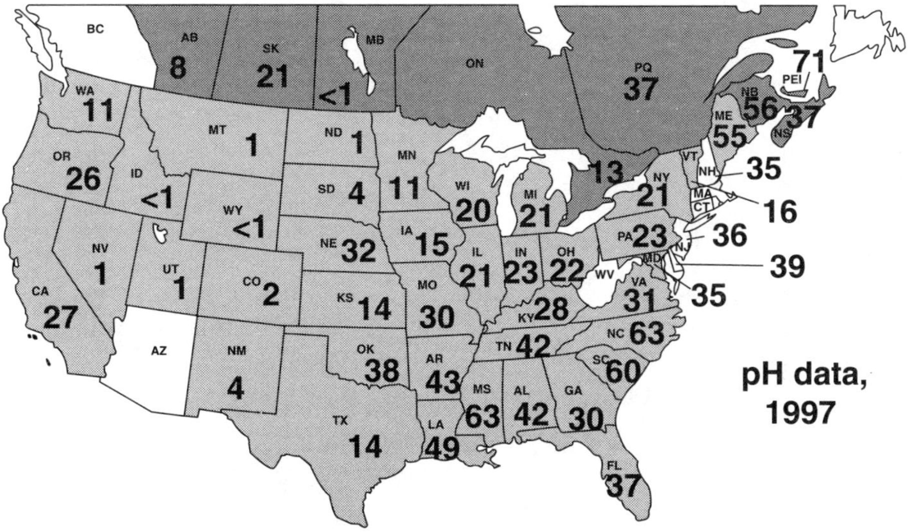

Figure 2.5
Percent of soils having water pHs of less than 6.0 in the United States and Canada, 1997. (From Potash & Phosphate Technical Bulletin 1998-3, Norcross, GA.)

pHs less than 6.0 as shown in Figure 2.5. It stated that "historically, soil pH values have tended to be more acid where rainfall is higher and where larger amounts of vegetation have helped to acidify the soil. Those conditions have been associated with areas east of the Mississippi River in the United States. The highest frequency of soil acidification continues to be found in the southeast where in some states over 60 percent of the soils test below pH 6.0."

Mullenax et al. (2001) collected data from 1968 through 1997 that showed a globally synchronous annual increase in soil acidity during the period June through August, which was highly correlated with a progressively decreasing geomagnetic field strength despite periodic local alterations in electromagnetic activity at the planetary boundary layer. Monthly determinations of soil pH for periods of 1 or more years at sites in Australia, South America, North America, Scotland, and Germany showed pH values to be consistently and significantly correlated with the monthly average geomagnetic field strength and the auroral electrojet values recorded during the time period covered by the data supporting the cyclic nature of the pH change. Gascho et al. (1996) measured water pH monthly in a Tifton loamy sand and found that the minimum pH occurred in September (pH 5.42) and the highest pH in January (pH 5.92).

By definition, soils with a pH less than 7.0 are generally defined as "acidic," with a pH above 7.0, "alkaline," and with pH at 7.0, "neutral." Two categorized classification systems based on water pH are as follows:

Category	pH	Category	pH
Very acid	4.5–5.5	Acid	<4.5
Acid	5.6–6.0	Weakly acid	4.5–6.5
Slightly acid	6.1–6.8	Neutral	6.6–7.5
Neutral	6.9–7.6	Weakly basic	7.6–9.5
Alkaline	7.7–8.3	Basic	>9.5

Generally, more descriptive terms are used to define pH in terms of its probable effect on the soil itself as well as on plant growth and development (Adams, 1984; Woodruff, 1967; van Lierop, 1990).

In general, the pH levels in 0.01 *M* $CaCl_2$ may be categorized as follows:

Category	pH
Acid	<4.5
Weakly acid	4.5–6.5
Neutral	6.5–7.5
Weakly basic	7.5–9.5
Basic	>9.5

For an interprepation of a soil pH determination, refer to Adams (1984), Alley and Zelazny (1987), Black (1993b), Anonymous (1996a), Maynard and Hochmuth (1997), Reid (1998a), and Slattery et al. (1999). The percent of soils having a water pH of less than 6.0 in North America for 1997 has been prepared by the Potash & Phosphate Institute (Anonymous, 1998a) and is shown in Figure 2.5.

The pH of a soil significantly affects plant growth, primarily as a result of the change in availability of both the essential elements, such as P, and most of the micronutrients, Cu, Fe, Mn, Mo, and Zn, as well as nonessential elements such as Al, which can be toxic to plants at elevated concentrations (Woodruff, 1967; Foy, 1984; Black, 1993a; Slattery et al., 1999). The activities of microbial populations are also affected by pH as are the activities of some types of pest chemicals applied to soils.

The pH of a soil can have a significant effect on plant growth and development by increasing or decreasing the availability of elements found in the soil, as is illustrated in Figure 2.6 for mineral soils, and the same influence on elemental availability occurs in organic soils (Figure 2.7; Sartoretto, 1991).

The pH change in H^+ ion concentration of the soil per se is not the reason plants respond differently to a change in soil pH, but the response is due to the change in concentration of elements in the soil solution and/or the change in "availability" of an element for root absorption. Simple examples are the increasing concentration of Al^{3+} ions in the soil solution with decreasing pH

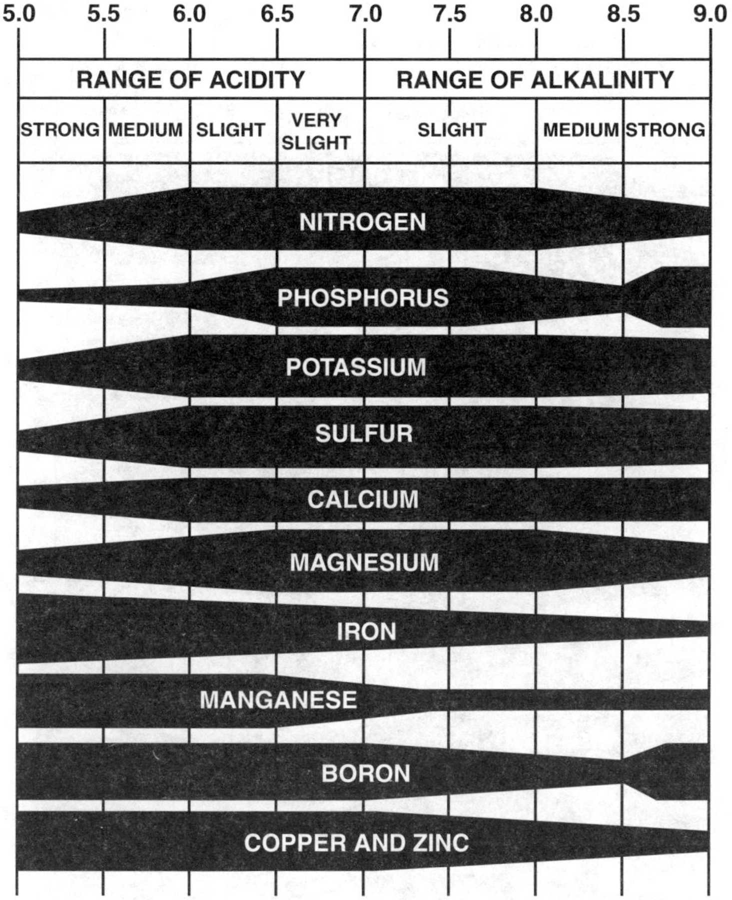

Figure 2.6
Availability of elements to plants at different pH levels for mineral soils. (From Sartoretto, 1991.)

(increased soil acidity), which adversely affects most plants, and the declining availability of Fe with increasing pH (increased soil alkalinity), resulting in Fe deficiency for many plants.

pH Effect on Elemental Availability and/or Soil Solution Composition

Element	pH decreasing	pH increasing
Aluminum (Al)	Increases	Decreases
Copper (Cu)	Increases	Decreases
Iron (Fe)	Increases	Decreases
Magnesium (Mg)	Decreases	Increases
Manganese (Mn)	Increases	Decreases
Zinc (Zn)	Increases	Decreases

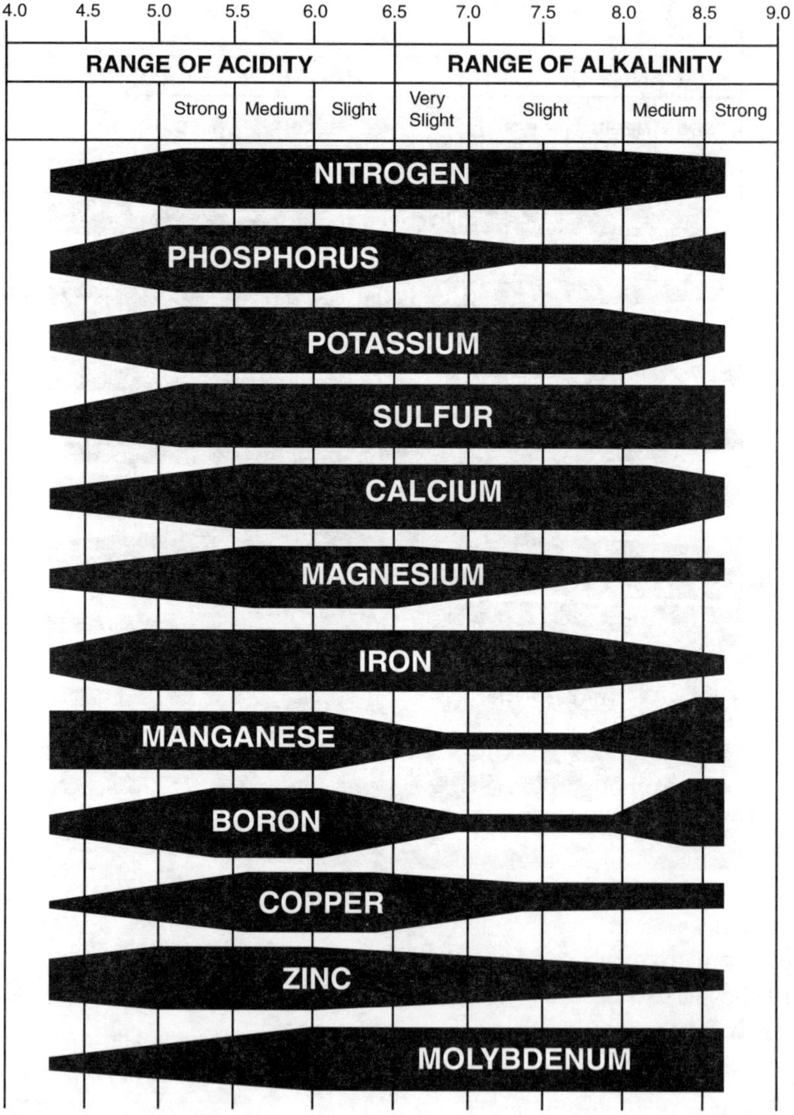

Figure 2.7
Availability of elements to plants at different pH levels for organic soils. (From Sartoretto, 1991.)

Fertilizer efficiency can also be affected by soil pH, as has been suggested and is shown in the following table.

Fertilizer Efficiency (%) at Varying Soil pH Levels

Soil acidity	pH	Nitrogen	Phosphate	Potash
Extreme	4.5	30	23	33
Very strong	5.0	53	34	52
Strong	5.5	77	46	77
Medium	6.0	89	52	100
Neutral	7.0	100	100	100

5. Soil Storage

The effects of long-term storage on the pH of a soil are not known, although Bates (1993) found that one laboratory reported that there was a decline in pH after just 1 month of storage.

J. Soil Buffer pH (Lime Requirement)

1. Introduction

Exchangeable acidity, a measure of the H ions on the soil colloids, can be measured by interacting a soil sample with a specially prepared buffer solution. Although exchangeable acidity can be determined accurately by titration (Yuan, 1959), the buffer method is quickly and easily performed in the laboratory, making it well suited for routine use in soil testing laboratories (McLean, 1978). The change in the pH of the buffer when interacted with soil gives an estimate of the quantity of exchangeable H ions, plus that acidity derived from Al ions in solution (Evans and Kamprath, 1970; Wolf, 1982; Sims, 1996) or that absorbed on the soil colloids (Schwertmann and Jackson, 1964; Kamprath, 1970), thus estimating the amount of calcium carbonate (in the form of liming products) needed to neutralize the total acidity, forming the basis for determining the lime requirement.

The lime requirement is that quantity of agricultural limestone required to raise the soil–water pH from an acidic level to a pH near neutrality. By using the buffer pH and referring to suitable tables the lime requirement (the quantity of agricultural limestone needed to raise the soil–water pH to a specific value) can be obtained.

Comparisons of various lime requirement methods have been made by McLean et al. (1966), Yuan (1975), Fox (1980), and Doerge and Gardner (1988).

There are several buffer techniques for determining the lime requirement, but only two are commonly used:

1. The Shoemaker, McLean, and Pratt (SMP) buffer for use on soils with a lime requirement greater than 2 tons/acre (4480 kg/ha) for acid soils with a pH below 5.8, containing less than 10% organic matter, and that have appreciable quantities of Al (Shoemaker et al., 1962).
2. The Adams–Evans buffer for use on soils with relatively small amounts of exchangeable acidity, 8 meq/100 g or less, sandy soils low in organic matter content (Adams and Evans, 1962).

Mehlich (1976) developed a buffer procedure relating the lime requirement to the buffer pH and nature of soil acidity that has more universal application, but it has not been widely adopted (Mehlich et al., 1976).

2. SMP Buffer

The SMP buffer method for estimating the lime requirement is reliable for soils whose characteristics were described above (Shoemaker et al., 1962). The original method was never considered to be very accurate for soils with lime requirements less than 2 tons/acre (4480 kg/ha) because of random variation between buffer indicated vs. actual lime requirement within this range. Also, on mineral soils of high organic matter content and high levels of exchangeable Al, the original SMP method gave lime requirements that were lower than the actual amounts required (McLean et al., 1977). More recently, the double-buffer adaptation (McLean et al., 1978), originally suggested by Yuan (1974), has been developed to alleviate some of the shortcomings of the original method (McLean, 1978). The SMP method is described by van Lierop (1990), Eckert (1988; 1991), Eckert and Sims (1995), Anonymous (1994c), and Watson and Brown (1998).

A sensitivity of 0.1 pH unit is needed for the determination of the lime requirement by the orginal SMP method (McLean, 1978), but the double-buffer adaptation calls for pH readings to the nearest 0.01 pH unit. A difference of 0.1 pH unit in the original method results in a difference of 0.4 to 0.6 tons/acre (896 to 1344 kg/ha) of lime for organic soils limed to pH 5.2, and 0.5 to 0.9 tons/acre (1120 to 2016 kg/ha) for mineral soils limed to near neutral pH. Similarly, a difference of 0.1 pH unit in one of the two buffers in a double-buffer adaptation may result in a difference in lime requirement of less than 0.1 ton/acre (224 kg/ha) for mineral soils of low lime requirement to more than 0.5 ton/acre (1120 kg/ha) for soils of high lime requirement.

SMP Buffer Solution

Weigh into an 18-L bottle 32.4 g paranitrophenol, 54.0 g potassium chromate ($KCrO_4$), and 955.8 g calcium chloride dihydrate ($CaCl_2{\cdot}2H_2O$).

Add approximately 9 L water. Shake vigorously as water is added and continue shaking for a few minutes to prevent formation of a crust over the salts.

Weigh 36.0 g calcium acetate [$Ca(C_2H_3O_2)_2{\cdot}H_2O$] into a separate container and dissolve in approximately 5 L water.

Add this latter solution to the former, shaking as they are combined.

Shake every 15 or 20 min for 2 or 3 h.

Add 45 mL triethanolamine, again shaking as the addition is made. Shake periodically until completely dissolved. This takes approximately 8 h.

Dilute to 18 L with water.

Adjust to pH 7.5 with 15% sodium hydroxide (NaOH) using a standardized pH meter.

Filter through a fiber glass sheet or cotton mat.

Connect an air inlet with 1- by 12-in. cylinder of Drierite, 1- by 12-in. cylinder of Ascarite, and 1- by 12-in. cylinder of Drierite in series to protect against contamination with carbon dioxide (CO_2) and water vapor.

Although less tedious procedures may be used for preparing small quantities of the buffer solution, the above procedure has been found to be most satisfactory for preparing bulk quantities of the buffer solution.

Buffer pH Determination

Weigh 5 g or scoop 4.25 cm^3 air-dried <10-mesh-sieved (2-mm) soil into a 50-mL cup in a tray designed for a mechanical shaker.

Add 5 mL water, shake or stir 1 min, let stand 10 min, and read pH in water with slight swirling of the electrodes.

Add 10 mL SMP buffer adjusted to pH 7.5 to the above soil suspension, shake on a mechanical shaker at >180 oscillations/min for 10 min, open the lid of the shaker, and let stand 30 min.

Read buffer pH (pH_1) using carefully adjusted pH meter to nearest 0.01 pH unit.

Note: *A 15-min stirring time and 15-min standing time may be used if more adaptable to the soil testing routine, since this gives essentially the same results as 10-min shaking plus 30-min standing times.*

If the original SMP (one-buffer) method is to be used, select the lime requirement from the table based on the buffer pH obtained. If the double-buffer adaptation is used, continue the procedure as follows:

Using an automatic pipette, add to the above soil–buffer suspension an aliquot of hydrochloric acid (HC1) equivalent to the amount required to decrease a 10-mL aliquot of pH 7.5 buffer to pH 6.0 (1 mL 0.206 *M* HC1 to 0.206 meg).

Repeat the 10-min shaking, 30-min standing (or 15 min of stirring plus 15 min standing), and reading of soil–buffer pH (pH_2).

The double-buffer adaptation involves the individual slope of the buffer-indicated vs. actual lime requirement curve for a given soil instead of a mean slope involved in the original method. The procedure for use of the double-buffer adaptation has not yet been worked out for liming organic soils to pH 5.2.

Interpretation. The regular (single-buffer) SMP method is probably the most satisfactory compromise between simplicity of measurement and reasonable accuracy for soils with a wide range of lime requirement. The table below is based on the SMP soil–buffer pH values and gives the lime requirement in terms of tons per acre of agricultural ground limestone of total neutralizing power (TNP) or $CaCO_3$ equivalent of 90% or above and an 8-in. plow depth (2,400,000 lb) to increase soil pH to selected levels.

The double-buffer adaptation is somewhat more accurate for all soils, but is especially so for soils having a relatively low lime requirement and probably so for acid mineral soils of relatively high organic matter content.

Amounts of limestone required to bring mineral and organic soils to the indicated pH according to soil–buffer pH (tons/acre 8-in. soil) are presented in the following table:

	Mineral soils				Organic soils
	7.0	7.0	6.5	6.0	5.2
Soil–buffer pH	Pure $CaCO_3$	Ag-ground limestone[a]			
6.8	1.1	1.4	1.2	1.0	0.7
6.7	1.8	2.4	2.1	1.7	1.3
6.6	2.4	3.4	2.9	2.4	1.8
6.5	3.1	4.5	3.8	3.1	2.4
6.4	4.0	5.5	4.7	3.8	2.9
6.3	4.7	6.5	5.5	4.5	3.5

Soil–buffer pH	Mineral soils				Organic soils
	7.0	7.0	6.5	6.0	5.2
	Pure $CaCO_3$	Ag-ground limestone[a]			
6.2	5.4	7.5	6.4	5.2	4.0
6.1	6.0	8.6	7.2	5.9	4.6
6.0	6.8	9.6	8.1	6.6	5.1
5.9	7.7	10.6	9.0	7.3	5.7
5.8	8.3	11.7	9.8	8.0	6.2
5.7	9.0	12.7	10.7	8.7	6.7
5.6	9.7	13.7	11.6	9.4	7.3
5.5	10.4	14.8	12.5	10.2	7.8
5.4	11.3	15.8	13.4	10.9	8.4
5.3	11.9	16.9	14.2	11.6	8.9
5.2	12.7	17.9	15.1	12.3	9.4
5.1	13.5	19.0	16.0	13.0	10.0
5.0	14.2	20.0	16.9	13.7	10.5
4.9	15.0	21.1	17.8	14.4	11.0
4.8	15.6	22.1	18.6	15.1	11.6

[a] Ag-ground limestone of 90% plus TNP or $CaCO_3$ equivalent, and fineness of 40% <100 mesh, 50% <60 mesh, 70% <20 mesh, and 95% <8 mesh.

Tons of Agricultural Limestone Needed to Raise the Soil pH to the Desired pH Level Based on the SMP Lime Test Index with an Incorporation Depth of 8 in.

Lime test Index[a]	Desired pH levels				
	Mineral soils, tons agricultural limestone/acre[b]			Organic soils	
	6.8	6.5	6.0	Soil pH	5.3, tons/acre
68	1.4	1.2	1.0	5.2	0.0
67	2.4	2.1	1.7	5.1	0.7
66	3.4	3.0	2.4	5.0	1.3
65	4.5	3.8	3.1	4.9	2.0
64	5.5	4.7	3.9	4.8	2.6
63	6.5	5.6	4.6	4.7	3.2
62	7.5	6.5	5.3	4.6	3.9

(continued)

Tons of Agricultural Limestone Needed to Raise the Soil pH to the Desired pH Level Based on the SMP Lime Test Index with an Incorporation Depth of 8 in. (continued)

	Desired pH levels				
Lime test	Mineral soils, tons agricultural limestone/acre[b]			Organic soils	
Index[a]	6.8	6.5	6.0	Soil pH	5.3, tons/acre
61	8.6	7.3	6.0	4.5	4.5
60	9.6	8.2	6.7	4.4	5.1

[a] Lime test index is the SMP buffer pH × 10.

[b] These values are based on agricultural limestone with a neutralizing value of 90% (Indiana RNV = 65. Ohio TNP = 90+). Adjustments in the application rate should be made for liming materials with different particle sizes, neutralizing values, and depth of incorporation.

3. Adams–Evans Buffer

This procedure for determining the lime requirement is for nonmontmorillonitic, low-organic-matter soils where the amounts of lime needed are small and the possibility of damage from overliming exists. The lime requirement of an acid soil is defined by this procedure as the amount of lime required to change an acid soil condition to a maximum water pH of 6.5 (Adams and Evans, 1962). The Adams–Evans method is described by van Lierop (1990) and Eckert and Sims (1995).

The determination and use of both the soil–water pH and buffer pH are required. Soil–water pH is used as a measure of acid saturation of the soil, designated H-sat$_1$, according to the following:

$$\text{Measured soil pH} = 7.79 - 5.5\ 5(\text{H-sat}_1) + 2.27\ (\text{H-sat}_2)^2$$

where H-saturation is expressed as a fraction of the cation exchange capacity (CEC). Buffer pH is used as a measure of soil acids, designated soil H below, according to the equation:

$$\text{Soil H} = 8\ (8.00 - \text{Buffer pH})$$

for a 10-cm^3 soil sample in 10 mL water plus 10 mL Adams–Evans buffer, where soil-H is in meq/100 g of soil. A pH change of 0.01 units in 20 mL of solution (10 mL water plus 10 mL buffer) is caused by 0.008 meq of acid anywhere between pH 7.0 and 8.0. CEC is calculated by using H-sat and soil H according to the equation:

$$CEC = \text{soil H/H-sat}_1$$

The desired soil pH (not to exceed 6.5) is expressed in terms of acid saturation (designated H-sat_2 below), according to the following:

$$\text{Desired soil pH} = 7.79 - 5.55\ (\text{H-sat}_2) + 2.27\ (\text{H-sat}_2)^2$$

The Adams–Evans buffer method is very reliable for soil with relatively small amounts of exchangeable acidity (maximum = 8 meq/100 g). The procedure provides a fairly high degree of accuracy for estimating lime requirements to reach pH 6.5 or less. Sensitivity for the lime requirement determination is within 500 lb/acre (560 kg/ha). A sensitivity of 0.01 in pH units of the buffer–soil slurry is needed for interpretation of this analysis.

Adams–Evans-based lime requirements in pounds per acre to raise the soil pH to 6.5 are given in the table on page 48.

Adams–Evans Lime Buffer Solution

Weigh 74 g potassium chloride (KCl) in 500 mL of water in a 1000-mL volumetric flask.
Add 10.5 g potassium hydroxide (KOH) and stir to bring into solution.
Add 20 g *p*-nitrophenol ($HO{\cdot}C_6H_4{\cdot}NO_2$) and continue to stir.
Add 15 g boric acid (H_3BO_3). Stir and heat, if necessary, to bring into solution.
Dilute to the mark with water when cool.

Buffer pH Determination

Scoop 10-cm^3 air-dried <10-mesh-sieved (2-mm) soil into a 50-mL cup.
Add 10 mL water and mix for 5 s.
Wait for 10 min and read the soil pH.
Only on soil samples with pH_w less than 6.4, add 10 mL Adams–Evans buffer solution to the cup.
Shake 10 min or stir intermittently for 10 min. Let stand for 30 min.
Read the soil–buffer pH on a standardized pH meter. Stir the soil suspension just prior to reading the pH and read the pH to the nearest 0.01 pH unit.

The lime requirement for low-CEC soils (and with a pH of about 4.5 when H-saturated) can be determined from the following lime requirement table. This table is based on the pH/Adams–Evans buffer values.

Limestone (Ag-Ground, TNP = in 1000 lb/acre) to Raise Soil pH to 6.5 to a Depth of 6⅔ in.

Buffer	Soil–water pH									
pH	6.2	6.0	5.8	5.6	5.4	5.2	5.0	4.8	4.6	4.4
7.95	0	0	0	0	2	2	2	2	2	2
7.90	0	0	0	0	2	2	2	2	2	2
7.85	0	0	2	2	2	2	2	2	2	2
7.80	0	1	2	2	2	2	2	2	2	2
7.75	0	1	2	2	2	2	2	2	2	2
7.70	1	1	2	2	2	2	2	2	2	3
7.65	1	1	2	2	2	2	3	3	3	3
7.60	1	2	2	2	2	3	3	3	3	4
7.55	1	2	2	2	3	3	3	4	4	4
7.50	1	2	3	3	3	3	4	4	4	5
7.45	2	2	3	3	3	4	4	4	5	5
7.40	2	2	3	3	4	4	4	5	5	5
7.35	2	2	3	4	4	5	5	5	5	6
7.30	2	3	4	4	4	5	5	5	6	6
7.25	2	3	4	4	5	5	5	6	6	7
7.20	2	3	4	5	5	6	6	6	7	7
7.15	2	3	4	5	5	6	6	7	7	8
7.10	2	3	5	5	4	7	7	7	8	8
7.05	3	4	5	5	6	7	7	7	8	8
7.00	3	4	5	6	7	7	8	8	8	9

4. Mehlich Buffer

The lime requirement is based first on the determination of the buffer pH acidity (AC) by means of the equation:

$$\text{AC (meq CaCO}_3\text{ per 100 cm}^3\text{ of soil)} = (6.6 - \text{BpH})/0.25$$

where BpH is the buffer pH reading and then the lime requirement (LR) is determined by the equations:

for mineral soils: $LR = 0.1(AC) + AC$
for Histosols: $LR = [-7.4 + 1.6(AC)]1.3$

The Mehlich buffer procedure is described by Mehlich (1976) and van Lierop (1990).

Sodium glycerophosphate

National Formulary (N.F.) mw 315.15.

Note: *The N.F. quality of sodium glycerophosphate [$Na_2C_3H_5(OH)_2PO_4 \cdot 5\ \frac{1}{2}H_2O$] is very satisfactory and considerably more economical than the crystal Beta form. (Source: Roussel Corporation, 155 E. 44th St., New York, NY 10017.)*

Buffer Solution

To about 1500 mL water in a 2000-mL volumetric flask or a 2000-mL calibrated bottle, add 5 mL *glacial* acetic acid (CH_3COOH) and 9 mL triethanolamine or, for ease of delivery, add 18 mL of a 1:1 aqueous mixture.

Add 86 g ammonium chloride (NH_4Cl) and 40 g barium chloride ($BaCl_2 \cdot 2H_2O$) and dissolve.

Dissolve separately 36 g sodium glycerophosphate in 400 mL water and transfer to the above 2000-mL volumetric flask or bottle.

Allow the endothermic reacted solution to reach room temperature and make up to volume with water and mix.

Dilute an aliquot of the buffer solution with an equal volume of water and determine the pH. The pH of the buffer reagent should be 6.6.

However, if it is above pH 6.64, add dropwise *glacial* CH_3COOH. If it is below pH 6.56, add dropwise 1:1 aqueous triethanolamine.

Check the concentration of the buffer by adding 10 mL 0.1 *M* $HCl \cdot AlCl_3$ mixture [dissolve 4.024 g aluminum chloride ($AlCl_3 \cdot 6H_2O$) in 0.05 *M* hydrochloric acid (HCl)] to 10 mL buffer + 10 mL water and determine the pH. The correct pH obtained should be 4.1 ± 0.05.

Buffer pH Determination

Scoop 10 cm^3 air-dried <10-mesh-sieved (2-mm) soil into a 50-mL cup.

To obtain weight per volume, weigh the measured 10 cm^3 soil to the nearest 0.1 g, divide by 10, and express the results in g/cm^3.

Add 10 mL water with sufficient force to mix.

After stirring for about 30 min, read soil pH while stirring (for poorly wettable Histosols, add 8 to 10 drops of ethanol).

Add to the soil from the pH determination 10 mL buffer solution with sufficient force to mix.

Read the buffer pH to the nearest 0.05 unit after 60 min while stirring.

If it is desired to extend buffer capacity below pH 4.0, add an additional 10 mL buffer solution, equilibrate with stirring, and measure pH_B.

Calculation for Exchangeable Acidity. Convert buffer pH (BpH) into buffer pH acidity (AC) as follows:

$$\text{AC (meq 100/cm}^3\text{) soil} = (6.6 - \text{BpH})/0.25 \tag{2.1}$$

(If a second 10-mL portion of buffer was used, multiply AC by 2.)

For unbuffered salt exchangeable acidity (ACe) based on AC of mineral soils, calculate:

$$\text{ACe (meq 100/cm}^3\text{)} = 0.54 + 0.96\text{ (AC)} \tag{2.2}$$

For ACe determination of Histosols and mineral soils having histic epipedon, calculate:

$$\text{ACe (meq 100/cm}^3\text{)} = -7.4 + 1.6\text{ (AC)} \tag{2.3}$$

Calculation for Lime Requirement. Convert BpH into AC. The lime requirement (LR) in the following equations may be expressed and is equivalent to meq $CaCO_3$ 100/cm^3 soil, metric tons (MT) ground limestone TNP = 90%/ha to a depth of 20 cm, or lb/acre (MT × 10^3).

Mineral Soils: For plants with slight to moderate tolerance for ACe and a soil reaction in water of pH 5.8 to 6.5,

$$\text{LR} = 0.1\text{ (AC)}^2 + \text{AC} \tag{2.4}$$

Mineral Soils: For plants with a low tolerance to ACe and a soil reaction in water of <6.5, multiply results with Equation 2.3 by 1.5 or 2.0.

Histosols or Mineral Soils with Histic Epipedon (OM 20% and above): For a soil reaction in water < pH 4.8 to 5.0 and 0.75 g W/V/cm^3, use Equation 2.2 × 1.3, that is,

$$\text{LR} = [-7.4 + 1.6\text{(AC)}]\ 1.3 \tag{2.5}$$

Mineral Soils High in Organic Matter (OM 10 to 19%): For a soil reaction in water < pH 5.3 to 5.5, and W/V within 0.75 to 0.95 g/cm^3, use Equation 2.4 with soils of sandy texture and Equation 2.3 with soils of silt and clay texture.

In all cases, when soil pH is below the indicated optimum, the use of 1 ton limestone/ha or its equivalent is suggested, even though the AC is less than 0.5 meq/100 cm^3.

Interpretation. The lime requirement equations based on the proposed BpH method may be used in a computerized soil testing program. For manual use, the calculated lime requirement values based on Equations 2.3 and 2.4 at 0.1 BpH intervals are recorded in the following table.

Buffer pH, AC, and Lime Requirement Conversion of Mineral and Organic Soils into MT/ha or lbs/acre (MT × 10^3) of Agriculture Ground Limestone with TNP = 90%

		Lime requirement for soils	
		Mineral	Organic
BpH	AC	Equation 2.4[a]	Equation 2.5[a]
6.6	0.0	0.0	0.0
6.5	0.4	0.4[b]	0.0
6.4	0.8	0.9	0.0
6.3	1.2	1.3	0.0
6.2	1.6	1.9	0.0
6.1	2.0	2.4	0.0
6.0	2.4	3.0	0.0
5.9	2.8	3.6	0.0
5.8	3.2	4.2	0.0
5.7	3.6	4.9	0.0
5.6	4.0	5.6	0.0
5.5	4.4	6.3	0.0
5.4	4.8	7.1	0.4[b]
5.3	5.2	7.9	1.2
5.2	5.6	8.7	2.0
5.1	6.0	9.6	2.9
5.0	6.4	10.5	3.7
4.9	6.8	11.4	4.5
4.8	7.2	12.4	5.4
4.7	7.6	13.4	6.2
4.6	8.0	14.4	7.0

(continued)

Buffer pH, AC, and Lime Requirement Conversion of Mineral and Organic Soils into MT/ha or lbs/acre (MT × 10^3) of Agriculture Ground Limestone with TNP = 90% (continued)

		Lime requirement for soils	
		Mineral	Organic
BpH	AC	Equation 2.4[a]	Equation 2.5[a]
4.5	8.4	15.5	7.9
4.4	8.8	16.5	8.7
4.3	9.2	17.7	9.5
4.2	9.6	18.8	10.3
4.1	10.0	20.0	11.2
4.0	10.4	21.2	12.0
3.9	10.8	22.5	12.8

[a] For Equations 2.4 and 2.5, see page 50. For crops with high lime requirements or very low tolerance to ACe, multiply the results of Equation 2.3 by a factor of 1.5 or 2.0.

[b] Using 1 ton limestone/ha or 1000 lbs/acre suggested when lime requirement based on pH is indicated.

Liming recommendations for crops using the Mehlich Buffer may be found in the circular by Tucker et al. (1997).

Interpretation. While liming needs are contingent on BpH, soil pH measured in a 1:1 soil/water ratio on a volume basis has been suggested as a criterion in the lime requirement decision-making process. Soil pH levels measured in 1 *N* potassium chloride (KCl) and 0.01 *M* calcium chloride ($CaCl_2{\cdot}2H_2O$) were found to deviate inconsistently from those measured in water. These deviations were largely related to the quantity and proportion of ACe to ACr, exchangeable Al^{3+} to H^+ and major soil components with respect to organic matter, layer silicates, and sesquioxides hydrates. Schofield and Taylor (1955) introduced the use of 0.01 *M* $CaCl_2{\cdot}2H_2O$ in a 1:2 soil/salt solution ratio on a weight-to-volume basis as a measure of "lime potential." The authors determined pH in the supernatant liquid. Jackson (1958) stirred the soil suspension just before immersing the electrodes, and Peech (1965) placed the glass electrode into the partly settled suspension and the calomel electrode into the clear supernatant solution. In the case of acid Ultisols, the relative decrease in pH from that obtained in a 1:1 soil/water suspension was, on the average, 1.0, 0.8, and 0.6 by the Jackson (1958), Peech (1965), and Schofield–Taylor (1955) procedures, respectively. With neutral to slightly acid soils, the total differences were in general less than one half of the above. In view of the variability of soil pH obtained with varying salts due to procedural differences and soil properties, and because of the importance of maintaining uniformity of soil test results, measurement of pH in a 1:1 soil/water suspension by volume in conjunction with the proposed BpH method for lime requirement is recommended.

5. Titratable and Exchangeable Acidity

The colloidial material in soil (clay and humus) contributes to the cation exchange capacity of a soil, and, therefore, the soil can be assayed for its acidity by titration (Yuan, 1959; Hesse, 1971; Tan, 1998). For determining the hydrogen ion content of a soil, a buffer procedure can be used as suggested by Mehlich (1939; 1953b; 1976), or it can be calculated as given in the Adams–Evans lime buffer procedure (see Section 3 above).

Two methods for the determination of exchangeable acidity are described below.

a. Determination of exchangeable acidity using barium chloride–TEA buffer

Principle of the Method. The method measures the acidity that is exchangeable by the barium chloride ($BaCl_2$)–TEA extractant that is buffered at pH 8.2. Thus, the exchangeable acidity measured comprises exchangeable aluminum (Al) and any hydrogen (H) that will dissociate when the soil is brought to a pH of 8.2 (potential acidity) (Sumner, 1992a; b). It is also a measure of the variable charge developed between the soil pH and pH 8.2. This method was developed by Mehlich (1953b) and is a modification of a previous Mehlich method (1939).

Reagents

Buffer Solution

Adjust 0.5 *N* barium chloride dihydrate ($BaCl_2 \cdot 2H_2O$) (61.07 g/L) and 0.2 *N* triethanolamine [$N(CH_2CH_2OH)_3$] (29.8 g/L) to pH 8.2 with hydrochloric acid (HCl).

Protect from carbon dioxide (CO_2) contamination by attaching a tube containing soda lime to the air intake.

Replacing Solution

Combine 0.5 *N* barium chloride dihydrate (61.07 g $BaCl_2 \cdot 2H_2O$ in 1000 mL water) with 0.4 mL Buffer Solution per 1000 mL.

Protect from carbon dioxide (CO_2) as with the Buffer Solution.

Hydrochloric Acid (HCl)

Approximately 0.2 *N* standard.

Bromocresol Green

0.1% aqueous solution.

Mixed Indicator

Weigh 1.250 g methyl red and 0.825 g methylene blue into 1000 mL 90% ethanol.

Procedure

Scoop 10 cm^3 air-dried <10-mesh-sieved soil into 100-mL beaker (use 5 cm^3 for very acid soils).
Add 25 mL Buffer Solution, mix well, and allow to stand for 1 h.
Transfer mixture to Buchner filtration system and add a further three aliquots (25 mL) Buffer Solution.
Continue with 100 mL Replacing Solution for a total of 200 mL.
Mix 100 mL Buffer Solution with 100 mL Replacing Solution to serve as a blank.
Add 2 drops bromocresol green and 10 drops mixed indicator.
Titrate with HCl to a green to purple end point.
Follow same method for soil filtrates.

Calculation

$$\frac{\text{exchangeable acidity}}{\text{meg/100 g}} = \frac{\text{mL HCl for blank} - \text{mL HC1 for soil filtrate}}{\text{sample, g}}$$

b. Determination of exchangeable acidity and exchangeable aluminum using 1 *N* potassium chloride

Principle of the Method. The acidity measured by the barium chloride ($BaCl_2$)–TEA method bears very little relationship to the acidity to which plant roots react. The 1 *N* potassium chloride (KCl) method extracts the acidity exchangeable at the existing soil pH and consists primarily of Al and some H (Sumner, 1992b). It is termed the *active* acidity in soil and determines to a substantial extent whether or not roots will grow in an acid soil.

Reagents

Replacing Solution (1 N KCl)

Weigh 74.56 g potassium chloride (KCl) in 1000 mL water.

Aluminum Complexing Solution (1 N KF)

Titrate 58.1 g potassium fluoride (KF) per 1000 mL to a phenolphathalein end point (clear to pink color) with 0.1 *N* sodium hydroxide (NaOH).

Hydrochloric Acid (HCl)

Approximately 0.1 *N* standardized.

Sodium Hydroxide (NaOH)

Approximately 0.1 *N* standardized.

Phenolphthalein Solution

Weigh 1 g phenolphthalein into 100 mL ethanol.

Procedure

Scoop 10 cm^3 air-dried <10-mesh-sieved soil into 100-mL beaker.
Add 25 mL 1 *N* KCl, mix, and allow to stand for 30 min.
Transfer mixture to Buchner filtration system and add 5 × 25-mL aliquots of 1 *N* KCl to give a total volume of 150 mL.
Titrate filtrate after adding 4 to 5 drops of phenolphthalein with 0.1 *N* NaOH to the first permanent pink end point. This titer gives exchangeable acidity.
Add 10 mL 1 *N* KF and titrate with 0.1 *N* HCl until pink color disappears.
Wait 30 min and add additional HCl to a clear end point. This titer gives exchangeable Al.

Calculation

$$\text{meq KCl acidity} = \frac{(\text{mL NaOH sample} - \text{mL NaOH blank}) \times \text{N} \times 100}{\text{sample, g}}$$

$$\text{meq KCl exchangeable Al} = \frac{\text{mL HCl} \times \text{N} \times 100}{\text{sample, g}}$$

$$\text{meq H} = \text{KCl exchangeable acidity} - \text{KCl exchangeable Al}$$

6. Interpretation

The concepts of liming are based on more than just a lime requirement value obtained by a particular test procedure, such as those described in this section. Aluminum is also a factor affecting a lime requirement determination (Schwertmann and Jackson, 1964; Kamprath, 1970; Evans and Kamprath, 1970; Wolf, 1982; Sims, 1996). Wolf (1982) suggests that for each ppm of Al (determined by the Wolf-modified Morgan method) above 10 ppm, add 45 lb limestone/acre above the lime requirement. Black (1993a) provides details on factors that would modify the lime requirement based on the obtained laboratory test value; these factors include limestone quality (purity, particle size, reactivity) and biological (crop differences) and soil factors (subsoil acidity). Depth of application and frequency are other considerations. For example, more frequent applications of small limestone doses may be preferable to less frequent larger doses, with the objective to maintain the soil pH within a narrower range of fluctuation by the former technique. The reader may refer to the book edited by Adams (1984) and the articles by van Lierop (1990), Sen Tran and van Lierop (1993), and Mikkelsen and Camberato (1995) for concepts related to liming as well as the PPI *Soil Fertility Manual* (Anonymous, 1996a).

Liming materials vary widely in their chemical and physical properties, which can significantly influence their effectiveness in changing soil pH (Barber, 1984; Black, 1993b). The next sections provide useful information on terms and on the characteristics of liming materials.

Wolf (1982) has prepared lime requirements based on water pH and soil texture as shown in the following table.

Quantity of Ag-Ground Limestone Required (in 1000 lb/acre) to Raise an Acid Soil pH to 6.5 to a Soil Depth of 6⅔ in. Based on Soil Water pH and Soil Texture

	Lime requirement (1000 lb/acre)[a]				
Soil water pH	S	SL	L	SiCl	O[b]
3.5–3.9	4.0	6.5	9.0	12.0	20.0
4.0–4.4	3.0	5.0	7.0	9.0	15.0
4.5–4.9	2.0	3.5	5.5	6.5	10.0
5.0–5.4	1.5	2.5	4.0	5.0	5.0

Quantity of Ag-Ground Limestone Required (in 1000 lb/acre) to Raise an Acid Soil pH to 6.5 to a Soil Depth of 6⅔ in. Based on Soil Water pH and Soil Texture (continued)

	Lime requirement (1000 lb/acre)[a]				
Soil water pH	S	SL	L	SiCl	O[b]
5.5–5.9	1.0	1.5	2.5	3.5	nr[c]
6.0–6.4	0.5	1.0	1.5	2.5	nr

[a] Ag-ground limestone with a TNP of 90 and fineness of 40% <100 mesh, 50% <60 mesh, and 95% <8 mesh. Rate adjustment is required for other liming materials or different mixing depths.

[b] S = sand; SL = sandy and gravely loam; L = loam; SiCl = silt and clay loams; O = organic soils and soilless mixes.

[c] nr = none required.

Adjustment of the above lime requirement must be made based on the amount of extractable Al (for each ppm of Al determined by the Wolf-modified Morgan method, above 10 ppm, add 45 lb limestone/acre above the lime requirement) (Wolf, 1982).

7. Definition of Liming Materials

Acid-forming fertilizer. A fertilizer that is capable of increasing the acidity (lowering the pH) of the soil following application.

Soil Acidification by N Fertilizers

Nitrogen fertilizer	Amount of CaO to compensate the soil acidification induced by 2.2 lb (1 kg) N[a]
Calcium ammonium nitrate (27% N)	1.32 lb (0.6 kg)
Ammonia, urea, and ammonium nitrate	2.2 lb (1 kg)
Diammonium phosphate and ammonium nitrate	4.4 lb (2 kg)
Ammonium sulfate	6.6 lb (3 kg)

[a] On the basis of 50% utilization rate.

Source: IFA World Fertilizer Use Manual, 1992.

Agricultural liming material. Any material that contains Ca and Mg in forms that are capable of reducing soil acidity.

AgLime. Synonymous with "agstolie." Produced by crushing and grinding calcitic or dolomitic limestone to a gradation or fineness that will enable it to neutralize soil acidity. Usually ground to pass sieves in the 8- to 100-mesh range or finer.

Calcite. The crystalline form of calcium carbonate ($CaCO_3$). Pure calcite contains 100% $CaCO_3$ (40% Ca) and 100% $CaCO_3$. Although calcite occurs in nature, limestones of this purity are not commercially available. It may be colorless, but it is usually variously tinted by impurities.

Calcitic limestone. Term widely used by agronomists when referring to agricultural limestone with high Ca content. Mainly contains calcium carbonate ($CaCO_3$), but may also contain small amounts of Mg. Term is not restrictive in definition as is calcite (see above), with which it is frequently confused.

Calcium carbonate ($CaCO_3$). A compound consisting of Ca combined with carbonate. It occurs in nature as limestone, marble, chalk, marl, shells, and similar substances.

Calcium carbonate equivalent (CCE). An expression of the acid-neutralizing capacity of a carbonate rock relative to that for pure calcium carbonate ($CaCO_3$), e.g., calcite. It is expressed as a percentage. For pure calcite, the value is 100%; for pure dolomite, 108.5%. The actual CCE of most limestones will vary from these percentages because of impurities in the rock, and the fact that commercially available limestones are most frequently composed of mixtures of calcite and dolomite rather than either in its pure form.

Calcium oxide (CaO). The chemical compound composed of Ca and oxygen. It is formed from calcium carbonate ($CaCO_3$) by heating limestone to drive off the carbon dioxide. Also known as quick lime, unslaked lime, burnt lime, caustic lime, etc., this compound does not occur in nature.

Calcium oxide equivalent. The percentage of calcium oxide (CaO) in a liming material plus 1.39 times the magnesium oxide (MgO) content. For pure calcite, the value is 56.0%; for pure dolomite, 60.8%. Used by some states as a measure of AgLime quality.

Dolomite. Limestone containing magnesium carbonate ($MgCO_3$) in an amount approximately equivalent to the calcium carbonate ($CaCO_3$) content of the stone. Limestone containing $MgCO_3$ in lesser proportions is referred to as Mg limestone or dolomitic limestone. Pure dolomite is 54.3% $CaCO_3$ and 45.7% $MgCO_3$ or, expressed another way, is composed of 30.4% CaO, 21.8% magnesia (MgO), and 47.8% CO_3.

Dolomitic limestone. Limestone that contains from 10% up to, but less than, 50% dolomite, and from 50 to 90% calcite. The $MgCO_3$ content of dolomitic limestone may range from approximately 4.4 to 22.6%.

Effective calcium carbonate equivalent (ECCE). An expression of AgLime effectiveness based on the combined effect of chemical purity (CCE) and fineness. Required for labeling in some states. Determined by multiplying CCE by a set of factors based on the fineness of grind of the limestone. Also referred to as effective neutralizing value (ENV), total neutralizing power (TNP), effective neutralizing power (ENP), effective neutralizing material (ENM), and, in one state, as the "lime score." Particle sizes are divided into effectiveness ratings based on sieve analysis: (1) smaller than 50 mesh, (2) between 10 and 50 mesh, (3) larger than 10 mesh. By using these size groups, a reliable fineness factor (percent of AgLime available based on fineness) can be determined. The formula is ECCE = % CCE × ½ (% passing 10 mesh + % passing 50 mesh).

Gypsum. A hydrated form of calcium sulfate ($CaSO_4$). It supplies Ca to the soil, but is a neutral substance and does not correct soil acidity; hence, it is not a liming material.

Lime. Chemically, calcium and magnesium oxide. Produced by calcining calcitic or dolomitic limestone, that is, replacing the carbonate (CO_3^{2-}) ion in limestone with oxygen under heat. However, the term is also broadly applied in agriculture to any material containing Ca and Mg in forms capable of correcting soil acidity.

Lime requirement. The quantity of agricultural limestone required to bring an acid soil to neutrality or to some other desired degree of acidity or pH. It is usually stated in pounds per acre of $CaCO_3$ needed to bring the soil to the desired pH under field conditions.

Magnesian limestone. Limestone that contains from 5 to 10% dolomite, and 90 to 95% calcite. The $MgCO_3$ content of magnesian limestone may range from 2.3 to 4.4%.

Magnesium carbonate. A compound consisting of Mg combined with carbonate. It occurs in nature as the mineral magnesite and as a constituent of dolomitic limestones and dolomite.

Magnesium oxide (MgO). The chemical compound composed of Mg and oxygen. It is formed from $MgCO_3$ by heating to drive off carbon dioxide, or in mixture with CaO by heating magnesium limestone or dolomite. Also known as magnesia, it occurs in nature as periclase.

Marble. A compact hard, polishable form of limestone.

Marl. A granular or loosely consolidated, earthly material composed largely of calcium carbonate ($CaCO_3$) as seashell fragments. It contains varying amounts of silt and organic matter.

Neutralizing value. For limestone to be effective as a liming material it must be finely ground before application; the finer the source, the quicker it will react with the soil. Limestone fragments over 5 mm in diameter will persist for so long that they are of no real neutralizing value; particles of lime (ground limestone) fine enough to pass through a 60-mesh sieve (60 wires/60 openings/in. with each opening about 0.21 mm across) will completely react with the soil within 3 years. Most limestone contains a variety of particle sizes that determine its relative effectiveness:

percent material coarser than 4 mesh × 0.0 = % effectiveness
percent material between 4 and 8 mesh × 0.1 = % effectiveness
percent material between 8 and 60 mesh × 0.4 = % effectiveness
percent material finer than 60 mesh × 1.0 = % effectiveness

Total percent effective during first 3 years = % effectiveness

A sample of lime that totally passes through the 4-mesh sieve but that leaves 10% on the 8-mesh sieve, 35% on a 60-mesh sieve yields 70% effectiveness during the first 3 years:

coarser than 4 mesh 0% × 0.0 = 0%
between 4 and 8 mesh 10% × 0.1 = 1%
between 8 and 60 mesh 35% × 0.4 = 14%
finer than 60 mesh 55% × 1.0 = 55%

Total percent effective first 3 years = 70%

Peletized lime. Produced by binding or compressing fine lime particles into large granules or pellets.

Suspension lime. Suspending finely ground AgLime, 100 to 200 mesh, in water, 50:50 lime/water suspension.

8. Acid-Neutralizing Values for AgLime Materials

The CCE of AgLime materials are given in the following table:

AgLime material	CCE, %
Calcium carbonate	100
Calcitic limestone	85–100
Dolomitic limestone	95–108
Marl (Selma chalk)	50–90
Calcium hydroxide (slaked lime)	120–135
Calcium oxide (burnt or quick lime)	150–175

AgLime material	CCE, %
Calcium silicate	86
Basic slag	50–70
Ground oyster shells	90–100
Cement kiln dusts	40–100
Wood ashes	40–50
Power plant ashes	25–50
Gypsum (land plaster)	None
By-products	Variable

Amounts of AgLime materials at different $CaCO_3$ equivalents required to equal 1 ton 100% $CaCO_3$ are listed in the following table:

$CaCO_3$ equivalent of liming material, %	Pounds needed to equal 1 ton of pure $CaCO_3$
60	3333
70	2857
80	2500
90	2222
100	2000
105	1905
110	1818
120	1667

9. Adjusting the AgLime Rate for Different Depths of Incorporation

Most lime requirement recommendations are based on either a 6⅔- or 8-in. depth of incorporation. If the depth of incorporation is less or greater than that specified, then the liming rate must be adjusted. The following table indicates such adjustments:

Depth of incorporation (in.)	Multiplying factor
3	0.43
4	0.57
5	0.71
6	0.86

(continued)

Depth of incorporation (in.)	Multiplying factor
7	1.00
8	1.14
9	1.29
10	1.43
11	1.57
12	171

10. Effect of Fineness on Availability of AgLime

Mesh size will have an effect on the rate of reaction as shown in this table:

	Years after application	
Mesh size	1	4
Coarser than 8	5	15
8–20	20	45
20–50	50	100
50–100	100	100

K. Extractable Phosphorus (P)

1. Introduction

Phosphorus (P) exists in various forms in mineral soils, about equally divided between that in soil organic matter and that in various inorganic forms. The inorganic P forms are primarily mixtures of aluminum (Al-P), iron (Fe-P), and calcium (Ca-P) phosphates; the relative percentage among these three forms is a function of soil pH, with higher percentages of Al-P and Fe-P occurring in acid soils, and a higher percentage as Ca-P in neutral to alkaline soils, as is shown in Figure 2.8 (Anonymous, 1996b). Therefore, the extraction procedure for the measurement of plant-available P is governed to a large degree by soil pH.

The current soil P test level in North American soils for 1997 was published by the Potash & Phosphate Institute (Anonymous, 1998a), which also included a graph of the changing trend in soil P test levels from 1975 to 1996. The levels remained fairly constant during this time period, although

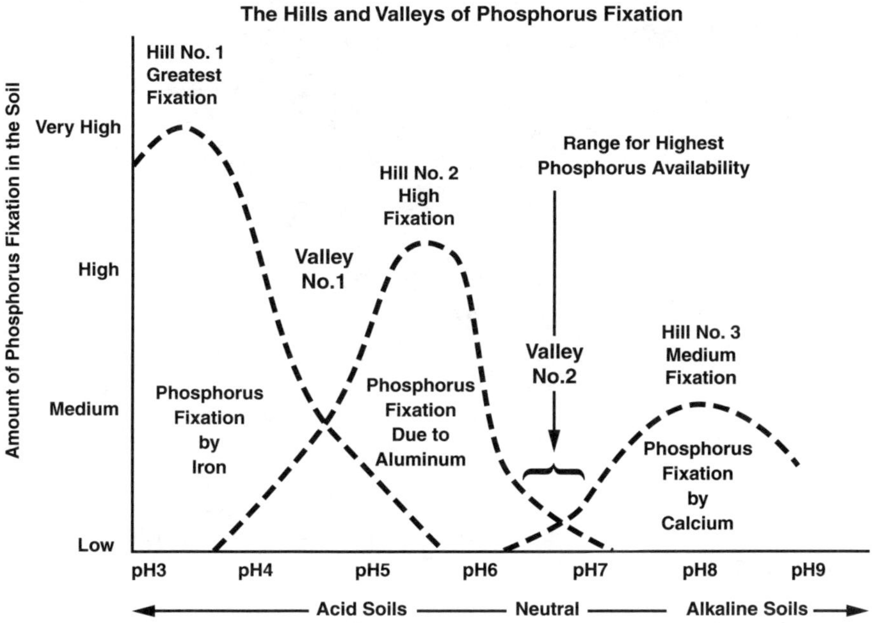

Figure 2.8
Form of P found in soils varies with soil pH. (From *Soil Fertility Manual*, Potash & Phosphate Institute, Norcross, GA.)

the percentage of soils testing "medium or below" varied from a high of 80% to a low of 30% depending on state, as shown in Figure 2.9 (Anonymous, 1998a).

Phosphorus soil and plant chemistry has been discussed by Lindsay and Moreno (1960), Khasawneh et al. (1980), Anonymous (1996b), Jones (1998b), and Moody and Bolland (1999), and reviews on P soil testing procedures have been published by Fixen and Grove (1990), Kuo (1996), Mallarino and Luellen (1998), and Radojevic and Bashkin (1999).

Today, there are eight soil testing procedures in common use for determining soil P; each method was designed for a specific soil situation. Frequently, a P soil test procedure is applied to a soil for which the test was not designed. Also, the interpretation ranges for these various test methods vary considerably, and frequently the levels of extractable P do not correlate well among P soil test methods unless the soil characteristics are similar. For example, Bray P1 and Mehlich No. 1 P values will track each other fairly closely for acid soils, and Mehlich No. 3 and Olsen track each other over a wide range of soil pHs, whereas, when compared with Bray P1, soils with a pH >7.4 do not.

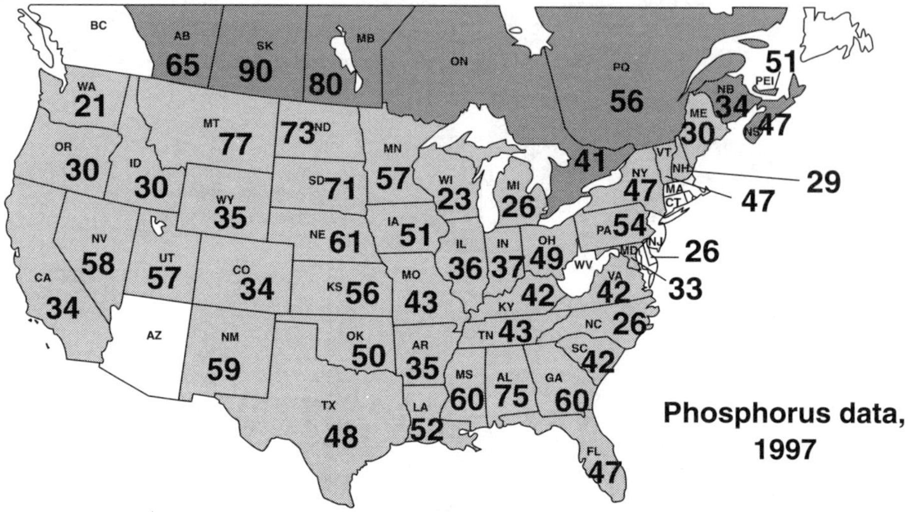

Figure 2.9
Percent of soils testing "medium" or lower in P, 1997. (From Potash & Phosphate Technical Bulletin 1998-3, Norcross, GA.)

Extraction reagents vary in their composition depending on their application for extracting a particular form of P found in a soil as well as other soil properties (mainly pH). Dilute acids solubilize Ca-P, Al-P, and to a lesser degree Fe-P, and F is included to complex Al and prevent readsorption of P by Fe oxides. The chelate EDTA and the $C_2H_3O_2^-$ and SO_4^{2-} anions serve somewhat the same function as the F^- ion, although the two anions are less effective complexers as compared with F. The HCO_3^--based extraction reagents apply particularly to alkaline soils in which the major portion of P exists as Ca-P.

Extraction time is based on time necessary to reach equilibrium. For most acid extraction reagents, equilibrium is reached quickly, and the time of the extraction is based more on ease of mechanical handling. For example, for the Bray P extraction reagent procedures, equilibrium is reached in 60 s for the P1 extraction reagent and in 40 s for the P2 extraction reagent, which were the initial recommended shaking times specified by Bray (Bray and Kurtz, 1945). Long periods of contact between soil and extraction reagent, by overextending the extraction time or delaying separation by either filtration or centrifugation, allow P reabsorption to take place. EDTA and DPTA are added to some P extraction reagents to extend their use to include other elements, mainly the micronutrients Cu, Fe, Mn, and Zn.

The following eight P soil test procedures are listed by name, date when first published, and adapted range of soil properties:

Test method	Date	Adapted range of soil properties
Morgan	1941	Acid soils with CEC of less than 10 meq/100 g
Bray P1	1945	Acid soils (pH_w < 6.8) of moderate texture
Bray P2	1945	Acid soils in which rock phosphate has been the primary P fertilizer source and/or the major portion of P exists in the soil as various forms of calcium phosphate
Mehlich No. 1	1953	Acid (pH_w < 6.5) coastal plain soils of low CEC (<10 meq/100 g) and low organic matter content (<5%)
Olsen	1954	Calcareous, alkaline, or neutral pH soils where soil P is mostly in various forms of calcium phosphate
AB–DTPA	1977	Calcareous, alkaline, or neutral pH soils where soil P is mostly in the various form of calcium phosphate; extractable P is highly correlated with Olsen P
Mehlich No. 3	1984	For a wide range of acid soils with extracted P correlating well with Bray P1 P for acid soils, and with Olsen P for calcareous, alkaline, or neutral pH soils
Calcium chloride	1999	All soils

The composition of the extraction reagent and the soil weight (volume), extractant volume, and shaking time are as follows:

Test method	Extraction reagent	Soil aliquot, g	Extractant volume, mL	Shaking time, min
Morgan	0.7 N $NaC_2H_3O_2$ + 0.54 N CH_3COOH, pH 4.8	5	25	30
Bray P1	0.03 N NH_4F + 0.025 N HCl	2	20	5
Bray P2	0.03 N NH_4F + 0.1 N HCl	2	20	5
Mehlich No. 1	0.05 N HCl + 0.025 N H_2SO_4	5	25	5
Olsen	0.5 N $NaHCO_3$, pH 8.5	2.5	50	30
AB–DTPA	1 M NH_4HCO_3 + 0.005 M DTPA, pH 7.6	10	20	15
Mehlich No. 3	0.2 N CH_3COOH + 0.015 N NH_4F + 0.25 N NH_4NO_3 + 0.013 N HNO_3 + 0.000 M EDTA	2.5 cm^3	25	5
Calcium chloride	0.01 M $CaCl_2 \cdot H_2O$	10	100	120

Eight extraction reagent procedures for P determination are described in this section.

2. Extraction Reagents and Procedures

a. Morgan

Principle of the Method. The method is used primarily for determining P content in acid soils with CECs of less than 20 meq/100 g. This method was initially proposed by Morgan (1932; 1941), then by Lunt et al. (1950), and later by Greweling and Peech (1965). The extracting reagent is well buffered at pH 4.8 and, when used in conjunction with "activated" carbon, yields clear and colorless extracts. The Morgan method was used by several state soil testing laboratories in the northeastern and northwestern United States (Jones, 1973; 1998a; Anonymous, 1995), but it is not in common use today. The method is described by Wolf and Beegle (1995).

Reagent

Extraction Reagent

Weigh 100 g sodium acetate ($NaC_2H_3O_2{\cdot}3H_2O$) in about 900 mL water in a 1000-mL volumetric flask.
Add 30 mL *glacial* acetic acid (CH_3COOH).
Adjust the pH to 4.8 and dilute to volume with water.

Extraction Procedure

Scoop 5 cm^3 air-dried <10-mesh-sieved (2-mm) soil into a 50-mL extraction vessel.
Add 25 mL Extraction Reagent and shake for 5 min on a reciprocating shaker at a minimum of 180 oscillations/min.
Immediately filter and collect the extract for P concentration determination.

Interpretation

Morgan soil P levels less than 3.5 lb P/acre would be considered deficient, and levels greater than 6.5 lb P/acre sufficient. The Wolf (1982) modification of the Morgan extraction reagent includes the chelate DTPA, which results in more P being extracted because of the complexing action of the chelate. Therefore, using the Wolf modification, the interpretative values would be as follows: less than 10 lb P/acre, deficient; 11 to 20 lb P/acre, medium; and 21 to 30 lb P/acre, high. Accurate fertilizer recommendations for P must be founded on known field responses based on local soil–climate–crop conditions (Peck, 1977; Cottenie, 1980; Brown, 1987; Dahnke and Olson, 1990).

b. Bray P1

Principle of Method. The extraction of P by the Bray P1 (Bray and Kurtz, 1945) method is based upon the solubilization effect of the H^+ ions on soil P and the ability of the F^- ion to lower the activity of Al^{3+} and, to a lesser extent, that of Ca^{2+} and Fe^{3+} ions in the extraction system. The extraction time and the solution-to-soil ratio in their procedure were 1 min and 7 mL extraction reagent to 1.0 g soil, respectively. To simplify adaptation to routine laboratory work and to extend the range of soils for which the extraction reagent is suitable, both the extraction time and the solution-to-soil ratio have been altered to 5 min and a 1:10 soil:extraction reagent ratio. This modification is in wide use in laboratories of the mideastern, midsouthern, and north-central areas of the United States (Jones, 1973; 1998a).

Clay soils with a moderately high degree of base saturation or silty clay loam soils that are calcareous or have a very high degree of base saturation will lessen the solubilizing ability of the extractant. Consequently, the method should normally be limited to soils with water pH values less than 6.8 when the texture is silty clay loam or finer. This method is described by Anonymous (1983a; 1994d), Hanlon and Johnson (1984), and Frank et al. (1998). Calcareous soils or high pH, fine-textured soils may be tested by this method, but higher ratios of extractant to soil are often used for such soils (Smith et al., 1957). The Bray P1 method is also suitable for organic soils. Other alternatives are the Olsen P (Olsen et al., 1954) and Mehlich No. 3 (Mehlich, 1984a) procedures. McLean and Mostaghimi (1983) recommend a two-step procedure, particularly for soils with concretions.

Extraction Reagent

Ammonium Fluoride (1 N NH_4F)

Weigh 37 g ammonium fluoride (NH_4F) into a 1000-mL volumetric flask and bring to volume with water.

Store in a polyethylene container and avoid prolonged contact with glass.

Hydrochloric Acid (0.5 N HCl)

Dilute 20.4 mL concentrated hydrochloric acid (HCl) to 500 mL with water.

Extraction Reagent Preparation

Mix 30 mL 1 *N* NH_4F with 50 mL 0.5 *N* HCl in a 1000-mL volumetric flask and dilute to volume with water.

Store in polyethylene.

Note: *This solution is 0.03 N in NH_4F and 0.025 N in HCl, has a pH of 2.6, and is stable for more than 1 year.*

Extraction Procedure

Weigh 2.0 g or scoop 1.70 cm^3 air-dried <10-mesh-sieved (2-mm) soil into a 50-mL extraction vessel.

Add 20 mL Extraction Reagent and shake for 5 min on a reciprocating shaker.

Immediately filter through Whatman No. 2 filter paper, limiting the filtration time to 10 min, and save the extract for P concentration determination.

Interpretation. In general, the extractable P levels may be categorized as follows:

	Extractable P in soil	
Category	kg/ha	lb/acre
Low	<34	<30
Medium	34–68	30–60
High	>68	>60

Beegle and Oravec (1990) compared P levels obtained by the Bray P1 and Mehlich No. 3 extraction reagents and found the critical P levels of 43 and 45 kg P/ha, respectively. Wolf and Baker (1989) compared results among four P soil test methods — Olsen, Bray P1, and Mehlich 1 and 2.

c. Bray P2

Principle of the Method. The extraction of P by the Bray P2 (Bray and Kurtz, 1945) method is based upon the solubilization effect of the H^+ ions on soil P and the ability of the F^- ion to lower the activity of Al^{3+} and, to a lesser extent, that of Ca^{2+} and Fe^{3+} ions in the extraction system. The extraction time and the solution-to-soil ratio in their procedure were 40 s and 7 mL extraction reagent to 1.0 g soil, respectively. To simplify adaptation to routine laboratory work and to extend the range of soils for which the extraction reagent is suitable, both the extraction time and the solution-to-soil ratio have been altered to 5 min and a 1:10 soil/extractant ratio. This modification is in wide use in laboratories of the mideastern, midsouthern, and north-central areas of the United States (Jones, 1973; 1998a).

As that for the Bray P1 procedure, the same criteria of soil characteristics apply. The acid (HCl) concentration was increased from 0.025 *N* in the Bray

P1 extraction reagent to 0.1 *N* to include P that exists in the soil as tricalcium phosphate. At the time this procedure was developed, farmers in the Midwest (Dr. Bray was a Professor at the University of Illinois) were using rock phosphate, which is tricalcium phosphate, as a P fertilizer source. Therefore, P that exists in the soil in this form will be included in the extract by using the P2 extraction reagent.

Extraction Reagent

Ammonium Fluoride (1 N NH_4F)

Weigh 37 g ammonium fluoride (NH_4F) into a 1000-mL volumetric flask and bring to volume with water.

Store in a polyethylene container and avoid prolonged contact with glass.

Hydrochloric Acid (0.5 N HCl)

Dilute 20.4 mL concentrated hydrochloric acid (HCl) to 500 mL with water.

Extraction Reagent Preparation

Mix 30 mL 1 *N* NH_4F with 200 mL 0.5 *N* HCl in a 1000-mL volumetric flask and dilute to volume with water.

Store in polyethylene.

Note: *This solution is 0.03 N in NH_4F and 0.1 N in HCl, has a pH of 2.6, and is stable for more than 1 year.*

Extraction Procedure

Weigh 2.0 g or scoop 1.70 cm^3 air-dried <10-mesh-sieved (2-mm) soil into a 50-mL extraction vessel.

Add 20 mL Extraction Reagent and shake for 5 min on a reciprocating shaker.

Immediately filter through Whatman No. 2 filter paper, limiting the filtration time to 10 min, and save the extract for P concentration determination.

Interpretation. The same interpretative ranges in P soil content would apply for the P2 extraction reagent procedure, but normally both P1 and P2 tests are done and a comparison of test results made. If the P2 result is twice that of P1, no P fertilizer would be recommended.

d. Mehlich No. 1 (North Carolina Double Acid)

Principle of the Method. This method is primarily used to determine P in sandy soils that have exchange capacities of less than 10 meq/100 g, are acid (pH less than 6.5) in reaction, and are relatively low (less than 5%) in organic matter content. The method is not suited for alkaline soils. This method was first published by Mehlich (1953a) and then by Nelson et al. (1953) as the North Carolina Double Acid Method. The method, which has been renamed Mehlich No. 1, is adaptable to the coastal plain soils of the eastern United States. It is currently being used by a number of state soil testing laboratories in the United States (Alabama, Delaware, Florida, Georgia, Maryland, New Jersey, South Carolina, and Virginia) (Anonymous, 1995; Jones, 1973; 1998a). This method is described by Donohue (1988), Anonymous (1983b), and Wolf and Beegle (1995). Wolf and Baker (1989) compared results among four P soil test methods — Olsen, Bray P1, and Mehlich 1 and 2.

Extraction Reagent

0.05 N HCl in 0.025 N H_2SO_4

Pipette 4 mL concentrated hydrochloric acid (HCl) and 0.7 mL concentrated sulfuric acid (H_2SO_4) into a 1000-mL volumetric flask and bring to volume with water.

Extraction Procedure

Weigh 5 g or scoop 4 cm^3 air-dried <10-mesh-sieved (2-mm) soil into a 50-mL extraction vessel.

Add 25 mL Extraction Reagent and shake for 5 min on a reciprocating shaker at a minimum of 180 oscillations/minute.

Immediately filter and collect the extract for P concentration determination.

Interpretation. For most soils and crops, the amount of P extracted is to be interpreted as follows:

	Extractable P in soil	
Category	kg/ha	lb/acre
Very low	<11	<10
Low	11–33	10–30
Medium	34–67	31–60
High	68–112	61–100
Very high	>112	>100

Gascho et al. (1990) found that although there was a significant correlation (0.85***) between Mehlich No. 1 and No. 3 extracted P, the amount of P extracted by the Mehlich No. 1 extraction reagent was about half that extracted by the Mehlich No. 3 extraction reagent.

e. Olsen's sodium bicarbonate

Principle of the Method. The extraction reagent is a 0.5 *M* sodium bicarbonate ($NaHCO_3$) solution at a pH of 8.5, which was first developed and described by Olsen et al. (1954). The original procedure required that 5 g soil be shaken for 30 min in 100 mL extraction reagent containing 1 teaspoon of carbon black (Darco G-60). The use of carbon black eliminated the color in the extract. This procedure was recently modified so that the use of carbon black was eliminated (Watanabe and Olsen, 1965). In the modified method, a single solution reagent, which consists of an acidified solution of ammonium molybdate containing ascorbic acid and a small amount of Sb, is used (Murphy and Riley, 1962; Watanabe and Olsen, 1965). The method is described by Anonymous (1994e) and Frank et al. (1998). The solubility of calcium phosphate in calcareous, alkaline, or neutral soils is increased because of the precipitation of Ca^{2+} as calcium carbonate ($CaCO_3$). In acid soils, P concentration in solution increases when aluminum and iron phosphates, such as variscite and strengite, are present (Lindsay and Moreno, 1960). Secondary precipitation reactions are reduced in acid and calcareous soils because Fe^{3+}, Al^{3+}, and Ca^{2+} ion concentrations remain low in the extract (Olsen and Dean, 1965). Recent studies have shown that precise maintenance of the pH of the extractant at 8.5 is essential to obtain reliable results. Wolf and Baker (1989) compared results among four P soil test methods — Olsen, Bray P1, and Mehlich 1 and 2.

Extraction Reagent

0.5 N $NaHCO_3$

Weigh 42.0 g sodium bicarbonate ($NaHCO_3$) into a 1000-mL volumetric flask and bring to volume with water.

Adjust the pH to 8.5 using either 50% sodium hydroxide (NaOH) or 0.5 *N* hydrochloric acid (HCl).

Add several drops of mineral oil to avoid exposure of the solution to air.

Store in a polyethylene container; check the pH of the solution before use and adjust if necessary.

Note: *Maintenance of the pH at 8.5 is essential.*

Extraction Procedure

Weigh 2.5 g or scoop 2 cm^3 air-dried <10-mesh-sieved (2-mm) soil into a 250-mL extraction vessel.
Add 50 mL Extraction Reagent and shake for 30 min on a reciprocating shaker.
Immediately filter and collect the filtrate for P concentration determination.

Caution: *Soil extraction is sensitive to temperature, changing 0.43 mg P/kg for each degree C for soils containing 5 to 40 mg P/kg.*

Interpretation. It has been shown by several workers (Olsen and Dean, 1965) that a P content of <12 kg P/ha in soil indicates a crop response to P fertilizers, between 12 and 24 kg P/ha indicates a probable response, and >24 kg P/ha indicates a crop response is unlikely. Ludwick (1998) suggests highly P-responsive soils at <18 kg P/ha, probable responsive soils in the range 18–34 kg P/ha, and nonresponsive soils at >34 kg P/ha. However, differences in climatic conditions and crop species may make the general guidelines given above not applicable to all conditions.

f. Ammonium bicarbonate–DTPA

Principle of the Method. The extraction reagent, 1 *M* ammonium bicarbonate (NH_4HCO_3) in 0.005 *M* DTPA adjusted to a pH of 7.6, was first proposed by Soltanpour and Schwab (1977), and later described by Soltanpour and Workman (1979) and Soltanpour (1991). Upon mixing the soil and extraction reagent, the pH rises as a result of the evolution of CO_2. As the pH rises, a fraction of the HCO_3 changes to CO_3. The CO_3^{2-} ions precipitate Ca from labile calcium phosphates, thus dissolving labile P in the 15 min of shaking. This method is highly correlated with the Olsen ($NaHCO_3$) method for P (Olsen et al., 1954). The method is described by Hanlon and Johnson (1984) and Anonymous (1994f).

Extraction Reagent

Ammonium Bicarbonate (AB)–DTPA

Obtain 0.005 *M* DTPA (diethylenetriaminepentaacetic acid) solution by adding 9.85 g DTPA (acid form) to 4500 mL water in a 5000-mL volumetric flask.
Shake for 5 h constantly to dissolve the DTPA.
Bring to 5000 mL with water. This solution is stable with regard to pH.

To 900 mL of the 0.005 *M* DTPA solution, add 79.06 g NH_4HCO_3 gradually and stir gently with a rod to facilitate dissolution and to prevent effervescence when bicarbonate is added. Dilute the solution to 1000 mL with the 0.005 *M* DTPA solution and mix gently with a rod.
Adjust the pH to 7.6 with dropwise 2 *M* hydrochloric acid (HCl) solution addition by slow agitation with a rod. The AB–DTPA solution must be stored under mineral oil.
Check the pH after storage and adjust it with a 2 *M* HCl solution dropwise, if necessary.

Note: *The cumulative volume of HCl added should not exceed 1 mL/L limit, after which a fresh solution should be prepared.*

Extraction Procedure

Weigh 10 g air-dried <10-mesh-sieved (2-mm) soil in a 125-mL conical flask.
Add 20 mL Extracting Reagent, and shake on an Eberbach reciprocal shaker or an equivalent shaker for exactly 15 min at 180 cycles/min with flasks kept open.
Immediately filter the mixture through Whatman 42 filter paper (Soltanpour and Workman, 1979) and collect the filtrate for P concentration determination.

Interpretation. The following table gives an interpretation of the index values for P for the AB–DTPA soil test. These are general guidelines and should be verified under different soil–climate–crop–management combinations.

Index Values for P in Soil

Category	P (mg/kg)
Low	0–4
Marginal	4–7
Adequate	> 7

g. Mehlich No. 3

Principle of the Method. Extraction of P by this procedure was designed to be applicable across a wide range of soil properties ranging in reaction from acid to basic as described by Mehlich (1984a). Phosphorus extracted by this method correlates ($r^2 = 0.966$) well with Bray P1 (Bray and Kurtz, 1954) on acid to neutral soil. It does not correlate with Bray P1 on calcareous soils. The Mehlich No. 3 method correlates ($r^2 = 0.918$) with the

Olsen extraction reagent (Olsen et al., 1954) on calcareous soils, even though the quantity of Mehlich No. 3–extractable P is considerably higher. This procedure was developed on a 1:10 soil/solution ratio (2.5 cm^3 soil + 25 cm^3 extraction reagent) for a 5-min shaking period at 200 4-cm reciprocations/min. The method is described by Hanlon and Johnson (1984), Tucker (1992a), Anonymous (1994g), Wolf and Beegle (1995), and Frank et al. (1998). Wolf and Baker (1989) compared results among four P soil test methods — Olsen, Bray P1, and Mehlich 1 and 2.

The Mehlich No. 3 method is being widely adopted in the United States because of its multielement capability and correlation with other extraction methods (Anonymous, 1995; Jones, 1998a).

Extraction Reagent

0.2 *N* acetic acid (CH_3COOH); 0.25 *N* ammonium nitrate (NH_4NO_3); 0.015 *N* ammonium fluoride (NH_4F); 0.13 *N* nitric acid (HNO_3); 0.001 *M* EDTA

Ammonium Fluoride–EDTA Stock Reagent

Add approximately 600 mL water to a 1000-mL volumetric flask.
Add 138.9 g ammonium fluoride (NH_4F) and dissolve.
Then add 73.05 g EDTA.
Dissolve the mixture and bring to volume with water.
Store in a plastic container.

Final Extraction Reagent Mixture

Add approximately 3000 mL water to a 4000-mL volumetric flask.
Add 80 g ammonium nitrate (NH_4NO_3) and dissolve.
Add 16 mL NH_4F–EDTA stock reagent (above) and mix well.
Add 46 mL *glacial* acetic acid (CH_3COOH) and 3.28 mL concentrated nitric acid (HNO_3).
Then bring to volume with water and mix thoroughly.
Achieve a final pH of 2.5 ± 0.1.
Store in a plastic container.

Extraction Procedure

Scoop 5 cm^3 air-dried <10-mesh-screened (2-mm) soil into an acid-washed 100-mL extraction vessel.
Add 50 mL Extraction Reagent and shake for 5 min on a reciprocating shaker.
Immediately filter and collect the filtrate and save for elemental content determination.
Store in a plastic container.

Note: *For the rationale of using a volume soil measure, refer to Mehlich (1973).*

Interpretation. Critical P levels proposed by Mehlich (1984a) are listed below.

Category	mg P/dm^3	kg P/ha	Expected crop response
Very low	<20	<40	Definite
Low	20–30	40–60	Probable
Medium	31–50	62–100	Less likely
High	>50	>100	Unlikely

Gascho et al. (1990) found that although there was a significant correlation (0.85***) between Mehlich No. 3 and No. 1 extracted P, the amount of P extracted by the Mehlich No. 3 extraction reagent was about twice that extracted by the Mehlich No. 1 extraction reagent.

Beegle and Oravec (1990) compared P levels obtained by the Mehlich No. 3 and Bray P1 extraction reagents and found critical P levels of 45 and 43 kg P/ha, respectively.

Crop fertilization based on the Mehlich No. 3 extraction reagent may be found in the circular by Tucker et al. (1997).

h. 0.01 *M* calcium chloride

Principle of the Method. The reagent has more or less the same ionic strength (0.3 *M*) as the average salt concentration in many soil solutions, and is able to extract P. The electrolyte concentration stays practically constant, the metal concentration reflects the differences in binding strength and/or solubility among various soils, and the measured elements reflect their availability at the pH of the soil since the extractant is an unbuffered solution. The method is described in detail by Houba et al. (1990; 2000).

Extraction Reagent

0.01 M $CaCl_2{\cdot}2H_2O$

Weigh 1.47 g calcium chloride ($CaCl_2{\cdot}2H_2O$) into a 1000-mL volumetric flask and dilute to volume with water.

Comment: *$CaCl_2{\cdot}2H_2O$ may absorb water on standing. The reagent should be standardized by titration with EDTA at pH = 10.0 with Eriochrome Black T as an indicator.*

Extraction Procedure

Weigh 10 g dry soil into a 250-mL polythene bottle.
Add 100 mL Extraction Reagent at 20°C (68°F) and shake mechanically for at least 2 h at room temperature (20°C; 68°F).
Either filter to collect the extract or collect the supernatant after centrifugation, and save for P concentration determination.

Interpretation. Comparison of results obtained by this method vs. other procedures has been made by Houba et al. (1986), and Simonis and Setatou (1996) using a group of northern Greece soils found that 0.01 *M* $CaCl_2$–extractable P was highly correlated with Olsen P (0.85***) and P uptake by ryegrass (0.75***).

3. Methods of Phosphorus Determination

The commonly used methods for P determination in soil extracts are by either ultraviolet–visible (UV-VIS) spectrophotometry or by plasma emission spectrometry (ICP-AES). Details on these methods of analysis are given in Chapter 5.

a. UV-VIS spectrophotometry

Three UV-VIS spectrophotometric procedures, vanadomolybdophosphoric acid (Kuo, 1996), chlorostannous acid or ascorbic acid molybdenum-blue (Rodriquez et al., 1994; Kuo, 1996), and malachite green (Novozamsky et al., 1993), can be used to determine P concentration in an obtained extractant. The ascorbic acid molybdenum-blue method is described below.

Reagents

Ascorbic Acid Solution

Weigh 176.0 g ascorbic acid into a 2000-mL volumetric flask and bring to volume with water.
Store in a dark glass bottle in a refrigerated compartment.

Sulfuric–Molybdate Solution

Weigh 100 g ammonium molybdate $[(NH_4)Mo_7O_{24}{\cdot}4H_2O]$ into 500 mL water in a 2000-mL volumetric flask and dissolve.
Weigh in 2.425 g antimony potassium tartrate $[K(SbO)C_4H_4O_6{\cdot}\frac{1}{2}H_2O]$ and stir to dissolve.

Add slowly 1400 mL concentrated sulfuric acid (H_2SO_4) and mix well.
Let it cool and bring to volume with water.
Store in a polyethylene or Pyrex bottle in a dark, refrigerated compartment.

Working Solution

Pipette 10 mL Ascorbic Acid Solution and 20 mL Sulfuric–Molybdate Solution into a 1000-mL volumetric flask and bring to volume with water.
Allow it to stand at least 1 h before using. The solution is stable for 2 to 3 days.

Phosphorus Standard (1000 mg P/L)

Use commercially prepared standard, or weigh 3.85 g ammonium dihydrogen phosphate ($NH_4H_2PO_4$) into a 1000-mL volumetric flask and bring to volume with Extraction Reagent.
Prepare working standards containing 1, 2, 5, 10, 15, and 20 mg P/L by diluting aliquots of the 1000 mg P/L standard with Extracting Reagent (Jones, 1990).

Color Development

Pipette 2 mL extractant into a spectrophotometer cuvet.
Add 23 mL Working Solution, mix well, and let it stand for 20 min.
Read the absorbance at 880 nm with a UV-VIS spectrophotometer zeroed against a blank consisting of the Extraction Reagent.

Caution: *Some variation may be required depending on the characteristics of the extraction reagent.*

b. Plasma emission spectrometry

Depending on the spectrometer design, several different spectral lines can be used to determine P in the aspirated filtrate. The most commonly used spectral line is 214.9 nm, which can be read also in the second order. Some modification of the filtrate may be required for delivery through the nebulizer system (see page 268).

4. Conversion Factors

Phosphorus results can be expressed in more than one set of units, and below are the those units; the initial number was selected for illustrative purposes only.

lb/acre	ppm (mg/kg)	kg/ha	mol_c/100 g
30	15	34	0.048

5. Interpretation

The critical values that distinquish deficiency from sufficiency and sufficiency from excess are listed below. The values are somewhat general and vary within a narrow range depending on soil type and crop; therefore, these values should be used appropriately.

Test method	Critical value, lb/acre Deficient	Excess
Morgan	<3.5	>6.5
Bray P1	<30	>60
Bray P2	<30	>60[a]
Mehlich No. 1	<30	>100
Olsen	<11	>22
AB–DTPA	<15	>30
Mehlich No. 3	<36	>90
0.01 *M* $CaCl_2 \cdot H_2O$		not known

[a] Frequently both Bray P1 and P2 tests are performed and the test results compared.

Interpretations may vary somewhat, depending on soil characteristics and different crops. Interpretative P data are given in Thomas and Peaslee (1973) and Fixen and Gove (1990).

6. Fertilizer Recommendations

Reviews on the use of P fertilizers can be found in Black (1993b), Withers and Sharpley (1995), Bolland and Gilkes (1998), and P fertilizer recommendations for some crops in the books by Halliday and Trenkel (1992), Maynard and Hochmuth (1997), Reid (1998b), and Ludwick (1998).

Accurate fertilizer recommendations for P must be based on field response data conducted under local soil–climate–crop conditions (Peck, 1977; Cottenie, 1980; Jones, 1985b; Brown, 1987; Dahnke and Olson, 1990; Tucker et al., 1997; Helyar and Price, 1999).

7. Effect of Fertilizer Phosphorus on Phosphorus Soil Test Level

In general, an increase of 1 ppm in soil P occurs when 12 to 28 lb of P_2O_5 per acre are applied above crop removal.

L. Major Cations (K, Ca, Mg, and Na)

1. Introduction

Potassium exists in the structure of soil minerals, such as mica, as nonexchangeable K that is trapped in the lattice of 2:1 clays, as an exchangeable cation (K^+) held on soil colloids, and as a K^+ ion in the soil solution. Calcium exists in several moderately insoluble chemical forms, such as calcium carbonate, calcium sulfate, and calcium phosphate, as an exchangeable cation (Ca^{2+}) held on soil colloids, and as a Ca^{2+} ion in the soil solution. Magnesium exists in the structure of soil minerals, such as biotite, serpentine, hornblende, and olivine, as an exchangeable cation (Mg^{2+}) held on soil colloids, and as Mg^{2+} ions in the soil solution. Sodium exists as an exchangeable cation (Na^+) held on soil colloids and as a Na^+ ion in the soil solution.

K, Ca, Mg, and Na are extracted from the soil by the use of an "exchange" cation, such as NH_4^+, H^+, or Na^+, in the extraction reagent of sufficient strength to remove most, if not all, of the cations on the collodial complex. One of the following is used: salt solutions, such as neutral normal ammonium acetate (Schollenberger and Simon, 1954), which use the NH_4^+ cation as the exchange cation; or the Morgan extractant (Morgan, 1932; 1941; Lunt et al., 1950) with its Wolf (1982) modification, which uses the Na^+ cation as the exchange cation; or the three Mehlich extractants (Mehlich, 1953a; 1978; 1984a; Nelson et al., 1953), which use the H^+ cation for Mehlich No. 1 and the NH_4^+ cation for Mehlich No. 2 and 3 as the exchange cation. For alkaline soils, water extraction (Yuan, 1974) is recommended, or the AB–DTPA extraction reagent (Soltanpour and Schwab, 1977), with the NH_4^+ cation the exchange cation, but this extraction reagent is suitable only for determining K and Na as Ca and Mg are partially precipitated by the HCO_3^- anion. The Mehlich No. 1 (Mehlich, 1953a; Nelson et al., 1953) extractant would have the same limitations for use as those specified for its use in determining extractable P.

There is an excellent review article on the exchangeable cations and their determination (Doll and Lucas, 1973) as well as specific procedures for exchangeable K (Carson, 1980a) and exchangeable Ca, K, Mg, and Na using neutral normal ammonium acetate (Warncke and Brown, 1998), Mehlich No.1 (Wolf and Beegle, 1995) and No. 3 (Tucker, 1992b; Gavlak et al., 1994; Wolf and Beegle, 1995), and Morgan (Wolf and Beegle, 1995). A two-step extraction procedure using neutral normal ammonium acetate has been suggested by McLean et al. (1982) for improving the interpretation of this extractant procedure for K.

Water (Bower and Wilcox, 1965; Hesse, 1971; Chapman and Pratt, 1982) and weak salt solutions, such as calcium chloride (0.01 *M* $CaCl_2$) (Baier and Baierova, 1998; Houba et al., 1990; 2000), extract those ions in the soil solution that are in equilibrium with that on the exchange complex.

The following lists commonly used soil test procedures for the exchangeable cations by name, date when first published, and adapted range of soil properties:

Test method	Date	Adapted range of soil properties
Morgan	1941	Acid soils with CEC of less than 10 meq/100 g
Ammonium acetate	1945	Acid to slightly alkaline soils
Mehlich No. 1	1953	Acid (pH_w < 6.5) coastal plain sandy soils of low CEC (<10 meq/100 g) and low (<5%) organic matter content
Water	1965	Primarily for alkaline soils
AB–DTPA	1977	Calcareous, alkaline, or neutral pH soils
Mehlich No. 3	1984	For a wide range of acid soils
0.01 *M* $CaCl_2$	2000	All soils

The composition of the extraction reagents and the soil weight (volume), extraction reagent volume, and shaking time are as follows:

Test method	Extraction reagent	Soil aliquot, g	Extraction reagent volume, mL	Shaking time, min
Morgan	0.7 *N* $NaC_2H_3O_2$ + 0.54 *N* CH_3COOH, pH 4.8	5	25	30
Ammonium acetate	1 *N* $NH_4C_2H_3O_2$, pH 7.0	5	25	5
Water	Water	5	25	15
Mehlich No. 1	0.05 *N* HCl + 0.025 *N* H_2SO_4	5	25	5
AB–DTPA	1 *M* NH_4HCO_3 + 0.005 *M* DTPA, pH 7.6	10	20	15

Test method	Extraction reagent	Soil aliquot, g	Extraction reagent volume, mL	Shaking time, min
Mehlich No. 3	0.2 *N* CH_3COOH + 0.015 *N* NH_4F + 0.25 *N* NH_4NO_3 + 0.013 *N* HNO_3 + 0.000 *M* EDTA	2.5 cm^3	25	5
Calcium chloride	0.01 *M* $CaCl_2{\cdot}H_2O$	10	100	120

Seven extracting reagents and procedures for the major cations Ca, K, Mg, and Na, frequently referred to as the exchangeable cations, are described.

2. Extraction Reagents and Procedures

a. Neutral normal ammonium acetate (K, Ca, Mg, and Na)

Principle of the Method. This method uses a neutral salt solution to replace the cations present on the soil exchange complex; therefore, the cation concentrations determined by this method are referred to as "exchangeable" for noncalcareous soils. For calcareous soils, the cations are referred to as "exchangeable plus soluble."

The use of neutral normal ammonium acetate ($NH_4C_2H_3O_2$) to determine exchangeable K was first described by Prianischnikov (1913). Schollenberger and Simon (1945) describe the advantages of this extracting reagent regarding its effectiveness in wetting soil, replacing exchangeable cations, ease of volatility during analysis, and suitability for use with flame emission spectrophotometry. More recently, this method has been described by Jackson (1958), Chapman and Pratt (1982), Hanlon and Johnson (1984), Haby et al. (1990), Bates and Richards (1993), Simard (1993), Anonymous (1994h), Helmeke and Sparks (1996), and Warncke and Brown (1998). The 1 *N* $NH_4C_2H_3O_2$, pH 7.0, extraction reagent is the most commonly used extraction reagent for determining K, Mg, Ca, and Na in soil testing laboratories in the United States (Jones, 1973; 1998a). Beegle and Oravec (1990) found that K determined by this extraction reagent is highly correlated (0.96***) with Mehlich No. 3–extractable K; the Cate–Nelson (1971) critical values for K are 0.24 and 0.20 cmol K/kg, respectively. Simonis and Setatou (1996), using northern Greece soils, obtained a high correlation (0.75***) between 1 *N* $NH_4C_2H_3O_2$, pH 7.0–extractable K and that extracted with 0.01 *M* $CaCl_2$.

Extraction Reagent

1 N $NH_4C_2H_3O_2$, pH 7.0

Dilute 57 mL *glacial* acetic acid (CH_3COOH) with water to a volume of approximately 500 mL.
Then add 69 mL concentrated ammonium hydroxide (NH_4OH).

Caution: *Use a fumehood.*

Add sufficient water to obtain a volume of 990 mL.
After thoroughly mixing the solution, adjust the pH to 7.0 using either NH_4OH or CH_3COOH.
Dilute to a final volume of 1000 mL with water.

Alternate Method

Weigh 77.1 g ammonium acetate ($NH_4C_2H_3O_2$) in about 900 mL water in a 1000-mL volumetric flask.
After thoroughly mixing the solution, adjust the pH to 7.0 using either 3 *N* acetic acid (CH_3COOH) or 3 *N* ammonium hydroxide (NH_4OH).
Bring to volume with water.

Extraction Procedure

Weigh 5 g or scoop 4.25 cm^3 air-dried <10-mesh-sieved (2-mm) soil into a 50-mL extraction vessel.
Add 25 mL Extraction Reagent and shake for 5 min on a reciprocating shaker.
Immediately filter and collect the filtrate for elemental determination.

b. Mehlich No. 1 (North Carolina Double Acid) (K, Ca, Mg, and Na)

Principle of the Method. This method is primarily used to determine K, Ca, Mg, and Na in soils that have exchange capacities of less than 10 meq/100 g, are acid (pH less than 6.5) in reaction, and are relatively low (less than 5%) in organic matter content. The use of Mehlich No. 1 as an extraction reagent was first described by Mehlich (1953a) and then published specifically as a P extraction reagent by Nelson et al. (1953) as the North Carolina Double Acid (now Mehlich No. 1) method. It is adaptable to the coastal plain sandy soils of the eastern United States. Determined values for K, Ca, and Mg using this extraction reagent were found to compare on a one-to-one basis with values obtained by the Mehlich No. 3 (Mehlich, 1984a) extraction reagent (Gascho et al., 1990). This extraction reagent is currently being used by a number of state soil testing laboratories in the United States (Alabama, Delaware, Florida, Georgia, Maryland, New Jersey, South Carolina, and Virginia) (Jones, 1973; 1998a; Anonymous, 1995). The method is not suited for alkaline soils. The method is described by Wolf and Beegle (1995).

Extraction Reagent

0.05 N HCl in 0.025 N H_2SO_4

Pipette 4 mL concentrated hydrochloric acid (HCl) and 0.7 mL concentrated sulfuric acid (H_2SO_4) into a 1000-mL volumetric flask and bring to volume with water.

Extraction Procedure

Weigh 5 g or scoop 5 cm^3 air-dried <10-mesh-sieved (2-mm) soil into a 50-mL extraction vessel.

Add 25 mL Extraction Reagent and shake for 5 min on a reciprocating shaker.

Immediately filter and collect the filtrate for elemental determination.

c. Mehlich No. 3 (K, Ca, Mg, and Na)

Principle of the Method. The extraction of K, Ca, Mg, and Na by this method is designed to be applicable across a wide range of soil properties ranging in reaction from acid to basic (Tucker, 1992b; Sen Tram and Simard, 1993; Warncke and Brown, 1998).

The Mehlich No. 3 method correlates well with Mehlich No. 1, Mehlich No. 2, and neutral normal ammonium acetate procedures (Hanlon and Johnson, 1984; Mehlich, 1984a; Sims, 1989; Schmisek et al., 1998). For specific extraction values and correlation coefficients, see Mehlich (1978; 1984a). The method is described by Tucker (1992b), Anonymous (1994g), and Wolf and Beegle (1995). The Mehlich No. 3 method is being widely adopted in the United States because of its multielement capability and correlation with other extraction methods (Sims, 1989; Wolf and Baker, 1989; Jones, 1998a).

Beegle and Oravec (1990) found that K determined by this extraction reagent is highly correlated (0.96***) with 1 *N* $NH_4C_2H_3O_2$, pH 7.0–extractable K, with the Cate–Nelson (1971) critical values for K 0.20 and 0.24 cmol K/kg, respectively. Determined values for K, Ca, and Mg using this extraction reagent were found to compare on a one-to-one basis with values obtained by the Mehlich No.1 extraction reagent (Gascho et al., 1990).

Crop fertilization based on the Mehlich No. 3 extraction reagent may be found in the circular by Tucker et al. (1997).

Extraction Reagent

0.2 *N* acetic acid (CH_3COOH); 0.25 *N* ammonium nitrate (NH_4NO_3); 0.015 *N* ammonium fluoride (NH_4F); 0.13 *N* nitric acid (HNO_3); 0.001 *M* EDTA

Ammonium Fluoride–EDTA Stock Reagent

Add approximately 600 mL water to a 1000-mL volumetric flask.
Add 138.9 g ammonium fluoride (NH_4F) and dissolve.
Then add 73.05 g EDTA.
Dissolve the mixture and bring to volume with water.
Store in a plastic container.

Final Extraction Reagent Mixture

Add approximately 3000 mL water to a 4000-mL volumetric flask.
Add 80 g ammonium nitrate (NH_4NO_3) and dissolve.
Add 16 mL NH_4F–EDTA stock reagent (above) and mix well.
Add 46 mL *glacial* acetic acid (CH_3COOH) and 3.28 mL concentrated nitric acid (HNO_3).
Then bring to volume with water and mix thoroughly.
Achieve a final pH of 2.5 ± 0.1.
Store in a plastic container.

Extraction Procedure

Scoop 5 cm^3 air-dried <10-mesh-screened (2-mm) soil into an acid-washed 100-mL extraction vessel.
Add 50 mL Extraction Reagent and shake for 5 min on a reciprocating shaker.
Immediately filter and collect the filtrate and save for elemental content determination.
Store in a plastic container.

Note: *For the rationale of using a volume soil measure, refer to Mehlich (1973).*

d. Morgan (K, Ca, and Mg)

Principle of the Method. This method is used primarily for determining K, Ca, and Mg in acid soils with cation exchange capacities of less than 20 meq/100 g. This method was initially proposed by Morgan (1932; 1941), and then by Lunt et al. (1950) and later by Greweling and Peech (1965). The Morgan method has been used by a number of soil testing laboratories in the northeastern and northwestern United States (Nelson et al., 1953; Jones, 1973; 1998a; Anonymous, 1995). The extraction reagent is well buffered at pH 4.8 and, when used in conjunction with *activated* carbon, yields clear and colorless extracts. The concentration of sodium acetate ($NaC_2H_3O_2{\cdot}3H_2O$) is sufficiently high to effect replacement of about 80% of the exchangeable cations. The method has been modified by Wolf (1982) to include additional elements. The method is described by Wolf and Beegle (1995).

Extraction Reagent

Weigh 100 g sodium acetate ($NaC_2H_3O_2{\cdot}3H_2O$) into a 1000-mL volumetric flask and add about 900 mL water.
Add 30 mL *glacial* acetic acid (CH_3COOH).
Adjust the pH to 4.8 and bring to volume with water.

Extraction Procedure

Scoop 5 cm^3 air-dried <10-mesh-sieved (2-mm) soil into a 50-mL extraction vessel.
Add 25 mL Extraction Reagent and shake for 5 min on a reciprocating shaker at a minimum of 180 oscillations/min.
Immediately filter and collect the filtrate for elemental determination.

Interpretation. The following table gives an interpretation of the index values for the Morgan soil test. These are general guidelines and should be verified under different soil–climate–crop–management combinations.

Index Values for K and Mg in Soil

	mg/kg in soil	
Category	K	Mg
Low	0–50	0–12
Marginal	51–100	13–40
Adequate	>100	>40

The optimum range in soil test values given by Wolf (1982) for K and Mg are 76 to 125 and 13 to 50 mg/kg, respectively. Additional Mg is recommended when the ratio between extractable Ca and Mg exceeds 20:1, particularly for Mg-sensitive crops.

e. Ammonium bicarbonate–DTPA (K)

Principle of the Method. The extraction reagent is 1 *M* ammonium bicarbonate (NH_4HCO_3) in 0.005 *M* DTPA adjusted to a pH of 7.6 (Soltanpour and Schwab, 1977; Soltanpour and Workman, 1979; Soltanpour, 1991). The NH_4^+ ion will exchange with the K^+ ion, and the colloidal complex will bring it into solution. This method is highly correlated with the 1 *N* ammonium acetate ($NH_4C_2H_3O_2$), pH 7.0 method for K. The method is described by Anonymous (1994f).

Extraction Reagent

NH_4HCO_3–DTPA

Obtain 0.005 *M* DTPA solution by adding 9.85 g DTPA (diethylenetriaminepentaacetic acid) (acid form) to 4500 mL water in a 5000-mL volumetric flask.
Shake for 5 h constantly to dissolve the DTPA.
Bring to 5000 mL with water. This solution is stable with regard to pH.
To 900 mL of the 0.005 *M* DTPA solution, add 79.06 g ammonium bicarbonate (NH_4HCO_3) gradually and stir gently with a rod to facilitate dissolution and to prevent effervescence when bicarbonate is added.
Dilute the solution to 1000 mL with the 0.005 *M* DTPA solution and mix gently with a rod.
Adjust the pH to 7.6 with 2 *M* hydrochloric acid (HCl) solution added dropwise and with slow agitation with a rod.

Note: *The AB–DTPA Extraction Reagent solution must be stored under mineral oil.*

Check the pH after storage and adjust it with a 2 *M* HCl solution dropwise, if necessary. Assure that the cumulative volume of HCl added does not exceed 1-mL/L limit, after which a fresh solution should be prepared.

Extraction Procedure

Weigh 10 g air-dried <10-mesh-sieved (2-mm) soil into a 125-mL conical flask.
Add 20 mL Extraction Reagent and shake on an Eberbach reciprocal shaker or an equivalent shaker for exactly 15 min at 180 cycles/minute with flasks kept open.

Immediately filter the extracts through Whatman 42 filter paper (Soltanpour and Workman, 1979) and save for elemental content determination.

Interpretation. The following table provides an interpretation of the index values for K for the AB–DTPA soil test. These are general guidelines and should be verified under different soil–climate–crop–management combinations.

Index Values for K in Soil

Category	K, mg/kg in soil
Low	0–60
Marginal	61–120
Adequate	>120

f. Water (K, Ca, Mg, and Na)

Principle of the Method. This method uses water to extract K, Ca, Mg, and Na from soil. A soil/water ratio of 1:5 (v:v) is the one adapted for routine analysis (Bower and Wilcox, 1965; Hesse, 1971; Chapman and Pratt, 1982). This method is relatively simple and can serve for quick, routine scanning. However, it suffers the disadvantage of yielding unrealistic figures for Ca and Na as a result of cation exchange equilibrium shifts (Chapman and Pratt, 1982).

Extraction Reagent

Pure water

Extraction Procedure

Weigh 5 g or scoop 4.25 cm^3 air-dried <10-mesh-sieved (2-mm) soil into a 50-mL extraction vessel, add 25 mL water, seal the bottle with a stopper, and shake for 30 min on a reciprocating shaker.

Allow to stand for 15 min to let the bulk of the soil settle.

Filter the supernatant liquid and discard the initial filtrate if it is turbid, and save for elemental content determination.

g. 0.01 *M* calcium chloride (K, Mg, and Na)

Principle of the Method. The reagent has more or less the same ionic strength (0.3 *M*) as the average salt concentration in many soil solutions, and is able to extract adsorbed cations. The electrolyte concentration stays practically constant, that metal concentration reflects the differences in binding strength and/or solubility among various soils, and the measured elements reflect their availability at the pH of the soil since the extractant is an unbuffered solution. Loch et al. (1998) found a high correlation between Mg extracted by this method and by other procedures commonly used in Europe, concluding that this extraction reagent could be successfully used to evaluate the Mg status of soils. This extraction reagent method is described in detail by Houba et al. (1990; 2000).

Extraction Reagent

0.01 M $CaCl_2 \cdot 2H_2O$

Weigh 1.47 g calcium chloride ($CaCl_2 \cdot 2H_2O$) into a 1000-mL volumetric flask and dilute to volume with water.

Comment: *$CaCl_2 \cdot 2H_2O$ may absorb water on standing. The reagent should be standardized by titration with EDTA at pH = 10.0 with Eriochrome Black T as an indicator.*

Extraction Procedure

Weigh 10 g dry soil into a 250-mL polythene bottle.
Add 100 mL 0.01 *M* $CaCl_2$ solution at 20°C (68°F) and shake mechanically for at least 2 h at room temperature (20°C; 68°F).
Either filter to collect the extract or collect the supernatant after centrifugation and save for elemental content determination.

Interpretation. Simonis and Setatou (1996) using northern Greece soils found that there was a significant correlation (0.75***) between 0.01 *M* $CaCl_2$–extractable K and neutral normal ammonium acetate–extractable K and a significant correlation (0.70***) between 0.01 *M* $CaCl_2$–extractable K and K uptake by ryegrass.

3. Methods of Cation Determination

Some form of spectrometry is the instrumental procedure for determining cation concentration in prepared extracts. Flame emission spectrophotometry is probably the best instrumental procedure for the determination of K and Na, whereas Ca and Mg are best determined by atomic absorption spectrophotometry. Unfortunately, both methods are single-element determination procedures, and they have a limited concentration range of no more than 2 decades, therefore requiring dilution for extracts containing high cation concentration levels. Calcium and Mg can be determined by flame emission spectrophotometry if interfering ions are removed or compensated for (known as a matrix effect), whereas K and Na can be determined in the absorption mode by atomic absorption spectrophotometry, although analytical performance may not be optimal.

Today, in most soil testing laboratories, the analytical procedure of choice is plasma emission spectrometry as all four cations can be determined simultaneously with minimum interference from matrix composition and without dilution for extracts having high cation concentrations (a 3- to 4-decade range in concentration is determinable). Details on these analytical methods of determination are given in Chapter 5.

Jones (1990) has suggested that working standards for Ca, K, Mg, and Na be 0 to 1000 mg/L, 0 to 100 mg/L, 0 to 500 mg/L, and 0 to 100 mg/L, respectively, prepared in the extraction reagent.

4. Methods of Expression

Cation results can be expressed in more than one set of units, and below are those units; the initial number was selected for illustrative purposes only.

Major cation	lb/acre	ppm (mg/kg)	kg/ha	mol_c/100 g (meq/100 g)
Calcium (Ca^{2+})	4000	2000	2240	10.0
Magnesium (Mg^{2+})	100	50	56	0.42
Potassium (K^+)	400	200	224	0.51
Sodium (Na^+)	100	50	56	0.11

Conversion to milliequivalents:

lb Ca/acre divided by 400 = meq Ca/100 g
lb Mg/acre divided by 240 = meq Mg/100 g
lb Na/acre divided by 460 = meq Na/100 g
lb K/acre divided by 780 = meq K/100 g

5. Interpretation

Evaluation of the analytical results for determination of fertilizer recommendations, particularly for the elements K and Mg, must be based on field response data conducted under local soil–climate–crop conditions (Peck, 1977; Cottenie, 1980; Brown, 1987; Dahnke and Olson, 1990; Haby et al., 1990; Black, 1993a; Mikkelsen and Camberato, 1995; Maynard and Hochmuth, 1997; Ludwick, 1998; Reid, 1998b). Mehlich proposed critical levels of K and Mg for the

Mehlich No. 3 extractant (Mehlich, 1984a) as well as interpretative guidelines for evaluating percentage of Ca and base saturation (Mehlich, 1978). The concept of ionic balance among the cations is discussed by Geraldson (1970), Baker (1973), Lindsay (1979), McLean (1982a), and Barber (1995).

The basic concepts of cation soil chemistry are given in the Potash & Phosphate Institute *Soil Fertility Manual* (Anonymous, 1996c), plant chemistry concepts by Jones (1998b), and soil testing procedures and their interpretation have been described by Haby et al. (1990), Helmke and Sparks (1996), and Suarez (1996). The major cations are discussed in a recent book (Peverill et al., 1999) in which the chapters on K, Ca, and Mg are written by Gourley (1999), Bruce (1999), and Aitken and Scott (1999), respectively.

Calcium (Ca)

It is generally assumed that if the soil pH is maintained within the proper range, there will be sufficient Ca to meet most crop requirements for this essential element. Some have based soil Ca sufficiency on percent base saturation, with the so-called ideal soil having 70 to 90% of the saturation percentage as Ca, although for highly weathered soils, percent base saturation Ca levels as low as 40% may be sufficient. Basing liming and additions of Ca to the soil on establishing or maintaining a desired percent soil saturation level is questionable (McLean, 1977; Dahnke and Olson, 1990). In absolute levels, soils in the silt loam to clay loam textural range are considered sufficient in Ca with exchangeable Ca levels between 2000 and 4000 lb/acre (2240 and 4480 kg/ha), sandy soils 200+ lb/acre (224+ kg/ha), and low (6 to 10 meq/100 g) CEC soils 400+ lb/acre (448+ kg/ha); crop requirement considerations and fertilizer treatments must also be factored in.

The basic concepts of Ca soil chemistry are given in the Potash & Phosphate Institute *Soil Fertility Manual* (Anonymous, 1996a), plant chemistry concepts by Jones (1998b), and soil testing procedures and their interpretation have been described by Haby et al. (1990), Helmke and Sparks (1996), Suarez (1996), and Bruce (1999).

Magnesium (Mg)

Magnesium soil chemistry is quite complex and generalizations are difficult to make. The Mg cation is the least competitive for plant uptake compared with either Ca or K, and the presence of the NH_4^+ cation can significantly reduce Mg uptake as well. Soil moisture is a factor affecting Mg plant availability, decreasing availability with decreasing soil moisture. In addition, some crop species are particularly sensitive to Mg. The so-called ideal soil has been defined as one that has 6 to 12% of the cation saturation percentage as Mg, with some suggesting that a saturation Mg percentage less than 6 would define a Mg-deficient soil (Bergmann, 1992; Mayland and Wilkinson,

1989). It is generally assumed that if the soil pH is maintained within the desired range using dolomitic limestone (Mg-containing limestone), there should be sufficient added Mg to ensure Mg sufficiency. Magnesium deficiency can be induced by heavy applications of K fertilizer as well as by the use of ammonium-form fertilizers, particularly when Mg-sensitive crops are being grown. In absolute levels, soils in the silt loam to clay loam textural range are considered sufficient in Mg with exchangeable Mg levels between 300 and 700 lb/acre (336 and 784 kg/ha), sandy soils 30+ lb/acre (33+ kg/ha), for high Mg-requirement crops 60+ lb/acre (67+ kg/ha), and low (6 to 10 meq/100 g) CEC soils 120+ lb/acre (134+ kg/ha).

The basic concepts of Mg soil chemistry are given in the Potash & Phosphate Institute *Soil Fertility Manual* (Anonymous, 1996a), plant chemistry concepts by Jones (1998b), and soil testing procedures and their interpretation have been described by Haby et al. (1990), Helmke and Sparks (1996), Suarez (1996), and Aitken and Scott (1999). In some instances, the availability and uptake of Mg are factors in plant and animal health, and Mayland and Wilkinson (1989) evaluated those soil factors that affected this availability, noting decreased availability with decreasing pH, low uptake by grass species, and availability decreasing with decreasing temperature.

Potassium (K)

The state-by-state summary of the percent of soils testing "medium" or lower in soil test K in North America for 1997 was published by the Potash & Phosphate Institute (Anonymous, 1998a) as is shown in Figure 2.10. The summary data show that the percentage of soils testing "medium" or lower declined from around 70 to 80% in the late 1960s to 40 to 60% in 1997.

The basic concepts of K soil chemistry are given in the *Soil Fertility Manual* (Anonymous, 1996c), plant chemistry concepts by Jones (1998b), and soil testing procedures and their interpretation have been described by Haby et al. (1990), Mikkelsen and Camberato (1995), Helmke and Sparks (1996), and Gourley (1999).

In absolute levels, soils in the silt loam to clay loam textural range are considered sufficient in K with exchangeable K levels between 180 and 280 lb/acre (202 and 314 kg/ha), sandy soils 60+ lb/acre (67+ kg/ha), and low (6 to 10 meq/100 g) CEC soils 100+ lb/acre (112+ kg/ha). At these K soil test levels, K fertilizer recommendations would be based on the crop requirement.

6. Effect of Fertilizer Potassium on the Potassium Soil Test Level

As a general rule of thumb, an increase in the K soil test level by 1 ppm (mg/kg) occurs when 8 to 16 lb K_2O/acre is applied above crop removal.

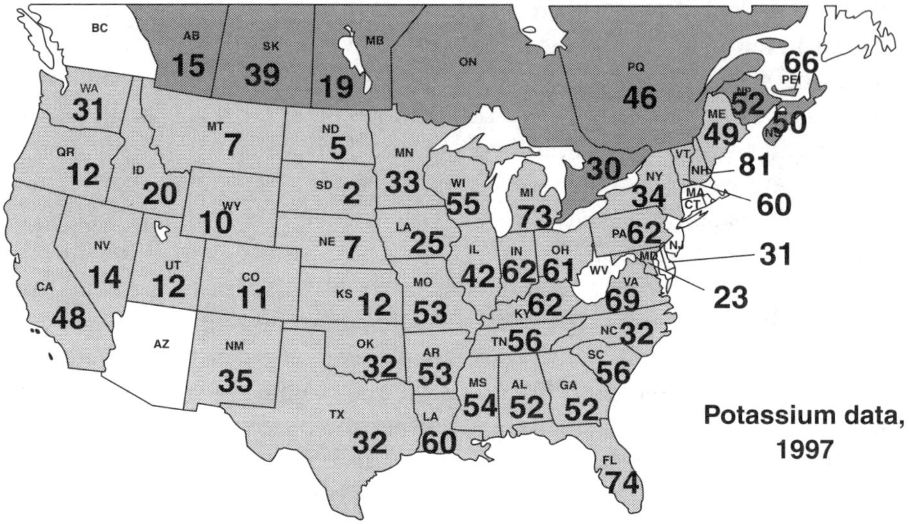

Figure 2.10
Percent of soils testing "medium" or lower in potassium, 1997. (From Potash & Phosphate Technical Bulletin 1998-3, Norcross, GA.)

7. Calculation of Cation Exchange Capacity (CEC)

The CEC of a soil is expressed as milliequivalents (meq)/100 g of soil. Soil colloidal components, clay and humus, contribute to the CEC of the soil, which depends on the percentage of clay and clay type (meq/100 g: vermiculite: 100 to 500; smectite: 70 to 95; illite: 10 to 40; kaolinite: 3 to 15) and level of humus (meq/100 g: 200) in the soil. Soils high in clay content and 2:1 lattice clays and/or high in organic matter content will have high (15 to 30 mg/100 g) CECs, as is shown in the following table:

Cation Exchange Capacity (CEC) of Colloids

Soil colloids	meq/100 g
Humus	200
Vermiculite	100–150
Smectite	70–95
Illite	10–40
Kaolinite	3–15
Sesquioxides	2–4

The CEC is also closely related to soil texture; i.e., sandy soils have CECs ranging from 1 to 8; loamy sands, 9 to 12; sandy or silty loam, 13 to 20; loam, 21 to 28; clay loam, 29 to 40; and clay soils >40 meq/100 g, respectively. The CEC is most commonly determined by summation of the exchangeable bases (Ca^{2+}, K^+, Mg^{2+}, Na^+) plus hydrogen (H^+). Normal neutral ammonium acetate (1 *N* $NH_4C_2H_3O_2$, pH 7.0) is the commonly used extraction reagent for cation determination and the procedure for exchangeable H^+ is by either titration (see Chapter 2, Section I) or calculation from a buffer pH determination (see Chapter 2, Section J). Hajek et al. (1972) describes a rapid method for determining the CEC of a soil. Sumner and Miller (1996) describe the various commonly used methods for determining the CEC of a soil and Gilliam et al. (1983) compared a number of these determining methods, concluding that most of the commonly used procedures yield comparable results.

8. Determination of Percent Base Saturation and Use

The percent base saturation is expressed as a percent of the total CEC that is contributed by the bases Ca, Mg, K, and Na. Base saturation percentages have been used for formulating lime and fertilizer recommendations (McLean, 1977; Eckert, 1987; Dahnke and Olson, 1990); an "ideal" soil has been defined by Prince et al. (1947) on the basis of ranges in base saturation percentages for the various cations and further defined by Graham et al. (1956). One example of the use of base saturation percent is the suggested need for amelioration of soils having Na percentages greater than 5% of the total CEC.

M. Micronutrients (B, Cl, Cu, Fe, Mn, and Zn)

1. Introduction

The micronutrients B, Cl, Cu, Fe, Mn, Mo, and Zn are the seven elements essential for plants (Römheld and Marschner, 1991) at requirement levels of less than 0.10% in the plant's dry matter (Epstein, 1972; Glass, 1989). General references on the micronutrients can be found in the books by Adriano (1996a, b), Pais and Jones (1997), and Kabata-Pendias (2000), and in the book edited by Mortvedt et al. (1991). Welsh et al. (1991) describe the geographic distribution of the micronutrients in the United States on both soil and plant factors, demonstrating heterogeneous distribution. Sillanpää (1972) described the general characteristics of the micronutrients (trace elements), and then later he (Sillanpää, 1990) made an assessment of the micronutrient status of soils on a countrywide basis. Since the micronutrients Cu, Mn, and Zn are also classed as heavy metals, some of these elements are included in Section N. A soil test procedure for Mo is not given in this guide.

In general, most of the micronutrient soil tests are limited in their application and are frequently associated with specific crop and soil characteristics (Martens and Lindsay, 1990; Johnson and Fixen, 1990; Adriano, 1996a, b; Pais and Jones, 1997; Jones, 1998), considerations that are essential for the interpretation of a micronutrient test (see interpretation section given later in this section). A general discussion of the micronutrients is given in the Potash & Phosphate Institute *Soil Fertility Manual* (Anonymous, 1996d).

Shuman (1991) discussed the chemical forms of the micronutrients in soil, Harter (1991) their adsorption–desorption reactions, Lindsay (1991) inorganic equilibria, Stevenson (1991) organic matter reactions, and Moraghan and Mascagni (1991) the environmental and soil factors affecting deficiencies and toxicities, factors that influence micronutrient uptake and utilization by crops.

Soil analysis (test) methods for all the micronutrients are given by Cox and Kamprath (1972), Sims and Johnson (1991), Sims (1995), and Whitney (1998a); for B by John (1973), Johnson and Fixen (1990), Gupta (1990), Isaac (1992), Anonymous (1994d), Keren (1996), Watson (1998b), and Bell (1999); for Cu by Makarim and Cox (1973), Martens and Lindsay (1990), Anonymous (1994j), and Brennan and Best (1999); for Cu and Zn by Johnson (1992a), Tucker (1992c), Reed and Martens (1996), and Anonymous (1994j); for Fe by Martens and Lindsay (1990), Anonymous (1994j), Loeppert and Inskeep (1996), and McFairlane (1999); for Mn by Mascagni and Cox (1983), Martens and Lindsay (1990), Tucker (1992c), Johnson (1992a), Anonymous (1994j), Gambrell (1996), and Uren (1999); and Zn by Martens and Lindsay (1990) and Armour and Brennan (1999). Micronutrient test procedures used in northeastern U.S. soil testing laboratories are given in the Northeastern Regional Bulletin 493 (Anonymous, 1995).

Extraction reagents, their interacting factors, and the range in critical level values have been summarized by Sims and Johnson (1991) as shown in the following table.

Soil Test Methods, Soil Factors Influencing Their Interpretation, and Typical Ranges in Critical Levels for the Micronutrients

Element	Interacting factors	Method	Range in critical level, mg/kg
Boron	Crop yield goal, pH, soil moisture, texture, organic matter, soil type	Hot-water soluble	0.1–2.0
Copper	Crop, organic matter, pH, percent $CaCO_3$	Mehlich No. 1	0.1–10.0
		Mehlich No. 3	—
		DTPA	0.1–2.5
		AB–DTPA	—
		0.1 *M* HC1	1.0–2.0
		Modified Olsen's	0.3–1.0
Iron	pH, percent $CaCO_3$, aeration, soil moisture, organic matter, CEC	DTPA	2.5–5.0
		AB–DTPA	4.0–5.0
		Modified Olsen's	10.0–16.0
Manganese	pH, texture, organic matter, percent $CaCO_3$	Mehlich No. 1	5.0 at pH 6.0
			10.0 at pH 7.0
		Mehlich No. 3	4.0 at pH 6.0
			8.0 at pH 7.0
		DTPA	1.0–5.0
		0.1 *M* HC1	1.0–4.0
		0.03 *M* H_3PO_4	10.0–20.0
		Modified Olsen's	2.0–5.0
Zinc	pH, percent $CaCO_3$, P, organic matter, percent clay, CEC	Mehlich No. 1	0.5–3.0
		Mehlich No. 3	1.0–2.0
		DTPA	0.2–2.0
		AB–DTPA	0.5–1.0
		Modified Olsen's	1.5–3.0
		0.1 *M* HC1	1.0–5.0

2. Extraction Reagents and Procedures

a. Hot water (B)

Principle of the Method. This method, which determines the amount of available soil B, was first proposed by Berger and Truog (1940; 1944) and later given in detail by Wear (1965), Bingham (1982), Johnson and Fixen (1990), Anonymous (1994i), Keren (1996), Isaac (1992), and Bell (1999). Various modifications of the technique have been studied by Gupta (1967) and Odom (1980). Wolf (1971; 1974) has developed a different extraction method using Azomethine-H as the color development reagent. However, hot water extraction is a method in common use, although other methods have been proposed (Johnson and Fixen, 1990). Shuman et al. (1992) found that B extracted by either the Mehlich No. 1 or No. 3 extraction reagents compared favorably with that extracted by hot water.

Extraction Procedure

Weigh 10 g or scoop 8.5 cm^3 air-dried <10-mesh-sieved (2-mm) soil into a refluxing flask and add 20 mL water.
Assemble the refluxing apparatus and place the flasks on the hot plate.
Bring to a boil and boil 10 min.
Filter through double filter paper and collect the filtrate for B determination.

Reagents

Azomethine-H Reagent

Weigh 0.9 g Azomethine-H and 2 g ascorbic acid into 10 mL water with gentle heating in a water bath.
When it is dissolved, dilute to 100 mL with water.
If the solution is turbid, reheat in the water bath until it is clear.
Store refrigerated for as long as 14 days.

Buffer Masking Reagent

Weigh 250 g ammonium acetate ($NH_4C_2H_3O_2$), 25 g tetrasodium salt of (ethylenedinitrillo) tetraacetic acid, and 10 g disodium salt of nitrilotriacetic acid in 400 mL water.
Slowly add 125 mL *glacial* acetic acid (CH_3COOH).

Color Development

Pipette 4 mL extractant into a spectrophotometer cuvet.
Add 1 mL Buffer Masking Reagent, 1 mL Azomethine-H Reagent, and mix immediately.
Let stand for 1 h.
Read the transmittance (% T) at 430 nm, with the blank water, using a UV-VIS spectrophotometer.

Interpretation. Accurate fertilizer recommendations for B must be based on known field responses based on local soil–climate–crop conditions (Berger and Truog, 1940; 1944; Gupta, 1967; Wear, 1968; Reisenauer et al., 1973; Johnson and Fixen, 1990; Jones, 1998b). For most soils and crops, the amount of B extracted should be interpreted as follows:

Category	mg B/kg in soil
Insufficient	<1.0
For normal growth	1.0–2.0
High	2.1–5.0
Excessive	>5.0

b. Mehlich No. 1 (Zn)

Principle of the Method. This method for determining extractable Zn has been evaluated only on soils that have CECs of less than 10 meq/100 g, are acid (pH less than 7.0) in reaction, and are relatively low (less than 5%) in organic matter content. Its suitability for use on alkaline or organic soils has not been determined. This method is described in some detail by Perkins (1970) for use with the sandy coastal plain soils of the southeastern United States. The use of Mehlich No. 1 as an extraction reagent for cations and P was first reported by Mehlich (1953a), and later classified as a P extraction reagent by Nelson et al. (1953) as the North Carolina Double Acid Method (renamed Mehlich No. 1). Wear and Evans (1968) compared Mehlich No.1 with 0.1 *N* hydrochloric acid (HCl) and EDTA as a Zn extraction reagent on 12 soils and found Mehlich No. 1–extractable Zn to correlate more closely with Zn uptake by corn and sorghum plants. Alley et al. (1972) developed a prediction equation for field conditions using Mehlich No.1–extractable Zn. The equation was improved considerably by taking into consideration soil pH and Mehlich No.1–extractable P.

Extraction Reagent

0.05 N HCl in 0.025 NH_2SO_4

Pipette 4 mL concentrated hydrochloric acid (HCl) and 0.7 mL concentrated sulfuric acid (H_2SO_4) into a 1000-mL volumetric flask and bring to volume with water.

Extraction Procedure

Weigh 5 g or scoop 4 cm^3 air-dried <10-mesh-sieved (2-mm) soil into an acid-washed 50-mL extraction vessel.

Add 20 mL Extraction Reagent and shake for 5 min on a reciprocating shaker. Immediately filter and collect the filtrate for Zn concentration determination.

Interpretation. An evaluation of the results as well as Zn recommendations must be based on field response data conducted under local soil–climate–crop conditions (Viets et al., 1973; Jones, 1998). Interpretative data that would be applicable to the southeastern United States are given by Alley et al. (1972), Perkins (1970), Cox and Wear (1977), and Martens and Lindsay (1990). The critical soil test Zn level for corn as interpreted by Cox and Wear (1977) is 0.8 mg Zn/kg. The probability of a corn yield response to Zn fertilization on soils testing below this value would be high. This critical level may not apply to extremely high CEC soils, high P-content soils, or very acid soils.

c. 0.1 *N* hydrochloric acid (Zn)

Principle of the Method. This method is primarily intended for determining extractable Zn in acid soils (pH less than 7.0) by 0.1 *N* hydrochloric acid (HCl). The test is designed to divide soils into two groups: those that cannot supply the crop requirement and, therefore, will require Zn; and those that have an adequate supply of Zn to meet the crop requirement. The method is not suitable for alkaline soils unless additional measurements are made (Nelson et al., 1959; Viets and Boawn, 1965; Martens and Lindsay, 1990; Lindsay, 1991; Sims and Johnson, 1991). This procedure as presented is a modification of a method used by Wear and Evans (1968) with early calibration work by Gilroy (1969). Many variations of the method have been used. The main differences between the methods include modifications in shaking time and soil-to-extractant ratios. The procedure is based upon the assumption that all or a portion of the soil Zn that will become available for plant uptake during a growing season is acid soluble. The quantity of acid-soluble Zn serves as an index of availability (Tucker and Kurtz, 1955).

Extraction Reagent

0.1 N HCl

Dilute 16.7 mL redistilled 6 *N* hydrochloric acid (HCl) to 1000 mL with water.

Titrate with standard base to the phenolphthalein end point (clear to pink).

Add dilute acid or pure water to obtain a 0.1 *N* solution.

Extraction Procedure

Weigh 5 g or scoop 4.25 cm^3 of air-dried <10-mesh-sieved soil into a 50-mL extraction vessel.
Add 20 mL Extraction Reagent and shake for 30 min on a reciprocating shaker.
Immediately filter and collect the filtrate for Zn concentration determination.
Carry a blank through the entire procedure with each run.

Interpretation. An evaluation of the results, as well as accurate fertilizer recommendations, must be based upon field response data conducted under local soil–climate–crop conditions (Viets and Lindsay, 1973; Sims and Johnson, 1991; Mikkelsen and Camberato, 1995; Jones, 1998b). This procedure is a routine soil test used in Missouri. Field studies have shown that soils with less than 2 mg 0.1 *N* HCl–extractable Zn/kg will probably need Zn soil fertilization to obtain optimum Zn levels for corn and grain sorghum. Cox and Wear (1977) found 3.1 kg Zn/ha or better in the soil adequate for corn on sandy soils.

d. Mehlich No. 3 (B, Cu, Mn, and Zn)

Principle of the Method. The extraction reagent (Mehlich, 1984a) and determination of B, Cu, Mn, and Zn by this procedure are applicable across a wide range of soil properties, from acid to basic in reaction (Anonymous, 1994j). Although the method was correlated with established extraction reagents from several regions and critical levels were established, the specific critical levels should be based on local soil–crop–climate conditions.

Good correlations were obtained between Mehlich No. 1 and Mehlich No. 3 for the micronutrients, B (Shuman et al., 1992), Cu, Mn, and Zn (Sims, 1989), and Mehlich No. 2 (Mehlich, 1978) and Mehlich No. 3 for Mn and Zn, even though the mean values were not the same (Schmisek et al., 1989; Tucker, 1992c). Critical Zn levels for this extractant have been given by Mehlich (1984a).

Extraction Reagent

0.2 *N* acetic acid CH_3COOH; 0.25 *N* ammonium nitrate NH_4NO_3; 0.015 *N* ammonium fluoride NH_4F; 0.013 *N* nitric acid HNO_3; 0.001 *M* EDTA

Ammonium Fluoride–EDTA Stock Reagent

Add approximately 600 mL water to a 1000-mL volumetric flask.
Add 138.9 g ammonium fluoride (NH_4F) and dissolve.

Then add 73.05 g EDTA.
Dissolve the mixture and bring to volume with water.
Store in a plastic container.

Final Extraction Reagent Mixture

Add approximately 3000 mL water to a 4000-mL volumetric flask.
Add 80 g ammonium nitrate (NH_4NO_3) and dissolve.
Add 16 mL NH_4F–EDTA stock reagent (above) and mix well.
Add 46 mL *glacial* acetic acid (CH_3COOH) and 3.28 mL concentrated nitric acid (HNO_3).
Then bring to volume with water and mix thoroughly.
Achieve a final pH of 2.5 ± 0.1.
Store in a plastic container.

Extraction Procedure

Scoop 5 cm^3 air-dried <10-mesh-screened (2-mm) soil into an acid-washed 100-mL extraction vessel.
Add 50 mL Extraction Reagent and shake for 5 min on a reciprocating shaker.
Immediately filter and collect the filtrate and save for elemental content determination.
Store in a plastic container.

Note: *For the rationale of using a volume soil measure, refer to Mehlich (1973).*

Interpretation.

Boron: Critical B soil test levels are similar to that for hot water-extraction B (Shuman et al., 1992).

Copper: Critical Cu soil test level was established with the Mehlich No. 3 Extraction Reagent (Mehlich, 1984a). The critical level is 0.5 mg Cu/dm^3, which equates to a soil test index of 25.

Manganese: Calibration of the Mn soil test with this extraction reagent is based on extractable Mn and soil pH (Mascagni and Cox, 1983). Equations predicting the Mn availability index (MnAI) for soybeans and corn are as follows:

$$\text{Soybean: } \quad \text{MnAI} = 101.2 + 0.6\,(\text{MnI}) - 15.2\,(\text{pH})$$

$$\text{Corn: } \quad \text{MnAI} = 108.2 + 0.6\,(\text{MnI}) - 15.2\,(\text{pH})$$

The critical soil test MnAI = 4 mg Mn/dm^3, which is equal to a 25 soil test index. Because of the limited soil test calibration for other crops, calculation of the MnAI for these crops is based on their sensitivity to Mn, as compared with corn or soybeans. For example, the soybean MnAI is used to predict Mn needs for small grains, since their sensitivity is closely related to that of soybeans.

Zinc: Critical Zn soil test level by this procedure is 1.0 mg Zn/dm^3 which equates to a soil test index of 25. A Zn availability index (ZnAI) has been established for mineral, mineral–organic, and organic soils and is based on the relationship between extractable Zn and soil pH (Junus and Cox, 1987). These values are as follows:

$$\text{ZnAI (mineral soils)} = \text{ZnI} \times 1.0$$

$$\text{ZnAI (mineral–organic soils)} = \text{ZnI} \times 1.25$$

$$\text{ZnAI (organic soils)} = \text{ZnI} \times 1.66$$

Micronutrient crop fertilization based on the Mehlich No. 3 extraction reagent may be found in the circular by Tucker et al. (1997).

e. Ammonium Bicarbonate–DTPA (Cu, Fe, Mn, and Zn)

Principle of the Method. The Extraction Reagent is 1 *M* ammonium bicarbonate (NH_4HCO_3) in 0.005 *M* DTPA adjusted to a pH of 7.6. The method was first proposed by Soltanpour and Schwab (1977), and later by Soltanpour and Workman (1979), and Soltanpour (1991). The original pH of 7.6 allows DTPA to chelate and extract Fe and other metals. This method is highly correlated with the DTPA (Lindsay and Norvell, 1978) method for Zn, Fe, Mn, and Cu (Soltanpour and Schwab, 1977; Soltanpour et al., 1976; 1979). The method is described by Anonymous (1994f).

Extraction Reagent

NH_4HCO_3–DTPA

Obtain 0.005 *M* DTPA solution by adding 9.85 g DTPA (diethylenetriaminepentaacetic acid) (acid form) to 4500 mL water in a 5000-mL volumetric flask.

Shake constantly for 5 h to dissolve the DTPA, and bring to 5000 mL with water. This solution is stable with regard to pH.

To 900 mL of the 0.005 *M* DTPA solution, gradually add 79.06 g ammonium bicarbonate (NH_4HCO_3) and stir gently with a rod to facilitate dissolution and prevent effervescence when bicarbonate is added.

Dilute the solution to 1000 mL with the 0.005 *M* DTPA solution and mix gently with a rod.

Adjust the pH to 7.6 with dropwise additions of 2 *M* hydrochloric acid (HCl) and slow agitation with a rod.

Store under mineral oil.
Check the pH after storage and adjust it with a 2 *M* HCl solution dropwise, if necessary.

Note: *The cumulative volume of HCl added should not exceed 1 mL/L, after which a fresh solution should be prepared.*

Extraction Procedure

Weigh 10 g air-dried <10-mesh-screened (2-mm) soil in a 125-mL conical flask.
Add 20 mL Extraction Reagent and shake on an Eberbach reciprocal shaker or an equivalent shaker for exactly 15 min at 180 cycles/min with flasks kept open.
Immediately filter the extracts through Whatman 42 filter paper (Lindsay and Norvell, 1978) and save for micronutrient concentration determination.

Precautions: *The Extraction Reagent is 1 M ammonium bicarbonate (NH_4HCO_3) in 0.005 M DTPA (AB–DTPA) adjusted to a pH of 7.6 (Soltanpour and Schwab, 1977; Soltanpour and Workman, 1979; Soltanpour, 1991). The original pH of 7.6 allows DTPA to chelate and extract Fe and other metals. The range and sensitivity are the same as those for the DTPA-extractable micronutrients (Soltanpour et al., 1977). The AB–DTPA Extraction Reagent is unstable with regard to pH and should be kept under mineral oil to prevent a pH change. Stainless steel soil sampling tubes and polyvinyl chloride mixing buckets for field soil sampling should be used to prevent contamination with trace elements. Use high-density aluminum oxide grinders equipped with stainless steel sieves to prevent soil contamination with trace elements. If the above grinder is not available, test other grinders with pure sand to make sure they do not contaminate the soil being processed. Grinding force, time, and the amount of soil in the grinder should be adjusted so that the extracted levels of trace elements are equivalent to those obtained with a wooden roller (Soltanpour et al., 1979). The uneven distribution of Fe, Zn, and Cu in soil particle-sized fractions is discussed by Kahn (1979). A coefficient of variability ranging from 5 to 10% can be expected for different determinations. Accuracy is comparable to that obtained with the DTPA extractant for the micronutrients (Lindsay and Norvell, 1978).*

Interpretation. The following tables provide an interpretation of the index values for Cu, Fe, Mn, and Zn. These are general guidelines and should be verified under different soil–climate–crop–management combinations.

Index Values for Cu, Fe, Mn, and Zn

	mg/kg in soil			
Category	Cu	Fe	Mn	Zn
Low	0.0–0.2	0.0–3.0	0.0–0.5	0.0–0.9
Marginal	0.3–0.5	3.1–5.0	0.6–1.0	1.0–1.5
Adequate	>0.5	>5.0	>1.0	>1.5

f. DTPA (Cu, Fe, Mn, and Zn)

Principle of the Method. The theoretical basis for the DTPA extraction is the equilibrium of the metal in the soil with the chelating agent. A pH level of 7.3 enables DTPA to extract Fe and other metals. The use of DTPA as an extraction reagent was developed by Lindsay and Norvell (1978), and the method is described by Johnson (1992a), Liang and Karamanos (1993), and Anonymous (1994j).

Extraction Reagent

DTPA (diethylenetriaminepentaacetic acid)

Weigh 1.96 DTPA $\{[(HOCOCH_2)_2NCH_2]_2NCH_2COOH\}$ into a 1000-mL volumetric flask.

Add 14.92 g triethanolamine (TEA) and bring to volume of approximately 950 mL with water.

Add 1.47 g calcium chloride ($CaCl_2{\cdot}2H_2O$) and bring to 1000 mL with water while adjusting the pH to exactly 7.3 with 6 *N* hydrochloric acid (HCl).

The final concentration will be 0.005 *M* DTPA, 0.1 *M* TEA, and 0.01 *M* $CaCl_2{\cdot}2H_2O$.

Note: *The DTPA reagent should be the acid form.*

Extraction Procedure

Weigh 10 g or scoop 8.5 cm^3 air-dried <10-mesh-screened (2-mm) soil into a 125-mL extraction vessel.

Add 20 mL Extraction Reagent and shake on a reciprocating shaker for 2 h.

Note: *Samples shaken longer than 2 h will give high results because a final equilibrium of the metal and soil is not reached in 2 h.*

Immediately filter and collect the filtrate for micronutrient concentration determination.

Interpretation. An evaluation of the analysis results as well as accurate fertilizer recommendations for Cu, Fe, Mn, and Zn must be based on field response for each crop and local field condition. Interpretative data for critical levels as established by Viets and Lindsay (1973) for Colorado soil are available. Boawn (1971) did work with DTPA for Zn on Washington soil.

g. 0.01 *M* calcium chloride (B, Cu, Fe, Mn, and Zn)

Principle of the Method. The reagent has more or less the same ionic strength (0.3 *M*) as the average salt concentration in many soil solutions, and is able to extract adsorbed cations; the electrolyte concentration remains practically constant, metal concentration reflects the differences in binding strength and/or solubility among various soils, and the measured elements reflect their availability at the pH of the soil since the extractant is an unbuffered solution. The method was initially described by Houba et al. (1990) and later by Houba et al. (2000).

Extraction Reagent

0.01 M $\boldsymbol{CaCl_2{\cdot}2H_2O}$

Weigh 1.47 g calcium chloride ($CaCl_2{\cdot}2H_2O$) into a 1000-mL volumetric flask and dilute to volume with water.

Comment: *$CaCl_2{\cdot}2H_2O$ may absorb water on standing. The reagent should be standardized by titration with EDTA at pH = 10.0 with Eriochrome Black T as an indicator.*

Extraction Procedure

Weigh 10 g dry soil into a 250-mL polyethylene bottle.

Add 100 mL 0.01 *M* $CaCl_2$ solution at 20°C (68°F) and shake mechanically for at least 2 h at room temperature (20°C; 68°F).

Either filter to collect the extract or collect the supernatant after centrifugation, and save for micronutrient concentration determination.

h. Morgan (B, Cu, Fe, Mn, and Zn)

Principle of the Method. The method is primarily used for determining K, Ca, and Mg in acid soils with cation exchange capacities of less than 20 meq/100 g. This method was initially proposed by Morgan (1932; 1941), and then by Lunt et al. (1950) and later by Greweling and Peech

(1965). The Morgan method has been used by a number of soil testing laboratories in the northeastern and northwestern United States (Nelson et al., 1953; Jones, 1973; 1998a; Anonymous, 1995). The Extraction Reagent is well buffered at pH 4.8 and, when used in conjunction with *activated* carbon, yields clear and colorless extracts. The concentration of sodium acetate ($NaC_2H_3O_2 \cdot 3H_2O$) is sufficiently high to effect replacement of about 80% of the exchangeable cations. The method has been modified by Wolf (1982) to include more efficient extraction of the micronutrients. The method is described by Wolf and Beegle (1995).

Extraction Reagent

Weigh 100 g sodium acetate ($NaC_2H_3O_2 \cdot 3H_2O$) into a 1000-mL volumetric flask and add about 900 mL water.

Add 30 mL *glacial* acetic acid (CH_3COOH), adjust the pH to 4.8, and bring to volume with water.

Extraction Procedure

Scoop 5 cm^3 air-dried <10-mesh-screened (2-mm) soil into a 50-mL extraction vessel.

Add 25 mL Extraction Reagent and shake for 5 min on a reciprocating shaker at a minimum of 180 oscillations/min.

Immediately filter and collect the filtrate for micronutrient concentration determination.

Interpretation. According to Wolf (1982), critical soil test values (no crop response above) for B, Cu, Fe, Mn, and Zn are 2.0, 1.3, 2.5, 2.5, and 2.5 mg/kg, respectively. These are general guidelines and should be verified under different soil–climate–crop–management combinations.

i. 0.01 *M* calcium nitrate (Cl)

Principle of Method. This method is a modification of the procedure of Adriano and Doner (1982) for Cl determination and uses an extraction procedure similar to that suggested by Bolton (1971). In this spectrophotometric method, Cl displaces thiocyanate, which, in the presence of Fe^{3+}–Fe, forms a highly colored ferric thiocyanate complex:

$$2\ Cl^- + Hg(SCN)_2 + 2\ Fe^{3+} \rightarrow HgCl_2 + 2\ Fe(SCN)^{2+}$$

The color of the resulting solution is stable and proportional to the original Cl^- ion concentration.

The procedure is very sensitive and has a detection limit of approximately 1 µg Cl/g soil. Nitrate, S^-, CN^-, CSN^-, Br^-, and I^- can cause interference but are usually not present in sufficient amounts to be a problem. Similar procedures have been modified for use with an AutoAnalyzer. Frankenberger et al. (1996) describe the various procedures for determining Cl.

Precision varies with level in the soil with coefficients from 9 to 24% for Cl levels greater than 12 mg/kg, and from 15 to 25% for Cl levels less than 10 mg/kg (Gelderman et al., 1998).

Extraction Reagent

0.01 M $Ca(NO_3)_2{\cdot}4H_2O$

Weigh 4.72 g calcium nitrate [$Ca(NO_3)_2{\cdot}4H_2O$] into a 2000-mL volumetric flask and bring to volume with water.

Extraction Procedure

Scoop 10 g crushed soil into a 50-mL Erlenmeyer flask.
Add approximately 25 mg 0.01 *M* $Ca(NO_3)_2$–washed charcoal and 25 mL Extraction Reagent, and shake for 15 min at 180 or more opm.
Filter immediately following shaking using Whatman No. 42 filter paper or equivalent.
Transfer 10 mL aliquot to a 50-mL beaker.
Do duplicate or triplicate analyses and include a blank sample.

Interpretation. The calibration and implementation of the Cl soil test have been reviewed by Fixen et al. (1987), and yield responses have been obtained when the amount of extractable Cl in the 0- to 24-in. depth is less than 20 lb Cl/acre for winter wheat (Lamond et al., 1999), and less than 20 to 25 lb Cl/acre for corn and grain sorghum (Lamond et al., 2000).

j. 0.5 *M* potassium sulfate (Cl)

Principle of the Method. The basic approach of the method was reported by Burton (1971) for Cl and F determination of P and involves measuring the electrode potential before and after addition of a known quantity of Cl to a sample. The change in potential is then related to sample concentration by assuming a Nernst-type relationship and a theoretical electrode response of 59.1 mV per tenfold change in concentration. This electrode response should be verified by measuring the potential after successive additions of the standard.

Direct reading of soil extracts with the solid-state Cl electrode has not been reliable across diverse soils and may give high readings (Hipp and Langdale, 1971). The electrode has worked well when used as an end-point indicator in titrations. A more convenient alternative to potentiometric titrations is the potentiometric known addition method outlined here. It is particularly well suited for situations where occasional analysis for Cl concentration is needed since no calibration is necessary.

Precision varies with level in the soil with coefficients from 9 to 24% for Cl levels greater than 12 mg/kg, and from 15 to 25% for Cl levels less than 10 mg/kg (Gelderman et al., 1998). Frankenberger et al. (1996) describe the various procedures for determining Cl.

Extraction Reagent

0.5 M K_2SO_4

Weigh 87.0 g potassium sulfate (K_2SO_4) into a 2000-mL volumetric flask and bring to volume with water.

Extraction Procedure

Scoop 10 g crushed soil into a 50-mL Erlenmeyer flask.
Add 30 mL Extraction Reagent and shake for 15 min at 180 or more opm.
Immediately filter (Whatman No. 2 or equivalent), centrifuge, or leave to settle samples to produce clear solutions.

Note: *Do duplicate or triplicate analyses; include a blank sample.*

Interpretation. The calibration and implementation of the Cl soil test have been reviewed by Fixen et al. (1987), and yield responses have been obtained when the amount of extractable Cl in the 0- to 24-in. depth is less than 20 lb Cl/acre for winter wheat (Lamond et al., 1999) and less than 20 to 25 lb Cl/acre for corn and grain sorghum (Lamond et al., 2000).

k. Saturated calcium hydroxide (Cl)

Principle of the Method. Chemically suppressed ion chromatography was introduced by Smith et al. (1975). The main advantages of this method are high sensitivity, the ability to separate and quantify similar types of ions (i.e., F, Cl, and Br), the ability to perform multiple element analyses, and increased freedom from sample matrix effect. Mosko (1984) demonstrated some problems encountered in the analyses of a range of aqueous samples. The method of extraction of Cl from the soil is the same method used for NO_3 (Carson, 1980b). This allows the potential for multielement analyses.

The procedure is very sensitive and has a detection limit of approximately 1 µg Cl/g soil. Precision varies with level in the soil with coefficients from 9 to 24% for Cl levels greater than 12 mg/kg, and from 15 to 25% for Cl levels less than 10 mg/kg (Gelderman et al., 1998). Frankenberger et al. (1996) describe the various procedures for determining Cl.

Extraction Reagent

Saturated $Ca(OH)_2$

Weigh 3 g calcium oxide (CaO) into 1000 mL of water and shake thoroughly.
Filter solution if desired, but this is not necessary.

Extraction Procedure

Weigh 10.0 g crushed soil into a 50-mL Erlenmeyer flask.
Dispense 25 mL Extraction Reagent into the flask and shake for 5 min at 180 or more opm.
Filter sample into a filter tube through Whatman No. 2 filter paper that has been washed with water.
Set up mechanized vacuum extractor utilizing 0.2-µm filters.
Pour the filtered sample extract or standard into each syringe and allow it to equilibrate with the exchange resin about 5 min.
Extract samples and/or standards through 0.2-µm filters.

Interpretation. The calibration and implementation of the Cl soil test have been reviewed by Fixen et al. (1987), and yield responses have been obtained when the amount of extractable Cl in the 0- to 24-in. depth is less than 20 lb Cl/acre for winter wheat (Lamond et al., 1999), and less than 20 to 25 lb Cl/acre for corn and grain sorghum (Lamond et al., 2000).

3. Methods of Micronutrient Determination in Soil Extracts

The micronutrients B, Cu, Fe, Mn, and Zn can be determined by UV-VIS spectrophotometry, classical methods of analysis, Cl by specific ion electrode and ion chromatography (Mosko, 1984), Cu, Fe, Mn, and Zn by atomic absorption spectrophotometry (AAS), and B, Cu, Fe, Mn, and Zn by plasma emission spectrometry (ICP-AES). Details on these methods of analysis are given in Chapter 5.

Standard preparation is important, particularly when making working standards in the extraction reagents, with details given in Appendix B.

Since the interpretative critical values for most of the micronutrients are quite low in concentration, any addition due to contamination can significantly affect the interpretation; therefore, the use of a blank is highly recommended. A blank is obtained by carrying an extraction through the entire process without the sample. Contamination and/or alteration of the extraction reagent may occur in the extraction or filtering process. Reagents also may contain sufficient quantities of a determined micronutrient to affect the assay.

4. Methods of expression

Micronutrient concentrations are usually expressed as either parts per million (ppm) or pounds per acre (lb/acre), or as SI units, kilograms per hectare (kg/ha), and millimoles per kilogram (mmol/kg). Comparative values in various units for the micronutrients are as follows:

Micronutrient	ppm (mg/kg)	lb/acre	kg/ha	mmol/kg
Boron (B)	0.20	0.40	0.45	0.0185
Chlorine (Cl)	10.0	20.0	22.4	0.282
Copper (Cu)	0.12	0.24	0.27	0.0019
Iron (Fe)	11.1	22.2	24.8	0.198
Manganese (Mn)	0.55	1.10	1.23	0.010
Molybdenum (Mo)	0.01	0.02	0.022	0.0001
Zinc (Zn)	0.33	0.66	0.74	0.005

Note: Levels have been selected for illustrative purposes only.

5. Cleaning Laboratory Ware

For determination of the micronutrients, extraction vessels, funnels, and receiving vessels should be carefully washed to minimize the possibility of adding these elements to the obtained extract (Kammin et al., 1995). Two washing procedures are recommended:

a. Dilute acid, i.e., 1:10 hydrochloric acid (HCl) or nitric acid (HNO_3), wash:
   1. Rinse extraction vessels, funnels, and receiving vessels with water.
   2. Soak in a dilute acid bath (at least 30 min).
   3. Rinse three times with water.
b. 0.2% aluminum chloride ($AlCl_3 \cdot 6H_2O$) (for Cu, Fe, Mn, and Zn Mehlich No. 3 determination; Tucker, 1992c):
   1. Wash extraction vessels, funnels, and receiving vessels with hot tap water.
   2. Rinse with 0.2% $AlCl_3 \cdot 6H_2O$; then rinse with pure water.
   3. After placing filter paper into the funnels, rinse the paper with 0.2% $AlCl_3 \cdot 6H_2O$ followed with pure water. Allow to drain.

All washing apparatus should be constructed from stainless steel or plastic. Allow the extraction vessels, funnels, and receiving vessels to drain and do not wipe them dry; just washed and wet extraction vessels, funnels, and receiving vessels may be oven-dried.

6. Interpretation

The effects of varying soil (pH and organic matter content) and weather characteristics (soil temperature and leaching) on the interpretation of B, Cu, Fe, Mn, and Zn soil tests are discussed by Martens and Lindsay (1990). Shuman (1998) details micronutrient sources and their influence on crop production as affected by chemical form and soil characteristics. A factor that can determine how a micronutrient test is evaluated would be its "normal" level found in soil, as has been discussed by De Temmerman et al. (1984).

For most of the micronutrients, there is no simple relationship between test level and the recommended rate of application of a micronutrient as is shown in the following table:

Micronutrient Ratings and Recommendation Rates[a] (lb/acre) as Related to Soil Tests with Different Extraction Reagents on Several Soils

Micronutrient	Rating	Extraction reagent		Crop/soil parameter			
				Legume[b]		Nonlegume	
		Hot water		pH <6.8	pH >6.8	pH <6.8	pH >6.8
		ppm		Rate, lb/acre			
Boron	VL	<0.4		1.5	2.0	1.0	1.5
(B)	L	0.4–0.7		1.0	1.5	1.0	1.0
	M	0.8–1.2		1.0	1.0	0.5	0.5
	H	1.3–2.0		0.5	0.5	0	0.5
	VH	>2.0		0	0	0	0
		0.1 *N* HCl	**DTPA**	Mineral soil		Organic soil	
		ppm		Rate, lb/acre			
Copper	VL	<0.3	<0.3	2		4	
(Cu)	L	0.3–0.8	0.3–0.8	1		3	
	M	0.9–1.5	0.9–1.2	0–1		2	
	H	1.6–3.0	1.3–2.5	0		1	
	VH	>3.0	>2.5	0		0	
				pH <6.8		pH >6.8	
				Rate, lb/acre			
Iron	VL	0–3	0–5	3		3–5	
(Fe)	L	4–11	5–10	2		2–4	
	M	12–24	11–16	0–2		1–2	
	H	25–50	17–25	0		0	
	VH	>50	>25	0		0	
				pH <6.8		pH >6.8	
				Rate, lb/acre			
Manganese	VL	0–5	0–4	4–5		5–7	
(Mn)	L	6–14	4–8	2–4		3–5	
	M	15–29	9–12	0–2		3–5	
	H	30–50	13–30	0		0–1	
	VH	>50	>30	0		0	

(continued)

Micronutrient Ratings and Recommendation Rates[a] (lb/acre) as Related to Soil Tests with Different Extraction Reagents on Several Soils (continued)

Micronutrient	Rating	Extraction reagent		Crop/soil parameter: Legume[b] Low phosphorus pH <6.8	Legume[b] Low phosphorus pH >6.8	Nonlegume High phosphorus pH <6.8	Nonlegume High phosphorus pH >6.8
				Rate, lb/acre			
Zinc	VL	<1.0	<0.5	5	6	7	8
(Zn)	L	1.1–2.9	0.5–1.0	3–4	4–5	4–5	5–6
	M	3.0–5.0	1.1–3.0	1–2	2–3	2–3	2–4
	H	5.1–8.0	3.1–6.0	0	0–2	0–2	1–2
	VH	>8.0	>6.0	0	0	0	0

[a] Broadcast rates: Fe, Mn, and Zn are more efficient when applied in bands. Divide rates by three for band placement.

[b] For legumes or other crops with high B requirements.

Source: Courtesy of ARA Professional Dealer Manual Fluid Fertilizers, 1994 ed.

Testing any particular soil for its micronutrient status is not recommended unless there is justification based on established crop type and soil conditions as given in this table:

Soil Conditions and Crops Where Micronutrient Deficiencies Most Often Occur

Micronutrient	Sensitive crops	Soil conditions for deficiency
Boron (B)	Alfalfa, clover, cotton, peanut, sugar beet, *Brassica* (cabbage and relatives), cereals, potato, tomato, celery, grapes, cucumber, sunflower, fruit trees (apple and pear), and mustard	Acid sandy soils low in organic matter, overlimed soils, organic soils
Copper (Cu)	Corn, onions, small grains (oat), watermelon, legumes, sunflower, spinach, citrus seedlings, and gladiolus	Organic soils, mineral soil high in pH and organic matter
Iron (Fe)	Citrus, clover, pecan, sorghum, soybean, grape, several calcifuge species, rice, tobacco, and clover	Leached sandy soils low in organic matter, alkaline soils, soils high in P

Soil Conditions and Crops Where Micronutrient Deficiencies Most Often Occur (continued)

Micronutrient	Sensitive crops	Soil conditions for deficiency
Manganese (Mn)	Alfalfa, small grains (oats), soybean, sugar beet, fruit trees (apple, cherry, and citrus), legumes, potato, and cabbage	Leached acid soils, neutral to alkaline soil high in organic matter
Zinc (Zn)	Corn, field beans, pecan, sorghum, legumes, grasses, spinach, hops, fax, grape, fruit trees (citrus), and soybean	Leached acid sandy soils low in organic matter, neutral to alkaline soils, and/or high in P

Pais and Jones (1997) published a list on the sensitivity of crop plants to the micronutrients as shown in the following table:

Relative Sensitivities of Selected Crops to Micronutrient Deficiencies

	Micronutrient				
Crop	Boron	Copper	Iron	Manganese	Zinc
Alfalfa	High	High	Medium	Medium	Low
Asparagus	Low	Low	Medium	Low	Low
Barley	Low	Medium	High	Medium	Medium
Bean	Low	Low	High	High	High
Blueberry	Low	Medium	— [a]	Low	—
Broccoli	Medium	Medium	High	Medium	—
Cabbage	Medium	Medium	Medium	Medium	Medium
Carrot	Medium	Medium	—	Medium	Low
Cauliflower	High	Medium	High	Medium	—
Celery	High	Medium	—	Medium	—
Clover	Medium	Medium	—	Medium	Medium
Corn	Low	Medium	—	Medium	High
Cucumber	Low	Medium	—	Medium	—
Grass	Low	Low	High	Medium	Low
Lettuce	Medium	High	—	High	Medium
Oat	Low	High	Medium	High	Low
Onion	Low	High	—	High	High

(continued)

Relative Sensitivities of Selected Crops to Micronutrient Deficiencies (continued)

	Micronutrient				
Crop	Boron	Copper	Iron	Manganese	Zinc
Parsnip	Medium	Medium	—	Medium	—
Pea	Low	Low	—	High	Low
Peppermint	Low	Low	Low	Medium	Low
Potato	Low	Low	—	High	Medium
Radish	Medium	Medium	—	High	—
Rye	Low	Low	—	Low	Low
Sorghum	Low	Medium	High	High	High
Soybean	Low	Low	High	High	Medium
Spearmint	Low	Low	—	Medium	Low
Spinach	Medium	High	High	High	—
Sudan grass	Low	High	High	High	Medium
Sugar beet	High	Medium	High	Medium	Medium
Sweet corn	Medium	Medium	Medium	Medium	High
Table beet	High	High	High	High	Medium
Tomato	Medium	Medium	High	Medium	Medium
Turnip	High	Medium	—	Medium	—
Wheat	Low	High	Low	High	Low

[a] Inadequate data to categorize into low-, medium-, or high-sensitivity groups.

Normally, the primary focus on micronutrients relates to deficiency rather than their excess that would lead to toxicity. In most instances, crop species that are sensitive are also sensitive to excessive levels. Those crop species sensitive to excessive levels of a micronutrient are given in the following table:

Crop Species Sensitive to Excessive Levels of the Micronutrients

Micronutrient	Crop species
Boron (B)	Cereals, potato, tomato, cucumbers, sunflower, mustard
Chlorine (Cl)	Strawberry, navy bean, fruit trees, pea, onion
Copper (Cu)	Cereals, legumes, spinach, citrus seedlings, gladiolus
Iron (Fe)	Rice and tobacco
Manganese (Mn)	Cereals, legumes, potato, cabbage
Zinc (Zn)	Cereals, spinach

N. Trace Heavy Metals

1. Introduction

The heavy metals have been variously classified; one classification considers those elements that have atomic weights greater than 55 heavy metals. Under this classification, the micronutrients Cu (atomic weight 63.54), Fe (atomic weight 55.85), Mn (atomic weight 54.993), Mo (atomic weight 95.95), and Zn (atomic weight 65.38) would be identified as heavy metals, elements that are discussed in Section F.4. Elements discussed in this section are those that are considered toxic to plants and/or animals (Adriano, 1996b; Risser and Baker, 1990; Pais and Jones, 1997; Kabata-Pendias, 2000): the elements Cd, Cr, Hg, Ni, and Pb. Sillanpää and Jansson (1992) determined the Cd, Pb, Co, and Se status in soils and plants in 30 countries, finding high levels of Cd and Pb in many soils. Swaine (1969) published a treatise on the trace element content of soils and Alloway (1995) edited a book on the heavy metal content in soils.

Risser and Baker (1990) describe methods for the determination of Cd, Cr, Hg, Ni, and Pb using a variety of extraction reagents, some similar for the determination of the micronutrients, such as the DTPA procedure (see Section N.2.f). Beckett (1989) reviewed extraction procedures used to assess the trace element content of soils that have been treated with sewage sludges. Novozamsky et al. (1993a) described a single extraction procedure (0.01 *M* $CaCl_2$; see Houba et al., 2000) for the determination of heavy metals in soil.

Three extraction methods are described.

2. Extraction Reagents and Procedures

a. Ammonium bicarbonate–DTPA (AB–DTPA)

Principle of the Method. This method can be used for determination of bioavailability and biotoxicity of B, Mo, Ni, As, Cd, Pb, and Se in addition to the elements listed above (Baker and Amacher, 1981; Amacher, 1990; Huang and Fujii, 1990; Riser and Baker, 1990; Soon and Abbourd, 1993; Bartlett and James, 1996) in manure-amended, sludge-amended, mine spoil–contaminated, and nonamended soils (Rappaport et al., 1987; 1988; Soltanpour, 1991; Johnson and Donohue, 1992).

Extraction Reagent

NH_4 HCO_3–DTPA

Obtain 0.005 *M* DTPA (diethylenetriaminepentaacetic acid) solution by adding 9.85 g DTPA (acid form) to 4500 mL water in a 5000-mL volumetric flask.
Shake constantly for 5 h to dissolve the DTPA.
Bring to 5000 mL with water. This solution is stable with regard to pH.
To 900 mL of the 0.005 *M* DTPA solution, gradually add 79.06 g ammonium bicarbonate (NH_4HCO_3) and stir gently with a rod to facilitate dissolution but prevent effervescence when bicarbonate is added.
Dilute to 1000 mL with the 0.005 *M* DTPA solution and mix gently with a rod. Adjust the pH to 7.6 with 2 *M* hydrochloric acid (HCl) solution with slow agitation with a rod.
Store the AB–DTPA solution under mineral oil.
Check the pH after storage and adjust it with a 2 *M* HCI solution dropwise, if necessary.

Note: *The cumulative volume of HCl added should not exceed 1 mL/L, after which a fresh solution should be prepared.*

Extraction Procedure

Weigh 10 g air-dried <10-mesh-screened (2-mm) soil into a 125-mL conical flask.
Add 20 mL Extraction Reagent and shake on an Eberbach reciprocal shaker or an equivalent shaker for exactly 15 min at 180 cycles/min with flasks kept open.
Immediately filter the mixture through a Whatman 42 filter paper (Soltanpour and Workman, 1979) and save for heavy metal concentration determination.

Precautions: *The Extraction Reagent is 1 M ammonium bicarbonate (NH_4HCO_3) in 0.005 M DTPA (AB–DTPA) adjusted to a pH of 7.6 (Soltanpour and Schwab, 1977; Soltanpour and Workman, 1979; Soltanpour, 1991). The original pH of 7.6 allows DTPA to chelate and extract Fe and other metals. The range and sensitivity are the same as those for the DTPA-extractable micronutrients (Soltanpour et al., 1977). The AB–DTPA Extraction Reagent is unstable with regard to pH and should be kept under mineral oil to prevent a pH change. Stainless steel soil sampling tubes and polyvinyl chloride mixing buckets for field soil sampling should be used to prevent contamination with* *trace elements. Use high-density aluminum oxide grinders equipped with stainless steel sieves to prevent soil contamination with trace elements. If the above grinder is not available, test other grinders with pure sand to make sure they do not contaminate the soil being processed. Grinding force, time, and the amount of soil in the grinder should be adjusted so that the extracted levels of trace elements are equivalent to those obtained with a wooden roller (Soltanpour et al., 1979). The uneven distribution of Fe, Zn, and Cu in soil particle-sized fractions is discussed by Kahn (1979). A coefficient of variability ranging from 5 to 10% can be expected for different determinations. Accuracy is comparable to that obtained with the DTPA extractant for the micronutrients (Lindsay and Norvell, 1978).*

b. DTPA

Principle of the Method. Lindsay (1979) describes the favorable combination of stability constants for the simultaneous complexing of Cd and Ni that occurs by DTPA chelation. The theoretical basis for DTPA extraction is the equilibrium of the metal in the soil with the chelating agent. The 7.3 pH, which is buffered with triethanolamine (TEA), prevents excess dissolution of the trace (heavy) metals. The use of DTPA as an extracting agent was developed by Lindsay and Norvell (1978).

Extraction Reagent

DTPA (diethylenetriaminepentaacetic acid)

Weigh 1.96 DTPA {$[(HOCOCH_2)_2NCH_2]_2NCH_2COOH$} into a 1000-mL volumetric flask.

Add 14.92 g triethanolamine (TEA) and bring to approximately 950 mL with water.

Add 1.47 g calcium chloride ($CaCl_2 \cdot 2H_2O$) and bring to 1000-mL with water while adjusting the pH to exactly 7.3 with 6 *N* hydrochloric acid (HCl).

The final concentration will be 0.005 *M* DTPA (acid form), 0.1 *M* TEA, and 0.01 *M* $CaCl_2$.

Note: *The DTPA reagent should be in the acid form.*

Extraction Procedure

Weigh 10 g or measure 8.5 cm^3 air-dried <10-mesh-screened (2-mm) soil into a 125-mL extraction vessel.

Add 20 mL Extraction Reagent and shake on a reciprocating shaker for 2 h.

Note: *Samples that are shaken longer than 2 h will give high results because a final equilibrium of the metal and soil is reached in 2 h.*

Immediately filter and collect the filtrate for heavy metal concentration determination.

c. 0.01 *M* calcium chloride

Principle of the Method. The reagent has more or less the same ionic strength (0.3 *M*) as the average salt concentration in many soil solutions, and is able to extract adsorbed cations; the electrolyte concentration stays practically constant, the metal concentration reflects the differences in binding strength and/or solubility among various soils, and the measured elements reflect their availability at the pH of the soil since the extractant is an unbuffered solution. Novozamsky et al. (1993a) initially described this single extraction procedure, a method that is detailed by Houba et al. (2000).

Extraction Reagent

0.01 M $CaCl_2 \cdot 2H_2O$

Weigh 1.47 g calcium chloride ($CaCl_2 \cdot 2H_2O$) into a 1000-mL volumetric flask. Dilute to volume with water.

Comment: *$CaCl_2 \cdot 2H_2O$ may absorb water on standing. The reagent should be standardized by titration with EDTA at pH = 10.0 with Eriochrome Black T as an indicator.*

Extraction Procedure

Weigh 10 g air-dried <10-mesh-screened (2-mm) soil into a 250-mL polyethylene bottle.

Add 100 mL 0.01 *M* $CaCl_2$ solution of 20°C (68°F), and shake mechanically for at least 2 h at room temperature (20°C; 68°F).

Either filter to collect the extract or collect the supernatant after centrifugation and save for heavy metal concentration determination.

3. Method of Heavy Metal Determination

Commonly used elemental assay methods are either flame or flameless atomic absorption spectrophotometry (AAS) (Watson and Isaac, 1990; Wright and Stuczynski, 1996), inductively coupled plasma emission spectrometry (ICP-AES), or inductively coupled plasma emission–mass spectrometry (ICP-MS) (Soltanpour, 1991; Soltanpour et al., 1996; 1998). The sensitivity will vary with the type of instrument used and the wavelength selected. Since instruments vary in their operating conditions, no specific details are given. The standards should be prepared in the Extraction Reagent. Details on these methods of analysis are given in Chapter 5.

Standard preparation is important, particularly when making working standards in the extraction reagents, with details given in Appendix B.

Since the interpretative values for most of the heavy metals can be quite low in concentration, any addition due to contamination can significantly affect the interpretation; therefore, the use of a blank is highly recommended. A blank is obtained by carrying an extraction through the entire process without the sample. Contamination and/or alteration of the extraction reagent may occur in the extraction or filtering process. Reagents also may contain sufficient quantities of a determinated heavy metal to effect the assay.

Glassware and other constant substances can be a source of heavy metals; therefore, washing (Kammin et al., 1995) and cleaning techniques are critical to minimize possible contamination of the elements being assayed (see page 109).

Jones (1990) has suggested working standard ranges in concentration for the micronutrients, B, Cu, and Zn, 0 to 5 mg/L and Mn and Fl 0 to 20 mg/L make-up in the extraction reagent.

4. Interpretation

Heavy metal content of soils and plants and their effect on plant growth and development may be found in the books by Adriano (1996a, b), Pais and Jones (1997), and Kabata-Pendias (2000), and the article by Kabata-Pendias and Adriano (1995). Limited work has been done to evaluate soil test methods for sludge-amended soils. Rappaport et al. (1989) reported that the DTPA method correlated well with metals applied in sludge but found generally poor correlations with plant uptake. More research is needed in this area, particularly with heavy metal–sensitive crops (Kabata-Pendias and Adriano, 1995; Adriano, 1996a, b; Pais and Jones, 1997; Kabata-Pendias, 2000).

The total heavy metal content of agricultural soils in the United States was determined by Holmgren et al. (1993) and a summary table given in the Potash & Phosphate Institute Bulletin 1998-2 (Anonymous, 1998b), as shown in the following table:

Geometric Means of Selected Heavy Metals and Soil Properties for U.S. Mineral Soils

State	No. of samples	Metal, mg/kg					CEC, cmol/kg	Organic C, %	pH
		Cd	Zn	Cu	Ni	Pb			
AL	92	0.037	13.9	6.0	9.4	6.8	2.6	0.59	5.75
AR	62	0.113	37.6	13.5	15.0	13.8	12.5	1.03	5.72
AZ	14	0.233	70.5	38.1	27.9	13.3	13.9	0.35	7.68
CA	279	0.243	82.7	37.3	50.5	9.7	16.5	0.80	7.21
CO	85	0.309	76.1	18.0	14.4	12.8	12.7	0.74	7.67
FL	30	0.375	19.9	31.9	8.0	10.1	6.8	1.51	6.29
GA	136	0.037	11.4	5.3	6.8	6.7	3.2	0.68	5.88
IA	70	0.234	59.2	19.9	25.7	13.4	27.3	2.49	5.95
ID	54	0.338	64.3	20.9	24.4	10.4	16.9	1.07	7.35
IL	131	0.181	52.4	16.2	19.1	16.0	17.6	1.59	6.00
IN	72	0.196	43.4	14.3	14.1	12.0	11.8	1.27	5.71
KS	30	0.313	51.4	15.2	19.8	14.8	19.2	1.15	5.74
LA	113	0.120	39.5	15.1	17.0	14.5	16.6	1.26	5.61
MD	57	0.079	29.5	7.7	11.6	10.3	3.2	0.72	5.77
ME	27	0.165	71.8	64.8	41.2	12.6	13.3	2.23	4.47
MN	89	0.280	68.0	21.8	29.5	12.0	33.1	2.90	5.90
MO	28	0.268	59.5	18.4	24.3	19.8	19.9	1.69	6.62
MT	29	0.367	74.0	20.6	25.8	10.5	17.1	1.41	6.87
NC	163	0.068	12.9	7.0	6.2	9.6	5.3	1.14	5.17
ND	30	0.316	58.7	17.8	25.9	8.6	21.8	1.83	7.13
NE	64	0.332	50.8	15.2	18.4	13.1	19.1	1.43	6.43
NJ	114	0.090	29.2	11.0	8.4	13.0	4.6	0.60	5.92
NM	36	0.200	46.5	15.4	16.2	10.5	14.2	0.57	8.19
NY	74	0.173	60.9	27.0	19.7	15.3	8.1	1.19	5.48
OH	77	0.357	82.1	26.2	27.1	18.2	18.3	1.74	6.35
OK	94	0.083	21.0	9.7	11.1	6.7	7.8	0.61	6.38
OR	88	0.294	67.4	28.6	27.4	8.6	22.2	1.07	6.29
PA	40	0.190	83.8	28.3	24.6	10.6	8.7	1.26	6.00
SD	28	0.531	91.2	29.6	40.5	14.1	29.8	2.48	6.49
TX	349	0.123	30.4	9.5	12.5	7.4	10.9	0.65	7.08
WA	122	0.184	66.0	26.7	26.4	8.5	13.1	0.86	6.30
WI	94	0.207	53.5	17.1	17.5	10.1	12.5	1.58	6.14
U.S.	2771	0.155	41.1	15.5	17.1	10.4	11.1	1.01	6.33

Source: Holmgren, G.G.S. et al., *J. Environ. Qual.*, 22, 335, 1993.

Phosphorus fertilizer can be a source of the heavy metals As, Cr, Hg, Se, U, and V, minor constituents in phosphate rock, but the metals can be increased during manufacture into P fertilizers (Anonymous, 1998b).

The heavy metal content of soils and plants and their effect on plant growth and development may be found in the books by Adriano (1996a, b), Pais and Jones (1997), and Kabata-Pendias (2000), and in the articles by Bingham (1986) and Kabata-Pendias and Adriano (1995). Testing soils for the heavy metals has been discussed by Risser and Baker (1990), Amacher (1996), Bartlett and James (1996), Crock (1996), and Radojevic and Bashkin (1999), and waste-amended soil by Johnson and Donohue (1992).

In some instances, the effect of heavy metal content on plant growth and development is not the concern; rather, the transfer of these metals to food and feed products (Chaney et al., 1987) that will in turn affect human (van Campen, 1991) and animal health (Miller et al., 1991) is of concern. The primary focus has been on Cd (Wolnik et al., 1983; 1985) with the primary source sewage sludge and similar waste products (Chaney, 1980; Rappaport et al., 1987; 1988).

O. Extractable Nitrate–Nitrogen (NO_3–N)

1. Introduction

The presence and accumulation of N, primarily as NO_3–N, is of increasing concern, both because of the harmful effects of excess on crop production, yield, and product quality and because of the harmful effects of excess on the environment. Thus, determination of the NO_3–N that exists in the soil profile and in the plow layer is critical to regulation of fertilizer N additions. The goal is to avoid excess application of N that can pass through into water aquifers and run off onto the surface as NO_3, as well as negatively affect crop yield and quality.

Commonly used test procedures for determining the NO_3–N levels in soils have been described by Houba et al. (1987), Dahnke and Johnson (1990), Johnson (1992b), Maynard and Kalra (1993), Anonymous (1994l), Griffin et al. (1995), Mulvaney (1996), Gelderman and Beegle (1998), and Radojevic and Bashkin (1999). Strong and Mason (1999) discuss the N characteristics of Australian soils and methods of analysis and interpretation.

2. Extraction Reagents and Procedures

a. Ammonium bicarbonate–DTPA

Principle of the Method. The 1 *M* ammonium bicarbonate (NH_4HCO_3) in 0.005 *M* DTPA adjusted to a pH of 7.6 reagent, first proposed by Soltanpour and Schwab (1977) and later described by Soltanpour and Workman (1979) and Soltanpour (1991), is a multielement extraction reagent that extracts water-soluble NO_3.

Extraction Reagent

1 M NH_4HCO_3–DTPA

Obtain 0.005 *M* DTPA (diethylenetriaminepentaacetic acid) solution by adding 9.85 g DTPA (acid form) to 4500 mL water in a 5000-mL volumetric flask.
Shake for 5 h constantly to dissolve the DTPA.
Bring to 5000 mL with water. This solution is stable with regard to pH.
To 900 mL of the 0.005 *M* DTPA solution, add 79.06 g ammonium bicarbonate (NH_4HCO_3) gradually and stir gently with a rod to facilitate dissolution and to prevent effervescence when bicarbonate is added.
Dilute the solution to 1000 mL with the 0.005 *M* DTPA solution and mix gently with a rod.
Adjust the pH to 7.6 using a 2 *M* hydrochloric acid (HCl) solution with slow agitation with a rod. Store the AB–DTPA solution under mineral oil.
Check the pH after storage and adjust it with a 2 *M* HCl solution dropwise, if necessary.

Note: *The cumulative volume of HCl added should not exceed a 1-mL/L limit, after which a fresh solution should be prepared.*

Extraction Procedure

Weigh 10 g air-dried <10-mesh-screened (2-mm) soil into a 125-mL conical flask.
Add 20 mL Extraction Reagent and shake on an Eberbach reciprocal shaker or an equivalent shaker for exactly 15 min at 180 cycles/min with flasks kept open.

Immediately filter the extract through Whatman 42 filter paper and save the filtrate for NO_3 concentration determination.

b. 2 *M* potassium chloride

Principle of the Method. The 2 *M* potassium chloride (KCl) Extraction Reagent method is described by Dahnke and Johnson (1990) and Anonymous (19941), and Mulvaney (1996) uses a salt solution that extracts water-soluble NO_3.

Extraction Reagent

2 M KCl

Weigh 150 g potassium chloride (KCl) into a 1000-mL volumetric flask and bring to volume with water.

Extraction Procedure

Weigh 10 g air-dried <10-mesh-screened (2-mm) soil into a 125-mL conical flask.
Add 50 mL Extraction Reagent and shake on an Eberbach reciprocal shaker or an equivalent shaker for exactly 15 min at 180 cycles/min.
Immediately filter the extract through Whatman 42 filter paper and save the filtrate for NO_3 concentration determination.

c. 0.01 *M* calcium sulfate

Principle of the Method. The 0.01 *M* calcium sulfate ($CaSO_4$) Extraction Reagent, described by Griffin et al. (1995), is an equilibrium solution that extracts water-soluble NO_3.

Extraction Reagent

0.01 M $CaSO_4 \cdot 2H_2O$

Weigh 1.72 g calcium sulfate dihydrate ($CaSO_4 \cdot 2H_2O$) into a 1000-mL volumetric flask and bring to volume with water.

Extraction Procedure

Weigh 5 g (for spectrophotometric NO_3 determination) or 20 g (for determination by NO_3 specific-ion electrode) air-dried <10-mesh-screened (2-mm) soil into a 125-mL conical flask.

Add 50 mL Extraction Reagent and shake on an Eberbach reciprocal shaker or an equivalent shaker for exactly 15 min at 180 cycles/min.
Immediately filter the extract through Whatman 42 filter paper and save the filtrate for NO_3 determination.

d. 0.04 *M* ammonium sulfate

Principle of the Method. The 0.04 *M* ammonium sulfate [$(NH_4)_2SO_4$] Extraction Reagent, described by Griffin et al. (1995), is a salt solution that extracts water-soluble NO_3.

Extraction Reagent

0.04 M $(NH_4)_2SO_4$

Weigh 5.28 g ammonium sulfate [$(NH_4)_2SO_4$] into a 1000-mL volumetric flask and bring to volume with water.

Extraction Procedure

Weigh 5 g (for spectrophotometric NO_3 determination) or 20 g (for determination by NO_3 specific ion electrode determination) air-dried <10-mesh-screened (2-mm) soil into a 125-mL conical flask.
Add 50 mL Extraction Reagent and shake on an Eberbach reciprocal shaker or an equivalent shaker for exactly 15 min at 180 cycles/min.
Immediately filter the extract through Whatman 42 filter paper and save the filtrate for NO_3 determination.

e. 0.01 *M* calcium chloride

Principle of the Method. The extraction reagent is 0.01 *M* $CaCl_2$, which has more or less the same ionic strength (0.03 *M*) as the average salt concentration in many soil solutions. The method has been described in an article by Houba et al. (2000), which includes details on an automated NO_3–N determination.

Extraction Reagent

0.1 M $CaCl_2 \cdot 2H_2O$

Weigh 14.7 g calcium chloride dihydrate ($CaCl_2 \cdot 2H_2O$) into a 1000-mL volumetric flask and bring to volume with water.

Acidified Calcium Chloride Reagent (0.1 M)

Weigh 14.7 g calcium chloride dihydrate ($CaCl_2 \cdot 2H_2O$) into a 1000-mL volumetric flask and dissolve in some water.

Then add 8 mL hydrochloric acid (HCl) (37%) and bring to volume with water.

Extraction Procedure

Weigh 10 g air-dried <10-mesh-screened (2-mm) soil into a 250-mL polyethylene bottle.

Add 100 mL Extraction Reagent and shake mechanically for at least 2 h at room temperature (20°C; 68°F).

Decant about 60 mL of the soil/Extraction Reagent slurry into a centrifuge tube and centrifuge for 10 min at about 1800 *g*.

Use the clear centrifugate for analysis.

Prepare two blanks.

Comments: *(1) Filtration is not recommended because most filter papers either absorb analyte or are contaminated, or cause secondary reactions with soil suspensions, and (2) since only a small amount of clear solution is needed for the measurements, a 3 g-soil sample can be shaken with 30 mL Extraction Reagent directly in centrifuge tubes, the suspension centrifuged, and part of the centrifugate taken out with a pipette and used for the measurement.*

Save the filtrate or centrifugate for NO_3 determination noting the procedure for preparation of standards given below under the section of Standard Series for 0.01 *M* $CaCl_2$ extracts.

3. Nitrate Standards

Primary Nitrate–N Standard (1000 NO_3–N/L)

Obtain commercially prepared standard or weigh 7.22 g potassium nitrate (KNO_3) into a 1000-mL volumetric flask and bring to volume with water.

Working Nitrate–N Standards

Pipette 0.25, 0.5, 1.0, 1.5, and 2.5 mL aliquots of the Primary Nitrate–N Standard into 100-mL volumetric flasks and bring to volume with Extraction Reagent to obtain a series of standards containing 2.5, 5, 10, 15, and 25 mg NO_3–N/L, respectively.

Standard Series for 0.01 M $CaCl_2$ Extract Standards

Pipette 0, 0.5, 1.0, 1.5, and 2.0 mL Stock Solution [375 mg NO_3/L: weigh 2.7083 g potassium nitrate (KNO_3) into 1000-mL volumetric flask and bring to volume with water] into 250-mL volumetric flasks.

Add 25 mL Acidified Calcium Chloride Reagent and bring to volume with water. This standard series has concentrations of 0, 0.75, 1.50, 2.25, and 3.00 mg NO_3–N/L.

Notes: *This standard series has concentrations of 0, 0.75, 1.50, 2.25, and 3.00 mg NO_3–N/L. This standard series should be prepared fresh every week.*

4. Methods of Nitrate Determination

a. UV-VIS Spectrophotometric determination procedure

A number of spectrophotometric procedures can be used to determine the NO_3 concentration in a soil extract (Watson and Isaac, 1990; Mulvaney, 1996), one being by Cd reduction (Huffman and Barbarick, 1981; Keeney and Nelson, 1982; Dorich and Nelson, 1984; Houba et al., 1987; 2000) and the other the Chromotropic Acid Procedure, which is described below.

Reagents

Antimony Sulfate Solution

Weigh 0.5 g Sb metal in 80 mL concentrated sulfuric acid (H_2SO_4). Add 20 mL water to the acid carefully to prevent splattering.

Notes: *Heating will facilitate the dissolution of Sb metal in H_2SO_4. This solution is used for masking (complexing) Cl in the NO_3 determination.*

Chromotropic Acid Solution (CTA)

0.00137 *M* solution of CTA or 4,5-dihydroxy-2,7-naphthalene-disulfonic acid, disodium salt [$(HO)_2C_{10}H_4(SO_3Na)_2$]: weigh 0.5 g CTA in 4.0 kg concentrated sulfuric acid (H_2SO_4).

Note: *This solution is used to develop color with NO_3.*

Fisher G Carbon Black

Or comparable decolorizing carbon.

Determination Procedure

Mix 5 mL soil extract with one 1-mL scoop of Fisher G Carbon Black (used to eliminate organic matter interference).
Shake for 5 min or longer if required to decolorize the solution, and filter.
Place a 0.5-mL aliquot of the latter decolorized soil extract, standards, and Extraction Reagent into 2.5-cm matching spectrophotometric tubes.
Add 3.0 mL water to each tube using an automatic diluter.
Add 2.0 mL Antimony Sulfate Solution followed by 6.5 mL CTA in quick succession to each tube.
Mix thoroughly for consistent results.
After 2 h of cooling in tap water, set the UV-VIS spectrophotometer at zero absorbance at 420 nm with the 0.00 NO_3–N solution. Read the color intensity (absorbance) of soil extracts after 2 h.

Note: *The final concentrations in the standards will be 0.104, 0.208, 0.416, 0.624, and 1.04 mg NO_3–N/L.*

b. Specific-ion electrode determination

The operating characteristics of specific electrodes are covered in Chapter 5 of this guide. An Ionic Strength Adjusting Solution for the determination of NO_3 in an extract may be needed to partially mask the effect of nitrite (NO_2^-) and chloride (Cl^-) anions in the extract (Millham et al., 1970; Mills, 1980; Watson and Isaac, 1990; Gelderman and Beegle, 1998).

Ionic Strength Adjusting Solution {0.01 M aluminum sulfate [$(Al_2SO_4)_2$]; 0.02 M boric acid (H_3BO_3); 0.01 silver sulfate (Ag_2SO_4); 0.02 M sulfamic acid (NH_2HSO_3)}

Weigh 67 g $(Al_2SO_4)_2 \cdot 18H_2O$, 12 g H_3BO_3, 20 g Ag_2SO_4, and 19 g NH_2HSO_3 into a 1000-mL volumetric flask and bring to volume with water.

Determination Procedure

Weigh 20 g soil into a 100-mL vessel.
Add 50 mL Ionic Strength Adjusting Solution and shake for 5 min.
Read the potential while stirring the suspension with a magnetic stirrer.
Read the millivolt reading or read the NO_3–N concentration if a meter has been calibrated.

c. Kjeldahl distillation

The NO_3–N content in an obtained extract can be determined by Kjeldahl steam distillation (Anonymous, 1994m). If NH_4–N is present, it must be distilled off first by making the extract alkaline [normally by adding sodium hydroxide (NaOH) solution], and then Devada's alloy is added and the steam distillation continued. The following is the assay procedure.

Reagents

Magnesium Oxide

Oven-dry magnesium oxide (MgO) in a muffle furnace at 650°C (1202°F) for 2 h.

Sodium Hydroxide (40%)

Weigh 400 g sodium hydroxide (NaOH) pellets into a 1000-mL volumetric flask.

Add 500 mL to dissolve, let cool, and then bring to volume with water.

Devada's Alloy

Grind in a ball mill to pass 100-mesh sieve and 75% to pass a 200-mesh sieve.

Boric Acid Indicator

Weigh 20 g boric acid (H_3BO_3) into a 1000-mL volumetric flask.

Add 20 mL mixed indicator (weigh 0.3 g bromocresol green and 0.165 g methyl red indicators into 400 mL 95% ethanol, and bring to 500 mL with water).

Adjust indicator to a pH of about 5.0 and bring to volume with water.

Procedure

Transfer 50 mL extract into a 300-mL Kjeldahl flask.

Add 20 mL 40% NaOH solution and distill about 75 mL of distillate into an aliquot of the Boric Acid Indicator.

Titrate using a standard acid [0.01 *N* hydrochloric acid (HCl)] to obtain the amount of NH_4–N in the distillate (color change is from dark green to light pink).

Let the mixture in the Kjeldahl flask cool.

Add 0.8 g Devada's alloy, and steam-distill into another aliquot of Boric Acid Indicator about 75 mL of distillate.

Titrate using a standard acid (0.01 *N* HCl) to obtain the amount of NO_3–N in the distillate (color change is from dark green to light pink).

d. Ion chromatography

Nitrate can be determined by ion chromatography, a procedure described by Mosko (1984) and Watson and Issac (1990); more details are given in Chapter 5, Section E.

5. Effects of Storage

Air drying and storage should not have any significant effect on the level of NO_3 in the soil (Bates, 1993; Houba and Novozamsky, 1998).

6. Interpretation

Interpretation of the test results will depend on their use as described by Magdoff et al. (1984; 1990), Schmitt and Randall (1994), and Muchovej and Rechcigl (1995). Soil test interpretation methods have been described by Peck (1977) and Dahnke and Johnson (1990), and the basis and summarization of N fertilizer recommendations are given by Black (1993c), Muchovej and Rechcigl (1995), Ludwick (1998), Maynard and Hochmuth (1997), and Reid (1998b). The Pre-Plant Soil Nitrate Test (PPNT) and Pre-Sidedress Nitrate Test (PSNT) are discussed by Gelderman and Beegle (1998); these are test procedures that are being used extensively to regulate the application of N fertilizer, primarily for corn, with 20 to 25 mg NO_3–N/kg found in the 12-in. depth, the level at which no additional N fertilizer is needed.

P. Extractable Sulfate–Sulfur (SO_4–S)

1. Introduction

Sulfur exists in soil in the form of primary minerals, such as sulfide minerals, pyrite and marcasite (FeS_2), chalcopyrite ($CuFeS_2$), and pyrrhotite ($Fe_{1-x}S$), and as a component of soil organic matter. Over time, S is released from these sources that then contributes to the pool of S available to plants, such as the sulfate (SO_4^{2-}) anion. The SO_4^{2-} anion is easily absorbed by clay and iron and aluminum oxides, adsorption increasing with pH. The SO_4^{2-} anion is found in the soil solution with a major portion of this form existing primarily in the subsoil. Therefore, the depth of soil sampling will vary with method and those parameters used to interpret the analytical data (Lewis, 1999).

Three extraction reagents for determining SO_4–S are (1) monocalcium phosphate (500 mg P/L) and (2) 0.5 *M* ammonium acetate–0.25 *M* acetic acid (Rehm and Caldwell, 1968; Schulte and Eik, 1988; Johnson, 1992c; Singh et al., 1995; Combs et al., 1998) and 0.01 *M* calcium chloride extraction reagent/method for total and soluble S (Houba et al., 2000).

2. Extraction Reagent and Procedures

a. Monocalcium phosphate

Extraction Reagent

Calcium Phosphate Solution (500 mg/L)

Weigh 2.03 g calcium phosphate [$Ca(H_2PO_4)_2 \cdot 2H_2O$] into a 1000-mL volumetric flask and bring to volume with water.

Extraction Procedure

Weigh 10 g air-dried <10-mesh-screened (2-mm) soil into an extraction vessel.
Pipette 25 mL Extraction Reagent into the flask and shake for 30 min.
Add ¼ teaspoon (about 0.15 g) "Activated Carbon" (wash Darco G-60 Activated Carbon with Extraction Reagent until free of measureable S. Rinse with water and oven-dry. Store in a closed container) and shake for an additional 3 min.
Filter and transfer a 10-mL aliquot into another flask for SO_4 concentration determination.

b. 0.5 *M* ammonium acetate–0.25 *M* acetic acid

Extraction Reagent

0.5 M $NH_4C_2H_3O_2$–0.25 M CH_3COOH

Weigh 39 g ammonium acetate ($NH_4C_2H_3O_2$) into a 1000-mL volumetric flask and bring to volume with 0.25 *M* acetic acid (CH_3COOH) (dilute 14.31 *glacial* CH_3COOH in 1000 mL water).

Extraction Procedure

Weigh 10 g air-dried <10-mesh-screened (2-mm) soil into an extraction vessel.
Pipette 25 mL Extraction Reagent into the flask and shake for 30 min.
Add ¼ teaspoon (about 0.15 g) "Activated Carbon" (wash Darco G-60 Activated Carbon with Extraction Reagent until free of measureable S; rinse with water and oven-dry; store in a closed container) and shake for an additional 3 min.
Filter and transfer a 10-mL aliquot into another flask for SO_4 concentration determination.

c. 0.01 *M* calcium chloride

Principle of the Method. The extraction reagent is 0.01 *M* $CaCl_2$, which has more or less the same ionic strength (0.03 *M*) as the average salt concentration in many soil solutions. The method has been described in an article by Houba et al. (2000), which includes details on the determination of two forms of SO_4.

Extraction Reagent

0.01 M $CaCl_2 \cdot 2H_2O$

Weigh 1.47 g calcium chloride ($CaCl_2 \cdot 2H_2O$) into a 1000-mL volumetric flask and bring to volume with water.

Comment: *$CaCl_2 \cdot 2H_2O$ may absorb water on standing; therefore, the reagent should be standardardized by titration with EDTA at pH = 10.0 with Erichrome Black T as an indicator.*

Extraction Procedure

Weigh 10 g air-dried <10-mesh-screened (2-mm) soil into a 250-mL polyethylene bottle. Add 100 mL Extraction Reagent and shake mechanically for at least 2 h at room temperature (20°C; 68°F).
Decant about 60 mL of the extractant into a 100-mL centrifuge tube and centrifuge for 10 min at about 1800 *g*.
Use the clear centrifugate for analysis.
Prepare two blanks.

Comment: *Filtration is not recommended because most filter papers either absorb analyte or are contaminated, or cause secondary reactions with soil suspensions).*

3. Sulfate–Sulfur (SO_4–S) Standards

Stock Solution (1000 mg S/L)

Obtain commercially prepared standard, or weigh 5.434 g potassium sulfate (K_2SO_4) in about 400 mL EDTA Solution [weigh 5.84 g ethylenediaminetetraacetic acid (H_4EDTA) into 1000-mL volumetric flask, add 30 mL concentrated aqueous ammonia (NH_4OH, s.p. = 0.91 g/cm^3), and bring to volume with water] in a 1000-mL volumetric flask and mix to dissolve.

Bring to volume with EDTA Solution.

Standard Series Procedure

Pipette 0, 1, 2, 3, 4, and 5 mL Stock Solution into 100-mL volumetric flasks.

Bring to volume with EDTA Solution [weigh 5.84 g ethylenediaminetetraacetic acid (H_4EDTA) into 1000-mL volumetric flask, add 30 mL concentrated aqueous ammonia (NH_4OH, s.p. = 0.91 g/cm^3)].

Note: *This Standard Series has S concentrations of 0, 10, 20, 30, 40, and 50 mg/L, respectively.*

4. Determination Procedures

a. Turbidity

Reagents

Seed Solution

Dissolve 0.1087 g potassium sulfate (K_2SO_4) in 500 mL water and add 500 mL concentrated hydrochloric acid (HCl).

Stir to bring into solution and slowly add 2.0 g powdered Gum Acacia while stirring to bring into solution.

Store the prepared solution in a refrigerator.

Barium Chloride Crystals

Crush barium chloride ($BaCl_2 \cdot H_2O$) to pass a 20- to 30-mesh sieve.

Measurement Procedure

To a 10-mL aliquot of extractant, add 1 mL Seed Solution and immediately swirl the contents.
Place the flask on magnetic stirrer and add ¼ teaspoon (0.3 g) barium chloride crystals.
Stir for about 1 min.
Transfer an aliquot to a spectrophotometric tube or cuvet and then read the absorbance using a calibrated UV-VIS spectrophotometer at 420 nm.

b. Inductively coupled plasma emission spectrometry (ICP-AES)

This method of determination is described by Novozamsky et al. (1986), and the entire method is given in the 0.01 *M* $CaCl_2$ procedure detailed by Houba et al. (2000).

Reagents

Acidified Calcium Chloride Solution (0.1 M)

Dissolve 1.470 g calcium chloride dihydrate ($CaCl_2{\cdot}2H_2O$) in about 50 mL water in a 100-mL volumetric flask.
Add 8 mL concentrated hydrochloric acid (HCl) (37%) and bring to volume with water.

Barium Chloride Solution (1 M)

Weigh 244 g barium chloride dihydrate ($BaCl_2{\cdot}2H_2O$) into a 1000-mL volumetric flask.
Add about 600 mL water, and then 82 mL concentrated hydrochloric acid (HCl) (s.p. = 1.19 g/cm^3), and bring to volume with water.

Nitric Acid Solution (5 M)

Dilute 34.7 mL concentrated nitric acid (HNO_3) (65%) in about 30 mL water in a 100-mL volumetric flask.
Let cool and bring to volume with water.

EDTA Solution (0.02 M)

Weigh 5.84 g ethylenediaminetetraacetic acid (H_4EDTA) into a 1000-mL volumetric flask.
Add 30 mL concentrated aqueous ammonia (NH_4OH, s.p. = 0.91 g/cm^3) and bring to volume with water.

Determination. SO_4 ions in the calcium chloride ($CaCl_2$) extract are precipitated by Ba ions. After a clean-up, the barium sulfate ($BaSO_4$) precipitate is dissolved by EDTA. The solution is then nebulized into an argon plasma, where all components are vaporized. Sulphate ions decompose and the S ions thus formed emit radiation when returning to their ground state; the intensity of the emission is measured at wavelength 182.04 nm.

Separation of Sulfate

Weigh a series of 15-mL centrifuge tubes [empty weight gram per tube (*e*)].

Add to each tube 1.0 mL Barium Chloride Solution.

Then pipette 5.0 mL centrifuged extract and the Standard Series into these tubes and mix by shaking. Bring the tubes to the same weight by adding water, and centrifuge for 10 min at 1200 to 1500 *g*.

Carefully decant the supernatant so that the precipitate and only about 0.5 mL of liquid remain in the tube.

Next, add 5.0 mL water and repeat the shaking and the centrifugation.

Decant the supernatant again and weigh the tube (*c*) to establish the volume of liquid left behind.

Finally, add 5.0 mL EDTA Solution and shake to bring the precipitate into suspension.

Allow the precipitate to dissolve completely, preferably overnight but at least 2 h, swirling now and then.

Measurement Procedure

Using the Standard Series, pipette 0, 1.00, 2.00, 3.00, 4.00, and 5.00 mL Stock Solution (1000 mg S/L, see page 132) into 100-mL volumetric flasks.

Add 10 mL Acidified Calcium Chloride Solution and bring to volume with water (this Standard Series has concentrations of 0, 10, 20, 30, 40, and 50 mg S/L).

Centrifuge extract and blanks. Determine the concentration of S by ICP-AES at wavelength 182.04 nm.

Comments: *(1) To measure low concentrations of SO_4–S in the soil extracts a 1:1 dilution of the extract with EDTA can be used instead of the 1:5 dilution. The calculation should be amended accordingly. (2) A data management system and system controller are used. In this way the measurements are checked continuously and the data output is in concentration units in the extracts. (3) Because of the linearity of the ICP-AES, it is* *possible to calibrate the ICP-AES using only the highest and zero standard from the standard series. As a check, all standards can be measured as samples.*

Calculate the SO_4–S content of the soil material, expressed in mg SO_4–S/kg, by

$$(5 + c - e) \times (a - b) \times V/w$$

where
c = weight in grams of centrifuge tube with remaining liquid
e = weight in grams of empty centrifuge tube
a = the mg SO_4–S/L in the soil extract
b = the mg SO_4–S/L in the blank extract
V = the total volume in mL taken for extraction
w = weight in grams of soil sample

c. Determination by ion chromatography

Using the monocalcium phosphate extraction reagent procedure (see above), the SO_4–S content in the obtained extract can be determined by ion chromatography (Mosko, 1984; Anonymous, 1994n). Because of the sensitivity of the method, calibration standards containing 0.1, 0.3, 0.7, and 2 mg SO_4–S/L are prepared and 50 µL of the extract and standards is injected into the ion chromatograph; the assay time is from 6 to 8 min.

5. Interpretation

Soil test interpretation methods are described by Peck (1977) and Dahnke and Olson (1990), and the basis and a summary of S fertilizer recommendations are given by Johnson and Fixen (1990), Black (1993c), Mikkelsen and Camberato (1995), Ludwick (1998), Maynard and Hochmuth (1997), and Reid (1998b). The suggested critical range is 5 to 10 mg extractable S/kg (Lewis, 1999).

Q. Testing Organic Soils and Soilless Media

Organic soils and soilless media are tested to determine pH, soluble salts, NO_3–N, Cl, P, K, Ca, Mg, Na, B, Cu, Fe, Mn, and Zn in *Greenhouse Growth Media (Soilless Mixes)* by water saturation extraction.

1. Introduction

Growth media (soilless mixes) used for the production of plants in greenhouses provide relatively low nutrient-holding capacity. The soil solution is the primary source of nutrient elements for plant growth. A water saturation extract of the growth media, therefore, provides a good indication of the available nutrient element status. The medium is saturated with water without preliminary preparation. This procedure eliminates possible segregation of mix components and ensures analysis of the growth medium as the grower is actually using it. Soluble salt and nutrient element concentrations in the water saturation extract are related to the moisture-holding capacity of the growth medium. This process eliminates the need to consider bulk density as a factor in the analysis procedure. One set of guidelines can be used with all soilless mixes (Warncke and Krauskopf, 1983; Berghage et al., 1987; Kidder, 1992; Warncke, 1986; 1988; Whipker et al., 1994). The water-saturation procedure is described by Warncke (1990; 1995; 1998).

Water-saturation extraction for measuring the salt content of soil was adopted by the U.S. Salinity Laboratory (1954). Later, Geraldson (1957; 1970) used the saturated-soil extract approach to determine the nutrient element *intensity* and *balance* in the sandy soils of Florida. Lucas et al. (1972) studied the saturated soil-extract method for analyzing greenhouse growth media (soilless mixes) and found it provided more meaningful results and was more advantageous than the Spurway and Lawton (1949) method. Sonneveld and van den Ende (1971) and Sonneveld et al. (1974) describe procedures for extraction using 1:2 and 1:½ volume extracts for growing media, respectively.

Water-soluble levels of the key micronutrients in prepared growth media (soilless mixes) are quite low. Zinc and Mn concentrations in water saturation extracts of growth media rarely exceed 0.8 mg/L, and Fe concentrations rarely exceed 4.0 mg/L. Hence, it is difficult to distinguish between deficient and adequate levels. In peat and bark-based growth media, the micronutrients are complexed by organic compounds (Verloo, 1980). In evaluating 15 extractants, Berghage et al. (1987) found that extractable levels of Cu, Fe, Mn, and Zn could be increased greatly by using weak solutions of various salts, acids, or chelates in the saturating solution with the saturation-extract procedure. A 0.005 *M* DTPA reagent was found to increase extractable micronutrient levels most consistently while having only a minor effect on the other key test parameters: total soluble salts and extractable levels of NO_3–N, P, K, Ca, Mg, Na, B, and Cl.

2. Extraction with Water

Procedure

Fill a 600-mL beaker about two thirds full with the growth media.
Gradually add water while mixing until the sample is just saturated.

Note: *At saturation, the sample will flow slightly when the container is tipped and can be easily stirred with a spatula. Depending on the growth media composition, the saturated sample may glisten as it reflects light.*

After mixing, allow the sample to equilibrate for 1 h and then check the following criteria to ensure saturation. The saturated sample should have no appreciable free water on the surface, nor should it have stiffened. Adjust as necessary by adding either additional media or pure water.
Then allow to equilibrate for an additional ½ h.
Determine the pH using a calibrated pH meter (see Chapter 2, Section I.2), placing the electrodes into the saturated sample, and read the pH to the nearest 0.1 pH.
Separate the water from the medium by gravitation or vacuum filtration and save the filtrate for soluble salt determination (see Chapter 2, Section S.2) and nutrient element assay.

3. Extraction with 0.005 *M* DTPA to Improve Extraction of Micronutrients

Extraction Reagent

0.005 M DTPA

Weigh 1.97 g dry DTPA (diethylenetriaminepentaacetic acid) into a 1000-mL volumetric flask.
Bring to volume with 50°C (122°F) water to facilitate dissolution.
Allow to cool to room temperature and adjust the volume.

Extraction Procedure

Place 400 cm^3 of growth media in a 600-mL beaker.
Add 100 mL 0.005 *M* DTPA Reagent.
Mix, gradually adding pure water to bring the media just to the point of saturation.

After mixing, allow the sample to equilibrate for 1 h and then check the following criteria to ensure saturation. The saturated sample should have no appreciable free water on the surface, nor should it have stiffened. Adjust as necessary by adding either additional media or pure water.

Then allow to equilibrate for an additional ½ h.

Separate the water from the medium by gravitation or vacuum filtration and save the filtrate for nutrient element assay.

4. Nutrient Element Assay Procedures

Nitrate in the obtained filtrates can be determined by either UV-VIS spectrophotometry (see Chapter 2, Section O.4) or specific-ion electrode (see Chapter 2, Section O.5), and the mineral elements by either AAS or ICP-AES. All these analytical procedures are discussed in Chapter 5.

5. Interpretation

Optimum pH and soluble salt and nutrient levels vary with the greenhouse crop being grown and with management practices. The following general guidelines can be used in making preliminary assessments of the results (Warncke, 1988; 1990; 1995; Kidder, 1992; Whipker et al., 1994).

	Level of acceptance				
Analysis	Low	Acceptable	Optimum	High	Very high
Soluble salts, dS/m	0–0.75	0.75–2.0	2.0–3.5	3.5–5.0	5.0+
Nitrate–N, mg/L	0–39	40–99	100–199	200–299	300+
Phosphorus, mg/L	0–2	3–5	6–10	11–18	19+
Potassium, mg/L	0–59	69–149	150–249	250–349	350+
Calcium, mg/L	0–79	80–199	200+		
Magnesium, mg/L	0–29	30–69	70+		

In the desired nutrient balance, the total soluble salts comprise the following percentages of elements: 8% NO_3–N, 12% K, 15% Ca, and 5% Mg. If Cl and Na are determined, the percentage of each should be less than 10%. If NH_4–N is determined, its concentration should be less than 3%.

The following general interpretation guidelines should be used in the DTPA extraction procedure for B, Cu, Fe, Mn, and Zn. Specific desirable levels will vary with the crop being grown.

Micronutrient	Generally adequate range, mg/L
Boron (B)	0.7–2.5
Copper (Cu)	0.5–1.5
Iron (Fe)	15–40
Manganese (Mn)	5–30
Zinc (Zn)	5–30

The availability of elements in a soilless medium will be influenced by pH, as is shown in Figure 2.5 found in Chapter 2, Section I.4. For most elements, the maximum availability occurs in a pH between 5.5 to 6.0.

R. Organic Matter and Humic Matter Content Determinations

1. Introduction

Soil organic matter consists of two forms: (1) crop and microorganism residues and (2) humus. Crop and microorganism residues contain substantial quantities of several important essential elements, N, P, S, and B, elements that are released into the soil solution in the process of residue decomposition. For fertilizer N adjustments, for each percent organic matter, 10 to 60 lb N/acre can be released from soil organic matter. In addition, N credit (10 to 90 lb N/acre) can be given if the previous crop was a legume (i.e., alfalfa, clover, soybean), thereby reducing the amount of fertilizer N needed for a following grain crop, such as corn, as is shown below:

Organic matter, %	Range in N release, lb N/acre
1.0	40–75[a]
2.0	60–95
3.0	80–115
4.0	100–135
5.0	120–155
6.0	141–175
7.0	160–195

[a] Due to soil texture, there are lower values for clay soils and higher values for sandy soils.

Humus, the end product of residue decomposition, is a very stable complex of large molecular substances, primarily humic acid and fulvic acid. Humus has considerable impact on the physical (soil structure) and physio-chemical (adds considerably to the cation exchange capacity, or CEC) properties of the soil (Tan, 1998). Humus acts as a "glue" that binds soil particles together, which, in turn, forms water-stable aggregates that sustain a friable soil condition. Humus has a high CEC of approximately 200 meq/100 g. It is the humus portion of soil organic matter that deactivates herbicides (Weber and Peter, 1982).

Organic matter test procedures normally do not distinguish between these two forms of soil organic matter, undecomposed resdiues and humus; therefore, care should be used when interpreting an organic matter determination. However, there is a specific test for humus.

Soil organic matter exists in two forms, as crop and microbial residues that, depending on soil temperature and moisture, are continuously undergoing decomposition, and as humus, an end product of organic matter decomposition, which is very stable and contributes to soil structural stability, as well as the water-holding and CEC of the soil (Tan, 1998; Baldock and Skjemstad, 1999). Crop and microbial residues, upon decomposition, are the source for a number of essential plant nutrient elements, such as N, P, and B, whereas humus impacts the effectiveness of applied soil herbicides. The characterization of soil organic matter has been described in detail by Swift (1996).

The commonly used organic matter determination procedures, wet oxidation (Walkley, 1947; Graham, 1948; Nelson and Sommers, 1996; Combs and Nathan, 1998) and loss-on-ignition (Golden, 1987; Ben-Dor and Banin, 1989; Nelson and Sommers, 1996; Combs and Nathan, 1998), do not distinguish between these two forms of organic material in the soil. Therefore, those who use either of these organic matter determination procedures for predicting N release may either significantly under- or overestimate N contributions to a growing crop.

The loss-on-ignition procedure for organic matter determination is still undergoing study and modification (Golden, 1987).

2. Methods of Organic Matter Determination

a. Wet digestion

The total soil organic matter is routinely estimated by measuring organic C content. The procedure is described by Mebius (1960). The method described is a wet-oxidation procedure using potassium dichromate ($K_2Cr_2O_7$) with external heat and back-titration to measure the amount of unreacted dichromate.

This method and other methods are thoroughly discussed by Hesse (1971), Jackson (1958), and Allison (1965). The procedure is rapid and adapted for routine analysis in a soil testing laboratory. It is primarily used to determine the organic matter of mineral soils.

The method is useful for soils containing a range of organic C from very low organic C to as high as 12% organic C with a sensitivity of about 0.2 to 0.5% organic C. Soils containing large quantities of Cl, Mn^{2+}, and Fe^{2+} ions yield high results. The Cl interference can be eliminated by adding silver sulfate (Ag_2SO_4) to the Oxidizing Reagent. No known procedure is available to compensate for the other types of interference. The presence of $CaCO_3$ up to 50% causes no interference. This procedure is not recommended for high-organic-matter-content soils or organic soils. The method is an incomplete digestion, and a correction factor must be applied. The correction factor used is 1.15 (Allison, 1965).

Repeated analyses should produce results with a coefficient of variability no greater than 1 to 4%. Soil samples must be thoroughly ground and mixed before subsampling because heterogeneity is a serious problem in organic matter distribution within samples.

Titration Procedure

Reagents

Potassium Dichromate Reagent (0.267 N)

Weigh 13.072 g potassium dichromate ($K_2Cr_2O_7$) into a 1000-mL volumetric flask.
Add 400 mL water to dissolve and then add 550 mL concentrated sulfuric acid (H_2SO_4).
Let cool and bring to volume with water.

Mohr's Salt Solution (0.2 M)

Weigh 78.390 g ferrous ammonium sulfate [$Fe(NH_4)_2(SO_4)\cdot 6H_2O$] into a 1000-mL volumetric flask.
Add 500 mL water to dissolve and then add 50 mL concentrated H_2SO_4.
Let cool and bring to volume with water.
Prepare fresh for each use.

Indicator Solution

Weigh 200 mg *n*-phenylanthranilic acid into a 1000-mL volumetric flask containing 0.2% sodium carbonate (Na_2CO_3) solution.

Determination Procedure

Weigh 0.1 to 0.5 g (depending on estimated organic content) of air-dried <10-mesh-screened (2-mm) soil into a 500-mL Erlenmeyer flask and add 15 mL 0.267 *N* Potassium Dichromate Reagent.

Connect the flask to a reflux condenser and boil for 30 min.

Let cool.

Wash down the condenser and flush with pure water. Add 3 drops of the Indicator Solution and titrate with Mohr's Salt Solution at room temperature.

As the end point is approached, add a few more drops of the Indicator Solution. The color change is from violet to bright green.

A blank analysis with no soil added is carried through the procedure.

Calculation

$$\% \text{ organic carbon (C)} = \{[(\text{meq } K_2Cr_2O_7 - \text{meq } FeSO_4) \times 0.3]/\text{grams soil}\} \times 1.15$$

$$\% \text{ organic matter} = \% \text{ organic carbon} \times 1.724$$

Colorimetric Procedure

Reagents

Sodium Dichromate Reagent (0.67 M)

Weigh 4.000 g sodium dichromate ($Na_2Cr_2O_7$) into a 1000-mL volumetric flask and bring to volume with water.

Technical Grade Sulfuric Acid

H_2SO_4

Determination Procedure

Weigh 2.0 g or scoop 1.5 cm^3 air-dried <10-mesh-screened (2-mm) soil into a 200-mL test tube. Under a hood, add 20 mL 0.67 *M* Sodium Dichromate Reagent and then add 20 mL H_2SO_4.

Mix thoroughly but slowly in order to keep the soil and digestion mixture off the sides of the flask and allow it to cool at least 40 min.

After cooling, add 100 mL water, mix, and allow it to stand at least 8 h. Transfer an aliquot of the clarified solution to a spectrophotometer vial using a syringe pipette.

Measure absorbance (A) at 645 nm using a UV-VIS spectrophotometer.

Standard Curve. A standard curve should be established with several soils that have an adequate range of organic matter contents. The percentage of organic matter is determined by a standard method (see Titration Procedure). Absorbance values are determined for each known soil organic matter and a curve is constructed by plotting the percentage of organic matter vs. absorbance, including a reference sample with daily runs of the method aids in verifying equivalent conditions between the standard curve and daily runs.

b. Loss-on-ignition (LOI)

Total soil organic matter is estimated by loss-on-ignition (LOI). The procedure was initially described by Davies (1974), and the method described here is that given by Ben-Dor and Banin (1989). The method described is a procedure in which a soil sample is dried at 105°C (221°F) and then ashed at 400°C (752°F). The loss in weight between 105°C (221°F) and 400°C (752°F) constituents the organic matter content. The results obtained compare favorably with those obtained by the dichromate wet-oxidation method and by carbon analyzers (Gallardo and Saavedra, 1987; Golden, 1987; Ben-Dor and Banin, 1989; Lowther et al., 1990; Schulte et al., 1991). Others have used different ashing temperatures ranging from 360°C (680°F) (Schulte et al., 1991) to 600°C (1112°F) (Gallardo and Saavedra, 1987; Ben-Dor and Banin, 1989).

The method is useful for soils containing low to very high organic matter contents with a sensitivity of about 0.2 to 0.5% organic matter. However, the method is generally considered not suitable for organic matter determination for calcareous soils. The presence of calcium carbonate ($CaCO_3$) may interfere.

Consistent analytical results are obtainable under a range of sample sizes, ashing vessels, ashing temperatures, and length of ashing time. However, the mineral composition of the soil may be a factor in the determination and may require more than one calibration curve (Schulte et al., 1991). In addition, soil horizons may be another factor affecting LOI results (David, 1988).

An automated system for determining organic matter content by LOI has been described by Storer (1984). Repeated analyses should produce results with a coefficient of variability of no greater than 1 to 4%.

Determination Procedure

Weigh 5.00 to 10.00 g (weigh to the nearest 0.01 g) sieved 2-mm soil into an ashing vessel (50-mL beaker or other suitable vessel).

Place the ashing vessel and soil into the drying oven set at 105°C (221°F) and dry for 4 h.
Remove the ashing vessel from the drying oven and place in a dry atmosphere.
Once cool, weigh to the nearest 0.01 g.
Place the ashing vessel plus soil into a muffle furnace and bring the temperature to 400°C (752°F).
Ash in the furnace at 400°C (752°F) for 4 h.
Remove the ashing vessel from the muffle furnace, let cool in a dry atmosphere, and weigh to the nearest 0.01 g.

Calculation. Percent organic matter in the soil is determined by the formula:

$$\%\ \mathrm{OM} = [(W_{105} - W_{400}) \times 100]/W_{105}$$

where W_{105} is the weight of soil at 105°C (221°F) and W_{400} is the weight of soil at 400°C (752°F).

Standard Curve. A standard curve may be established with several soils that have a range of organic matter contents encompassing that in the unknowns. The percentage of organic matter in the standards will have been determined by other methods. More than one calibration curve may be required for varying soil types (David, 1988; Schulte et al., 1991).

c. Humic matter by 0.2 *N* sodium hydroxide extraction

This extraction method is designed to determine the sodium hydroxide (NaOH)-soluble humic matter, which consists of humic and fulvic acids. These components comprise approximately 85 to 90% of the soil humus and are responsible for the cation and anion exchange properties exhibited by the soil organic fraction. This method is based on the concept that humic matter compounds are soluble in dilute alkali solutions (Mortensen, 1965; Levesque and Schnitzer, 1967; Hayes et al., 1975; Tucker, 1992d). Acidic organic compounds are converted to ions with the subsequent formation of a physical solution of these ions in water (Mehlich, 1984b). The reaction of a dilute alkali with the humic matter results in a colored solution that is proportional to the soluble humic matter content within the soil. The color varies from shades of brown to black, depending on the type of soil from which the sample originates. Colorimetric determination of the humic matter content of soils by this method is based upon the color intensity of the solution following extraction with a dilute alkali extractant. The alkali used in the method is NaOH, which serves as the humic acid solvent (Tucker,

1992d). DTPA aids in the dispersion of some of the large molecular Ca–humate compounds, and ethanol aids in the solubility of hydrophobic lipid components of soil organic matter. Calibration data were generated from a standard humic matter source.

This method was designed to accomplish two major objectives: (1) to estimate the chemically reactive portion of the soil organic fraction for better prediction of herbicide rate requirements and (2) to remove Cr from the effluent of municipal waste systems. Experimental evidence has shown that this method can be used to predict herbicide rates (Strek and Weber, 1983; Strek, 1984).

As much as 10% of the humic matter content of soils can be determined by this method. Higher levels could be determined with a wider extraction ratio (Mehlich, 1984b). The method as described will encompass a majority of mineral soils. Saturation of the method is encountered on the organic and mineral organic soils where total organic matter is high. However, in some organic soils, the humic matter content is low even though the percentage of combustible organic content may be in excess of 90%. The sensitivity of this method depends on the quality and homogeneity of the field sample. See Appendix E for the ASI method of humus determination.

Extraction Reagents

0.2 *N* sodium hydroxide (NaOH); 0.0032 *M* DTPA; 2% ethanol (C_2H_5OH)

Preparation

Using a 4-L volumetric flask, add about 1000 mL water, 32 g sodium hydroxide (NaOH), and dissolve.

Then add 16 mL diethylenetriaminepentaacetic acid (DTPA), pentasodium salt, and 80 mL ethanol (C_2H_5OH).

Add water to volume and mix thoroughly.

Prepare larger volumes of extractant, depending on the numbers of samples to be analyzed.

Standard Humic Acid

Dry the Humic Acid Standard (Aldrich Chemical Co., 940 W. St. Paul Ave., Milwaukee, WI 53233) at 105°C (221°F) for approximately 4 h. Loss on ignition at 550°C (1022°F) shows that this humic matter standard contains 87% organic matter.

For calibration, weigh 0.115 g Standard Humic Acid ($0.1 \div 0.87 = 0.115$) and place into a 55-cm^3 polystyrene vial.

Standard Humic Matter Calibration Procedure

Weigh 0.115 g Standard Humic Acid into a 55-cm^3 polystyrene vial and add 20 mL Extracting Reagent with sufficient force to mix the Standard Humic Acid.

Allow the sample to set for 1 h and then add an additional 20 mL Extracting Reagent with sufficient force to mix well.

Add the two 20-mL portions of extractant separately to enhance dissolution of the humic matter.

Let the sample set overnight (16 to 18 h minimum), then pipette 5 mL of the supernatant and 35 mL water into 55-mL polystyrene vials.

Caution: *Care should be taken not to pipette colloidal precipitation from the bottom of vials.*

Achieve a final dilution of the sample of 1:8 ratio (5 mL sample + 35 mL water), which is required at this extraction ratio to fall within the UV-VIS spectrophotometer reading range.

Set the UV-VIS spectrophotometer at 100% T with 5 mL Extraction Reagent and 35 mL water.

Read the standard at 650 nm.

Note: *When a Brinkman probe spectrophotometer with a 2-cm light path is used, the standard humic acid standard should read 10% T. This equates to 10 g 100/cm^3 humic matter equivalent. A standard curve can be developed by sequential 1:1 dilutions of the 10% humic acid standard. To develop the factor for converting the UV-VIS spectrophotometer reading to g humic matter 100/cm^3, convert %T to absorbance, then divide g humic matter 100/cm^3 by the absorbance. Assuming linearity of the standard, the ratio of g humic matter absorbance should be a constant.*

If a larger volume of humic acid standard is required for calibration, use multiple quantities of standard humic acid and extractant.

Extraction Procedure

Scoop 1 cm^3 soil (screened 2-mm) into a 55-mL polystyrene vial and add 20 mL Extracting Reagent with sufficient force to mix well.

After 1 h, add another 20 mL Extracting Reagent with mixing force and allow the samples to set overnight.

Note: *In addition to allowing adequate reaction time of humic matter with the Extracting Reagent, setting allows soil particles to settle out, leaving a clear supernatant.*

Determination Procedure

Transfer 5 mL of undisturbed supernatant and 35 mL water into a 55-mL polystyrene vial.

Set the UV-VIS spectrophotometer to read 100% T with a blank (5 mL Extraction Reagent + 35 mL water).

Read the samples at 650 nm and record the %T.

Note: *A check sample whose humic matter content has been previously determined should be analyzed routinely with unknown samples. Samples that exceed 10% humic matter can be diluted with water and the appropriate dilution factor employed.*

Calculations. Humic matter (HM) content of a soil can be determined from a standard curve or by converting %T to absorbance (Abs) and multiplying by the factor developed in the calibration procedure. For this method the factor is 10; therefore, Abs × 10 = g humic matter equiv 100/cm^3 of soil. If the percentage of HM on a weight basis is desired, divide humic matter (g/100/cm^3) by the WV (weight/volume in g/cm^3) of each soil for specific values in the development of this procedure.

Calibration Procedure. Values shown below were developed to determine humic matter up to 10%, using an extraction ratio of 1:40 (1 cm^3 soil + 40 mL Extracting Reagent), with 0.115 g humic acid standard (87% organic matter).

Humic matter equiv,[a] g/100 cm^3	Abs	Humic matter, equiv/Abs	Factor[b]
10.000	—	—	
5.000	—	—	
2.500	—	—	
1.250[a]	1.000	1.25	10
0.625	0.509	1.23	
0.313	0.206	1.21	
0.156	0.131	1.19	
0.078	0.061	1.28	
0.036	0.027	1.33	
		Avg 1.25 × 8 = 10	

[a] The Standard Humic Acid (HA) sample is diluted sequentially (1:1) with water for the development of the standard curve.

[b] The factor is determined by taking the average of humic matter per absorbance multiplied by 8 (d.f.).

[c] The Standard HA sample is diluted 1:8 (5 cm^3 sample extract + 35 mL water). Read at 650 nm = 10% T or 1.0% absorbance. Unknown samples can be diluted in the same manner.

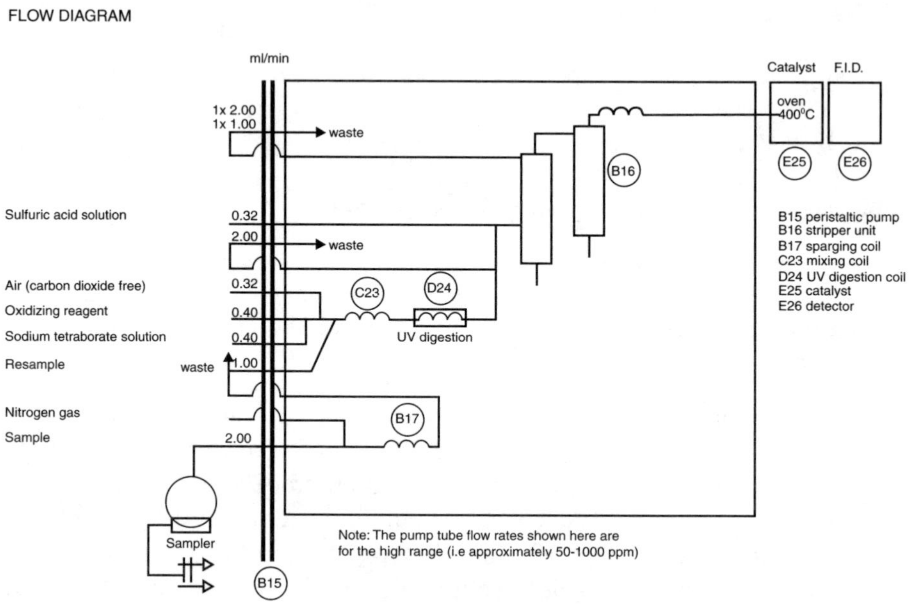

Figure 2.11
Flow diagram for the determination of soluble organic matter. (From Houba, V.J.G. et al., *Commun. Soil Sci. Plant Anal.*, 31:1299, 2000. With permission.)

d. Determination of soluble organic carbon

The soluble organic C is measured with an Organic Carbon Analyzer. Inorganic C is excluded from the measurement by acidification of the extract with sulfuric acid (H_2SO_4). The produced CO_2 is driven out of the system by N_2 gas. The extract is next mixed with a solution containing persulfate and tetraborate and digested under influence of UV light. Organic C is oxidized to CO_2, and this gas is expelled from the solution by acidification and next mixed with H_2 gas. This combination of gases is fed over a Ni catalyst at 400°C (752°F) by which the CO_2 is transformed into methane. The concentration of methane (CH_4) is next measured with a flame ionization detector. The flow diagram for the determination of soluble organic C is shown in Figure 2.11.

Concentrations between 0.5 and 50 mg C/L can be measured. The detection limit is approximately 1 mg C/L in the extract. The determination limit is approximately 3 mg C/L (30 mg C/kg in the soil). High salt levels can influence the peak height measurement. The reproducibility of determinations by this procedure should give, at reasonable control and thorough sample preparation, a coefficient of variation within 10%. The method is described in detail by Houba et al. (2000).

Reagents

Extraction Reagents (0.01 M $CaCl_2 \cdot 2H_2O$)

Weigh 14.7 g calcium chloride ($CaCl_2 \cdot 2H_2O$) into a 1000-mL volumetric flask and bring to volume with water.

Comment: *$CaCl_2 \cdot 2H_2O$ may absorb water on standing; therefore, the reagent should be standardized by titration with EDTA at pH = 10.0 with Erichrome Black T as an indicator.*

Stock Solution (1000 mg C/L)

Weigh 2.125 g potassium hydrogen phthalate ($KC_8H_5O_4$) in about 900 mL water in a 1000-mL volumetric flask and bring to volume with water.

Tetraborate Solution

Weigh 35 g sodium tetraborate decahydrate ($Na_2B_4O_7 \cdot 10H_2O$) in about 200 mL water in a 1000-mL volumetric flask and bring to volume with water.

Oxidizing Reagent

Weigh 35 g potassium persulfate ($K_2S_2O_8$) into 1000-mL volumetric flask and bring to volume with water.

Mix and leave this solution for 24 h in a refrigerator.

Thereafter, decant the clear supernatant solution.

Sulfuric Acid Solution (1 M)

Dilute carefully, in portions, 50 mL concentrated sulfuric acid (H_2SO_4) in about 500 mL water in a 1000-mL volumetric flask.

Allow the mixture to cool and bring to volume with water.

Procedure

Weigh 10 g air-dried <10-mesh-sieved (2-mm) soil into a 250-mL polystyrene bottle.

Add 100 mL Extraction Reagent and shake mechanically for at least 2 h at room temperature (20°C; 68°F).

Decant about 60 mL of extractant into a 100-mL centrifuge tube and centrifuge for 10 min at about 1800 *g*.

Use the clear centrifugate for analysis.

Prepare two blanks.

Comments: *Filtration is not recommended because most filter papers either absorb analyte or are contaminated, or cause secondary reactions with soil suspensions.*

In the Standard Series, soil extract, and blanks, determine the C concentration using the Organic Carbon Analyzer.

Comment: *A data management system and system controller are used. In this way, the measurements are checked continuously and the data output is in concentration units in the extracts.*

Standards

Standard Series

Pipette 0, 1, 2, 3, 4, and 5 mL Stock Solution (1000 mg C/L) in 100-mL volumetric flasks and bring to volume with water.

Note: *This standard series has C concentrations of 0, 10, 20, 30, 40, and 50 mg/L.*

Calibration Curve

Plot the output of the flame ionization detector vs. mg C/L in the standard series.

Calculation. Concentration soluble organic C in the soil material, expressed in mg C/kg, is calculated by: $(a - b) \times 10$, where a is the concentration of soluble organic C in the soil extract in mg/L and b is the concentration of soluble organic C in the blank extract in mg/L.

S. Soluble Salt Determination

1. Introduction

The level of soluble salts found in the soil solution can be classified by determining the electrical conductivity (EC) of the solution, or the soil solution can be assayed for its elemental content. The first procedure is more commonly performed. As the level of soluble salts increases, the usual effect is decreased plant growth; therefore, soluble salt determination has considerable significance. Soils affected by high soluble salt levels are also difficult to manage, particularly when Na is the major cation contributing to the high salt level (Richards, 1969; Shaw, 1999).

In the natural environment, soils with high soluble salt content are found in low-rainfall areas. In the managed environment, soluble salt problems occur as a result of the use of salt-laden irrigation waters, or as a result of improper fertilizer placement, high fertilizer application rates, or accumulation from repeated fertilizer applications.

Salinity, the result of high Na-salt presence, affects about 25% of the croplands in the world and is becoming an increasing problem in most irrigated croplands. Most field crops can be damaged by fertilizer placement too close to germinating seeds or young plants. High soluble salt is a common problem in container growing, the result of accumulating applied plant nutrient element ions. A soluble salt determination is recommended as an important procedure for the container and hydroponic grower, indicating when containers should be leached or hydroponic solutions replenished (Merkle and Dunkle, 1944; Geraldson, 1957; Kuehny and Morales, 1998).

The cations Na^+, K^+, Ca^{2+}, Mg^{2+}, and NH_4^+, and the anions Cl^-, NO_3^-, HCO_3^-, SO_4^{2-}, and CO_3^{2-} contribute to the conductivity of the soil solution or irrigation water.

Not only is the total ion content in the soil solution or irrigation water significant, but the ratio of Na to the cations Ca and Mg, known as the Sodium Adsorption Ratio (SAR), is also an indicator of soil or water quality. The SAR is calculated as follows:

$$\text{SAR} = \text{Na}\Big/\sqrt{(\text{Ca} + \text{Mg})/2}$$

Electrical resistance is defined by the equation $\mathbf{E} = IR$, where $\mathbf{E}$ is the electrical potential in volts, I is the current in amperes, and R is the resistance in ohms. Electrical conductance (C), or the conductivity of a solution, is the reciprocal of resistance:

$$C = I/R$$

C is expressed as mhos, but for convenience more commonly as millimhos, which is 1/1000 mhos.

Because the numbers for expressing specific conductance are relatively small, it has been found convenient to express specific conductance as millimhos per centimeter (mmhos/cm). In the older literature, it was common to see specific conductance expressed as mhos $\times$ 10^{-1}/cm. In International Units, mmhos/cm is equivalent to decisiemans per meter (dS/m).

The specific conductance or soluble salt level of a soil can be determined based on a water saturation extract or a 1:2 soil/water extraction. An appropriate quantity of extractant is obtained and its specific conductivity determined. The extraction procedure (soil/water ratio) chosen will yield different interpretative results; therefore, it is important that the interpretation values chosen are appropriate to the extraction procedure employed.

2. The Conductivity Meter

Specific conductance is determined using an electrical resistance (Wheatstone) bridge, commonly referred to as either a conductance or a conductivity meter. Specific conductance of a solution is the conductance that is measured at 25°C (77°F) between electrodes 1 cm^3 in surface area and placed 1 cm apart. For convenience, most cells for measuring the conductance of solutions are as described above so that the cell constant is 1. Most meters currently measure conductivity directly.

3. Standard Calibration Solution

It is not generally necessary to determine the cell constant unless there is some doubt about it. The cell constant can be determined by preparing a 0.01 *N* KCl solution (weigh 0.7456 g potassium chloride into 500 mL in a 1000-mL volumetric flask and bring to volume with water), which will give a specific conductance reading of 1.4118 dS/m at 1 at 25°C (77°F), which is the cell constant.

4. Procedures

The various methods for determining soluble salts or salinity have been described by Anonymous (1994k), Rhoades (1996), Whitney (1998b), and Shaw (1999).

2:1 Water/Soil Extraction

Scoop 10 cm^3 2-mm-sieved soil into a beaker, add 20 mL water, stir thoroughly, and allow the suspension to settle for at least 30 min or long enough for the solids to settle.

Draw the supernatant into the conductivity pipette and measure the conductivity (Anonymous, 1983c; Gartley, 1995; Rhoades, 1996).

1:1 Water/Soil Extraction

Scoop 20 cm^3 2-mm-sieved soil into a test tube or small container.

Add 20 mL water, stir thoroughly, and allow the suspension to stand 15 to 20 min.

Insert the conductivity cell into the suspension and read the electrical conductivity (Gartley, 1995; Rhoades, 1996; Whitney, 1998b).

Saturated Paste Method

Weigh 250 g air-dried <10-mesh-sieved (2-mm) soil into a 400-mL beaker. Add water while stirring with a spatula until the soil slides freely from the surface of the spatula.

Note: *At saturation, the soil paste will glisten as it reflects light. Let stand for 1 h.*

Transfer the saturated paste to a filter funnel and draw water from the soil by applying vacuum and determine the conductivity in the obtained filtrate (Anonymous, 1984k; Janzen, 1993; Gartley, 1995).

5. Interpretation

Interpretation of conductance readings (dS/m) for soils is as follows:

Saturated paste	2:1 Water/soil	Effects
<1.0	<0.40	*Nonsaline*: Salinity effects mostly negligible, except possibly beans and carrots
1.1–2.0	0.40–0.80	*Very Slightly Saline*: Yields of very salt-sensitive crops such as flax, clovers (alsike, red), carrots, onions, bell pepper, lettuce, and sweet potato may be reduced by 25 to 50%
2.1–4.0	0.81–1.20	*Moderately Saline*: Yield of salt-sensitive crops restricted; seedlings may be injured; satisfactory for well-drained greenhouse soils; crop yields reduced by 25 to 50% may include broccoli and potato plus the other plants listed above
4.1–8.0	1.21–1.60	*Saline Soils*: Crops tolerant include cotton, alfalfa, cereals, grain sorghum, sugar beets, bermuda grass, tall wheat grass, and Harding grass; salinity higher than desirable for greenhouse soils
8.1–16.0	1.61–3.20	*Strongly Saline*: Only salt-tolerant crops yield satisfactorily; for greenhouse crops leach soil with enough water so that 2 to 4 quarts (2 to 4 L) pass through each square foot (0.1 mL) of bench area, or 1 pint of water (0.5 L) per 6-in. (15-cm) pot; repeat after 1 h; repeat again if readings are still in the high range
>16.0	>3.2	*Very Strongly Saline*: Only salt-tolerant grasses, herbaceous plants, and certain shrubs and trees will grow

The relationship between conductivity (EC) and degree of salinity by the 1:1 saturated method and soil texture is as follows:

	Soil texture, dS/m			
Degree of salinity	Coarse to loamy sand	Loamy fine sand to loam	Silt loam to clay loam	Silty clay loam to clay
Nonsaline	0–1.1	0–1.2	0–1.31	0–1.4
Slightly saline	1.2–2.4	1.3–2.4	1.4–2.5	1.5–2.8
Moderately saline	2.5–4.4	2.5–4.7	2.6–5.0	2.0–5.7
Strongly saline	4.5–8.9	4.8–9.4	5.1–10.0	5.8–11.4
Very strongly saline	>9.0	>9.5	>10.1	>11.5

The relationship between conductivity (EC) and degree of salinity by the saturation method for all soils is as follows:

Degree of salinity	dS/m
Nonsaline	0.0–2.0
Slightly saline	2.2–4.0
Moderately saline	4.1–8.0
Strongly saline	8.1–16.0
Very strongly saline	>16.1

6. Electrical Conductivity Units and Conversions

1 millimho per centimeter = mmho/cm = EC × 10^{-3}
1 micromho per centimeter = µmho/cm = EC × 10^{-6}
1000 micromho per centimeter = 1 mmho/cm
1 milliseiman per centimeter = 1 mS/cm = 1 mmho/cm
1000 microseiman per centimeter = 1 µS/cm = 1 µmho/cm
1 deciseiman per meter = 1 dS/m = 1 mS/cm = 1 mmho/cm = 700 ppm
Acceptable media EC levels for most greenhouse crops range from 0.75 to 1.5 dS/m
Unacceptable alkalinity levels in irrigation water are approximately:
plug production = 1.5 meq/L or 75 ppm $CaCO_3$
container production = 2.0 meq/L or 100 ppm $CaCO_3$

7. Salinity (NaCl) and Alkalinity ($NaHCO_3$)

20–10–20 fertilizer (200 ppm) = EC = 1.5 dS/m and a pH of 5.6
100 ppm NaCl = EC of 3.0 dS/m
400 ppm NaCl = EC of 7.2 dS/m
100 ppm $NaHCO_3$ (pH = 6.06) = EC of 1.5 dS/m
200 ppm $NaHCO_3$ (pH = 7.14) = EC of 1.9 dS/m
300 ppm $NaHCO_3$ (pH = 7.43) = EC of 2.3 dS/m
400 ppm $NaHCO_3$ (pH = 7.62) = EC of 2.8 dS/m
500 ppm $NaHCO_3$ (pH = 7.88) = EC of 3.1 dS/m

T. Soil Texture (Mechanical Analysis)

1. Separate Characteristics

As with soil organic matter content, texture can tell much about the physico-chemical properties of a soil. Water-holding capacity, aeration, ease of handling, tendency to crust, and cation exchange capacity (CEC) are properties determined in some degree by particle size distribution. The three soil separates are sand (although this fraction may be separated into several particle size ranges), silt, and clay; designations are determined by particle diameter (size) based on the U.S. Department of Agriculture classification as follows:

Separates	Diameter limits, mm
Fine gravel	2.0–1.0
Coarse sand	1.0–0.5
Medium sand	0.05–0.25
Fine sand	0.25–0.10
Very fine sand	0.01–0.05
Silt	0.05–0.002
Clay	below 0.002

The international system for size limits of soil separates is as follows:

Fraction	Diameter range, mm
I	2.0–0.2
II	0.20–0.02
III	0.02–0.002
IV	below 0.002

2. Textural Classifications

By knowing the percentage of sand, silt, and clay present, a textural classification (there are 12 textural classes: sand, loamy sand, sandy loam, loam, silt loam, silt, sandy clay loam, clay loam, silty clay loam, sandy clay, silty clay, and clay) can be assigned to the soil, a system devised by the U.S. Department of Agriculture as shown in Figure 2.12. The U.S. Department of Agriculture Textural Classification Chart for Soil Material Less Than 2 mm is as follows.

U.S. Department of Agriculture Textural Classification Chart for Soil Material <2 mm

Basic soil class	Subclass	Soil separates: Very coarse sand, 2.0–1.0 mm	Coarse sand, 1.0–0.5 mm	Medium sand, 0.5–0.25 mm	Fine sand, 0.25–0.1 mm	Very fine sand, 0.1–0.05 mm
Sands	Coarse sand	≥ 25%		< 50%	< 50%	< 50%
	Sand	≥ 25%			< 50%	< 50%
	Fine sand		< 25%	*or*	≥ 50%	< 50%
	Very fine sand					≥ 50%
Loamy sands	Loamy coarse sand	≥ 25%		< 50%	< 50%	< 50%
	Loamy sand		≥ 25%	< 50%	< 50%	< 50%
	Loamy fine sand		< 25%	*or*	≥ 50%	< 50%
	Loamy very fine sand					> 50%
Sandy loams	Coarse sandy loam	≥ 25%		< 50%	< 50%	< 50%
	Sandy loam		≥ 30%			
					and	
		< 25%			< 30%	< 30%
	Fine sandy loam		*or*		≥ 30%	< 30%
			15 to 30%	*or*		≤ 30%
	Very fine sandy loam		< 15%			> 40%[a]

[a] Half of fine sand and very fine sand must be very fine sand.

Figure 2.13 shows a comparison of particle size limits in four systems of particle size classification.

3. Method of Determination

The amount of sand, silt, and clay in a soil can be determined by several different methods, although all are based on the same basic principle — varying settling velocities due to different particle size in a standing column

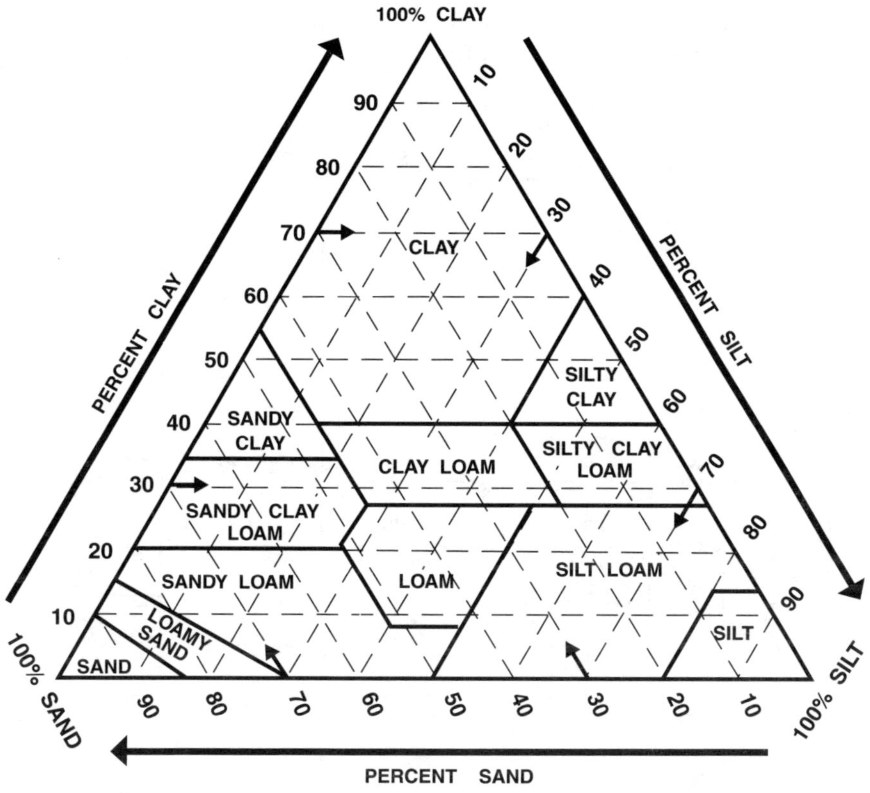

Figure 2.12
Soil textural classes based on percentage content of sand, silt, and clay.

of water (the principle of sedimentation is known as Stokes' law). This technique is frequently referred to as a "mechnical analysis." Particle separation by varying settling velocity assumes a consistent particle density (2.65 g/cm^3) for all the separates.

To separate the individual soil particles adequately, the soil sample is treated with a dispersing agent (a mixture of sodium heximetaphosphate and sodium carbonate — CALGON®) and vigorously mixed. Pretreatment of the soil is necessary to remove all particles greater than 2 mm in size (done by sieving), and treatment with hydrogen peroxide (H_2O_2) to oxidize (remove) the organic matter and an acid to remove carbonates may be necessary to enhance particle dispersion (Sheldrick and Wang, 1993).

Once the soil sample is dispersed and vigorously mixed, passing the dispersed soil through an ASTM Number 325 (USDA system) sieve will remove all the sand. Passing the dispersed soil through an ASTM Number 80

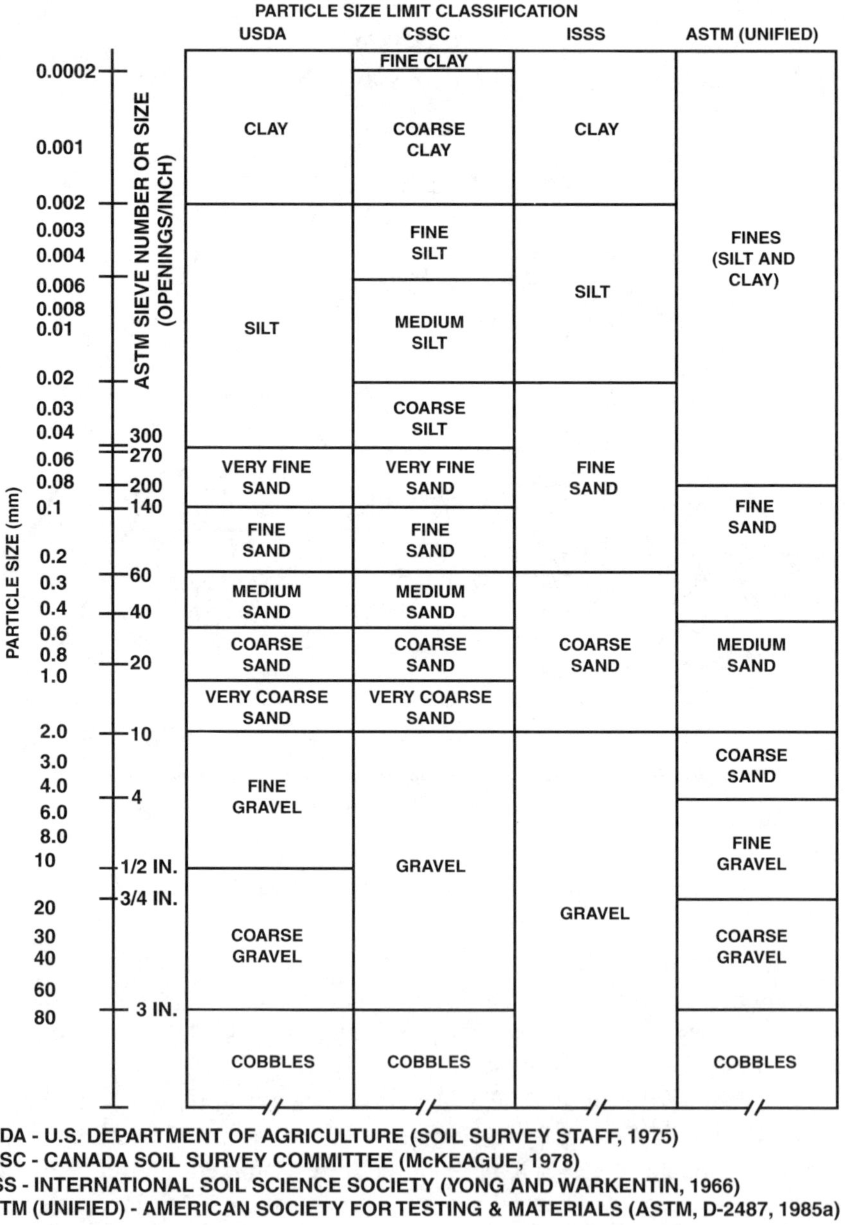

Figure 2.13
Comparison of particle size limits in four systems of particle size classification. (From McKeague, J.A., Ed., *Manual on Soil Sampling and Methods of Analysis*, 2nd ed., Canada Society of Soil Science, Ottawa, Ontario, Canada, 1978. With permission.)

sieve will remove the fine gravel and a coarse and medium sand. The particles retained on the sieves are oven-dried and further sieved for specific particle identification and quantification. These wet sieving procedures, which remove all or a portion of the sand fractions, are not common practices unless a careful detailed textural analysis of the coarser fractions in the soil is required.

The next step is to place the dispersed soil in a standing column of water (Genrich and Bremner, 1974) and, at specified time intervals, to remove a separate either by pouring or pipetting (Day, 1965; Gee and Bauder, 1986; Sheldrick and Wang, 1993) or by measuring the changing density of the column of water and dispersed soil with a hydrometer (Bouyoucos, 1962; Day, 1965). The hydrometer method, using the Bouyoucos hydrometer calibrated in g/L, is the most frequently used method because of its ease of operation and adequate results. In 2 h and with two hydrometer readings, the percentage of sand, silt, and clay in a dispersed soil can be determined and the textural class of the soil identified (Steinhardt, 1979).

4. Hydrometer Procedure

a. Soil preparation

Procedure

1. Weigh 50 g air-dried <10-mesh-sieved (2-mm) soil (100 g for sandy soils) into a beaker and add 100 mL 5% CALGON solution. Stir and let stand overnight (8 h). Transfer the entire contents of the beaker into a dispersion cup and fill the cup two thirds full with water. Add 1 drop of oil and place the cup on the Humbolt Mixer. Mix for 2 min. Remove the cup and cleanly transfer the entire contents of the cup into a 1000-mL cylinder.
2. Dilute to the mark (1000 mL for 50-g sample, 1120 mL for the 100-g sample) with water.

b. Hydrometer readings

Procedure

1. Cap the cylinder and invert 10 times with vigor. Put the cylinder down; uncap and place the Bouyoucos hydrometer (calibrated in g/L) into the water–soil slurry.
2. After 40 s of standing, make a hydrometer reading. Place a thermometer into the column and read the temperature. Record both readings.
3. Let the cylinder stand undisturbed for 2 h. At that time, make another hydrometer and temperature reading.

c. Blank hydrometer determination

Dilute 100 mL 5% CALGON solution to 1000 mL with water in a 1000-mL cylinder. Make a hydrometer reading and determine the temperature.

d. Calculation for % sand, silt, and clay

% Silt + % Clay: 40-second hydrometer reading:

hydrometer reading (sample – blank) + temp. corr.*/weight of soil, g

% Sand: 100 – (% Silt + % Clay)
% Clay: 2-h hydrometer reading:

hydrometer reading (sample – blank) + temp. corr./weight of soil, g

% Silt: 100 – % Sand – % Clay

References

Adams, F., Ed. 1984. *Soil Acidity and Liming*, 2nd ed., Agronomy Series Number 12, American Society of Agronomy, Madison, WI.

Adams, F. and C.E. Evans. 1962. A rapid method for measuring lime requirement of red-yellow podzolic soils. *Soil Sci. Soc. Am. Proc.*, 26:355–357.

Adriano, D.C., Ed. 1996a. *Biogeochemistry of Trace Elements*, Lewis Publishers, Boca Raton, FL.

Adriano, D.C. 1996b. *Trace Elements in the Terrestrial Environment*, Springer-Verlag, New York.

Adriano, D.C. and H.E. Doner. 1982. Bromine, chlorine, and fluoride, in A.L. Page, Ed., *Methods of Soil Analysis,* Part 2, 2nd ed., Agronomy Monograph Number 9, American Society of Agronomy, Madison, WI, 449–483.

Aitken, R.L. and B.J. Scott. 1999. Magnesium, in K.I. Peverill, L.A. Sparrow, and D.J. Reuter, Eds., *Soil Analysis: An Interpretation Manual*, CSIRO Publishing, Collingwood, Australia, 255–262.

Alley, M.M., D.C. Martens, M.G. Schnappinger, and G.W. Hawkins. 1972. Field calibration of soil tests for available zinc. *Soil Sci. Soc. Am. Proc.,* 36:621–624.

Alley, M.M. and L.W. Zelazny. 1987. Soil acidity: soil pH and lime needs, in J.R. Brown, Ed., *Soil Testing: Sampling, Correlation, Calibration, and Interpretation*, SSSA Special Publication 21, Soil Science Society of America, Madison, WI, 65–72.

---

* *Temperature Correction*: Degrees below 19.4°C (67°F), substract 0.2 units per degree from hydrometer reading. Degrees above 19.4°C (67°F), add 0.2 units per degree to hydrometer reading.

Allison, L.E. 1965. Organic carbon, in C.A. Black, Ed., *Methods of Soil Analysis*, Part 2, *Chemical and Microbiological Properties*, Agronomy Number 9, American Society of Agronomy, Madison, WI, 1367–1378.

Alloway, B.J., Ed. 1995. *Heavy Metals in Soils*, Blackie Academic and Professional, Glasgow, U.K.

Amacher, M.C. 1990. Nickel, cadmium, and lead, in D.L. Sparks, Ed., *Methods of Soil Analysis,* Part 3, *Chemical Methods*, SSSA Book Series Number 5, Soil Science Society of America, Madison, WI, 739–768.

Anonymous 1983a. Determination of phosphorus by Bray P1 extraction, in R.A. Isaac, Ed., Reference Soil Test Methods for the Southern Region of the United States, Southern Cooperative Series Bulletin 289, University of Georgia College of Agriculture Experiment Stations, Athens, 20–24.

Anonymous 1983b. Determination of phosphorus by Mehlich I (0.05 *N* HCl in 0.025 *N* H_2SO_4) extraction, in R.A. Isaac, Ed., Reference Soil Test Methods for the Southern Region of the United States, Southern Cooperative Series Bulletin 289, University of Georgia College of Agriculture Experiment Stations, Athens, 15–19.

Anonymous 1983c. Determination of specific conductance in supernatant 1:2 soil:water solution, in R.A. Isaac, Ed., Reference Soil Test Methods for the Southern Region of the United States, Southern Cooperative Series Bulletin 289, University of Georgia College of Agriculture Experiment Stations, Athens, 8–14.

Anonymous 1994a. Soil sample preparation, in R.G. Galvak, D.A. Horneck, and R.O. Miller, Eds., Plant, Soil, and Water Reference Methods for the Western Region, Western Regional Extension Publication WREP 125, University of Alaska, Fairbanks, 17.

Anonymous 1994b. pH: 1:2 soil to water ratio, in R.G. Gavlak, D.A. Horneck, and R.O. Miller, Eds., Plant, Soils, and Water Reference Methods for the Western Region, Western Regional Extension Publication WREP 125, University of Alaska, Fairbanks, 17–18.

Anonymous 1994c. Lime requirement: SMP buffer method, in R.G. Gavlak, D.A. Horneck, and R.O. Miller, Eds., Plant, Soils, and Water Reference Methods for the Western Region, Western Regional Extension Publication WREP 125, University of Alaska, Fairbanks, 19–20.

Anonymous 1994d. Extractable phosphorus: dilute acid-fluoride method (Bray-P1), in R.G. Gavlak, D.A. Horneck, and R.O. Miller, Eds., Plant, Soils, and Water Reference Methods for the Western Region, Western Regional Extension Publication WREP 125, University of Alaska, Fairbanks, 23–24.

Anonymous 1994e. Extractable phosphorus: sodium bicarbonate method (Olsen), in R.G. Gavlak, D.A. Horneck, and R.O. Miller, Eds., Plant, Soils, and Water Reference Methods for the Western Region, Western Regional Extension Publication WREP 125, University of Alaska, Fairbanks, 21–22.

Anonymous 1994f. Extractable phosphorus, potassium, sulfate–sulfur, zinc, iron, manganese, copper, nitrate: ammonium bicarbonate–DTPA method, in R.G. Gavlak, D.A. Horneck, and R.O. Miller, Eds., Plant, Soils, and Water Reference Methods for the Western Region, Western Regional Extension Publication WREP 125, University of Alaska, Fairbanks, 27–28.

Anonymous 1994g. Extractable phosphorus, potassium, calcium, magnesium, sodium, iron, manganese, and copper: Mehlich-3 method, in R.G. Gavlak, D.A. Horneck, and R.O. Miller, Eds., Plant, Soils, and Water Reference Methods for the Western Region, Western Regional Extension Publication WREP 125, University of Alaska, Fairbanks, 25–26.

Anonymous 1994h. Extractable calcium, magnesium, potassium, and sodium: ammonium acetate method, in R.G. Gavlak, D.A. Horneck, and R.O. Miller, Eds., Plant, Soils, and Water Reference Methods for the Western Region, Western Regional Extension Publication WREP 125, University of Alaska, Fairbanks, 31–32.

Anonymous 1994i. Hot-water extractable boron: Azomathione-H method, in R.G. Gavlak, D.A. Horneck, and R.O. Miller, Eds., Plant, Soils, and Water Reference Methods for the Western Region, Western Regional Extension Publication WREP 125, University of Alaska, Fairbanks, 33–34.

Anonymous 1994j. Extractable zinc, copper, iron, and manganese: DTPA method, in Plant, Soils, and Water Reference Methods for the Western Region, Western Regional Extension Publication WREP 125, R.G. Gavlak, D.A. Horneck, and R.O. Miller, Eds., University of Alaska, Fairbanks, 29–30.

Anonymous 1994k. Soluble salts, in R.G. Galvak, D.A. Horneck, and R.O. Miller, Eds., Plant, Soil, and Water Refererice Methods for the Western Region, Western Regional Extension Publication WREP 125, University of Alaska, Fairbanks, 38.

Anonymous 1994l. Ammonium- and nitrate-nitrogen: KCl extraction method, in R.G. Gavlak, D.A. Horneck, and R.O. Miller, Eds., Plant, Soils, and Water Reference Methods for the Western Region, Western Regional Extension Publication WREP 125, University of Alaska, Fairbanks, 48.

Anonymous 1994m. Kjeldahl distillation: CEC, TN, NH_4–N, NO_3–N, and mineralizable-N, in R.G. Gavlak, D.A. Horneck, and R.O. Miller, Eds., Plant, Soils, and Water Reference Methods for the Western Region, Western Regional Extension Publication WREP 125, University of Alaska, Fairbanks, 45–47.

Anonymous 1994n. Sulfate-sulfur (SO_4–S): ion chromatographic method, in R.G. Gavlak, D.A. Horneck, and R.O. Miller, Eds., Plant, Soils, and Water Reference Methods for the Western Region, Western Regional Extension Publication WREP 125, University of Alaska, Fairbanks, 49–50.

Anonymous 1995. Summary of soil testing methods in the northeast, in J.T. Sims and A.M. Wolf, Eds., Recommended Soil Testing Procedures for the Northeastern United States, 2nd ed., Northeast Regional Publication 493 (revised), Agricultural Experiment Station, University of Delaware, Newark, 93–94.

Anonymous 1996a. Soil reaction and liming, in *Soil Fertility Manual*, Potash & Phosphate Institute, Norcross, GA, 2:1–9.

Anonymous 1996b. Phosphorus, in *Soil Fertility Manual*, Potash & Phosphate Institute, Norcross, GA, 4:1–14.

Anonymous 1996c. Potassium, in *Soil Fertility Manual*, Potash & Phosphate Institute, Norcross, GA, 5:1–9.

Anonymous 1996d. The micronutrients, in *Soil Fertility Manual*, Potash & Phosphate Institute, Norcross, GA, 7:1–10.

Anonymous 1997. Water: what's in it and how to get it out. *Today's Chem.,* 6(1):16–19.

Anonymous 1998a. Soil Test Levels in North America: Summary Update, PPI/PPIC/FAR Technical Bulletin 1998-3, Potash & Phosphate Institute, Norcross, GA.

Anonymous 1998b. Heavy Metals in Soils and Phosphatic Fertilizers, PPI/PPIC/FAR Technical Bulletin 1998-2, Potash & Phosphate Institute, Norcross, GA.

Anonymous 1999a. Soil sampling strategies key. *Agric. Consultant,* 55:17–20.

Anonymous 1999b. *Soil Analysis Handbook of Reference Methods*, CRC Press, Boca Raton, FL.

Armour, J.D. and R.F. Brennan. 1999. Zinc, in K.I. Peverill, L.A. Sparrow, and D.J. Reuter, Eds., *Soil Analysis: An Interpretation Manual,* CSIRO Publishing, Collingwood, Australia, 281–285.

Baier, J. and V. Baierova. 1998. Hundredth molar calcium chloride extraction procedure. IV. Calibration with conventional soil testing methods for potassium. *Commun. Soil Sci. Plant Anal.*, 29:1641–1648.

Baker, D.E. 1971. A new approach to soil testing. *Soil Sci.,* 112:381–391.

Baker, D.E. 1973. A new approach to soil testing. II. Ionic equilibrium involving H, K, Ca, Mg, Mn, Fe, Cu, Zn, Na, P, and S. *Soil Sci. Soc. Am. Proc.,* 37:537–541.

Baker, D.E. 1990. Baker soil test theory and application. *Commun. Soil Sci. Plant Anal.*, 21:981–1008.

Baker, D.E. and M.C. Amacher. 1981. The Development and Interpretation of a Diagnostic Soil-Testing Program, Pennsylvania Agricultural Experiment Station Bulletin 826.

Baldock, J.A. and J.O. Skjemstad. 1999. Soil organic/soil organic matter, in K.I. Peverill, L.A. Sparrow, and D.J. Reuter, Eds., *Soil Analysis: An Interpretation Manual,* CSIRO Publishing, Collingwood, Australia, 159–170.

Barber, S.A. 1984. Liming materials and practices, in F. Adams, Ed., *Soil Acidity and Liming*, 2nd ed., American Society of Agronomy, Madison, WI, 171–209.

Barber, S.A. 1995. *Soil Nutrient Availability: A Mechanistic Approach,* 2nd ed., John Wiley & Sons, Chichester.

Bartlett, R.J. and B.R. James. 1996. Chromium, in D.L. Sparks, Ed., *Methods of Soil Analysis,* Part 3, *Chemical Methods*, SSSA Book Series Number 5, Soil Science Society of America, Madison, WI, 683–701.

Bates, R.G. 1973. *Determination of pH: Theory and Practice*, 2nd ed., John Wiley & Sons, New York.

Bates, T.E. 1993. Soil handling and preparation, in R.L. Carter, Ed., *Soil Sampling and Methods of Analysis*, CRC Press, Boca Raton, FL, 19–24.

Bates, T.E. and J.E. Richards. 1993. Available potassium, in R.L. Carter, Ed., *Soil Sampling and Methods of Analysis*, CRC Press, Boca Raton, FL, 59–64.

Beckett, P.H.T. 1989. The use of extractants in studies of trace metals in soils, sewage sludges, and sludge-treated soils. *Adv. Soil Sci.,* 9:143–176.

Beegle, D.B. and T.C. Oravec. 1990. Comparison of field calibrations for Mehlich-III P and K with Bray P1 and ammonium acetate K for corn. *Commun. Soil Sci. Plant Anal.,* 21:1025–1036.

Bell, R.W. 1999. Boron, in K.I. Peverill, L.A. Sparrow, and D.J. Reuter, Eds., *Soils Analysis: An Interpretation Manual,* CSIRO Publishing, Collingwood, Australia, 309–317.

Ben-Dor, E. and A. Banin. 1989. Determination of organic matter content in arid-zone soils using a simple "loss-on-ignition" method. *Commun. Soil Sci. Plant Anal.*, 20(15–16): 1675–1695.

Berger, K.C. and E. Truog. 1940. Boron deficiency as revealed by plant and soil tests. *J. Am. Soc. Agron.,* 32:297–301.

Berger, K.C. and E. Truog. 1944. Boron tests and determination for soils and plants. *Soil Sci.,* 57:25–36.

Berghage, R.D., D.M. Krauskopf, D.D. Warncke, and I. Widders. 1987. Micronutrient testing of plant growth media: extractant identification and evaluation. *Commun. Soil Sci. Plant Anal.*, 18(10):1089–1110.

Bergmann, W. 1992. *Nutritional Disorders of Plants — Developments, Visual and Analytical Diagnosis,* Gustav Fischer Verlag, New York.

Bingham, F.T. 1982. Boron, in A.L. Page, Ed., *Methods of Soil Analysis,* Part 2, *Chemical and Microbiological Properties*, 2nd ed., American Society of Agronomy, Madison, WI, 431–447.

Black, C.A. 1993a. *Soil Fertility Evaluation and Control*, Lewis Publishers, Boca Raton, FL, 647–728.

Black, C.A. 1993b. *Soil Fertility Evaluation and Control*, Lewis Publishers, Boca Raton, FL, 271–452.

Black, C.A. 1993c. *Soil Fertility Evaluation and Control*, Lewis Publishers, Boca Raton, FL.

Boawn, L.C. 1971. Evaluation of DTPA extractable zinc as a zinc soil test for Washington soils, in *Proceedings of the 22nd Annual Pacific North West Fertilizer Conference,* Pacific NW Plant Food Association, Bozeman, MT, 143–147.

Bolland, M.D.A. and R.J. Gilkes. 1998. The chemistry and agronomic effectiveness of phosphate fertilizers, in Z. Rengel, Ed., *Nutrient Use in Crop Production,* The Haworth Press, New York, 139–163.

Bolton, J. 1971. The chloride balance in a fertilizer experiment on sandy soil. *J. Sci. Food Agric.*, 22:292–294.

Bouyoucos, C.J. 1962. Hydrometer method improved for making particle-size analysis of soils. *Agron. J.,* 54:464–465.

Bower, C.A. and L.U. Wilcox. 1965. Soluble salts, in C.A. Black, Ed., *Methods of Soil Analysis,* Part 2, Agronomy Number 9, American Society of Agronomy, Madison, WI, 938–951.

Bray, R.H. and L.T. Kurtz. 1945. Determination of total, organic and available forms of phosphorus in soils. *Soil Sci.,* 59:39–45.

Brennan, R.F. and E. Best. 1999. Copper, in K.I. Peverill, L.A. Sparrow, and D.J. Reuter, Eds., *Soil Analysis: An Interpretation Manual,* CSIRO Publishing, Collingwood, Australia, 263–280.

Brown, A.J. 1999. Soil sampling and sample handling for chemical analysis, in K.I. Peverill, L.A. Sparrow, and D.J. Reuter, Eds., *Soil Analysis: An Interpretation Manual,* CSIRO Publishing, Collingwood, Australia, 35–53.

Brown, J.R., Ed. 1987. *Soil Testing: Sampling, Correlation, Calibration, and Interpretation.* SSSA Special Publication 21, Soil Science Society of America, Madison, WI.

Brown, L.R. 1997. *The Agricultrural Link: How Environmental Deterioration Could Disrupt Economic Progress*, WorldWatch Paper 136, WorldWatch Institute, Washington, D.C.

Bruce, R.C. 1999. Calcium, in K.I. Peverill, L.A. Sparrow, and D.J. Reuter, Eds., *Soil Analysis: An Interpretation Manual*, CSIRO Publishing, Collingwood, Australia, 247–254.

Burton, L.G. 1971. Known addition ion selective electrode technique for simultaneously determining fluoride and chloride in calcium halophosphate. *Anal. Chem.,* 43:479–481.

Camberato, J.J. and H.R. Deaton. 1994. Sampling for soil nitrogen, in W.O. Thom and W.E. Sabbe, Eds., Soil Sampling Procedures for the Southern Region of the United States, Southern Cooperative Series Bulletin Number 377, Kentucky Cooperative Extension Service, University of Kentucky, Lexington, 11–14.

Campbell, L.C. 1994. Beneficial impact of precision nutrient management on the environment and future needs. *Commun. Soil Sci. Plant Anal.,* 25:889–908.

Campbell, L.C. 1998. Managing soil fertility decline, in Z. Rengel, Ed., *Nutrient Use in Crop Production*, Haworth Press, Binghamton, New York, 29–52.

Carson, P.L. 1980a. Recommended potassium test, in W.C. Dahnke, Ed., Recommended Chemical Soil Test Procedures for the North Central Region, North Central Regional Publication 221, North Dakota Agricultural Experiment Station Bulletin 499, 17–18.

Carson, P.L. 1980b. Recommended nitrate-nitrogen tests, in W.C. Dahnke, Ed., Recommended Chemical Soil Test Procedures for the North Central Region, North Central Regional Publication 221, North Dakota Agricultural Experiment Station Bulletin 499, 12–13.

Carter, R.L., Ed. 1993. *Soil Sampling and Methods of Analysis,* CRC Press, Boca Raton, FL.

CAST. 2000. Relevance of Soil Testing to Agriculture and the Environment, Issue Paper Number 15, June 2000, Council for Agricultural Science and Technology, Ames, IA.

Cate, R.B., Jr. and L.A. Nelson. 1965. A Rapid Method for Correlation of Soil Test Analyses with Plant Response Data, International Soil Testing Series Technical Bulletin 1, North Carolina State University, Raleigh.

Chaney, R.L. 1980. Health risks associated with toxic metals in municipal sludge, in G. Bitton, B.L. Damron, G.T. Edds, and J.M. Davidson, Eds., *Health Risks of Land Application*, Ann Arbor Science, Ann Arbor, MI, 59–67.

Chaney, R.L., J.F. Bruins, D.E. Baker, R.F. Korcak, J.E. Smith, and D. Cole. 1987. Transfer of sludge-applied trace elements to the food chain, in A.L. Page, T. Logan, and J. Ryan, Eds., *Land Application of Sludge*, Lewis Publishers, Chelsea, MI, 67–93.

Chapman, H.D. and P. Pratt. 1982. *Method of Analysis for Soils, Plants and Water,* Priced Publication 4034, University of California, Division of Agricultural Sciences, Berkeley.

Combs, S.M. and M.V. Nathan. 1998. Soil organic matter, in J.R. Brown, Ed., Recommended Chemical Soil Test Procedures for the North Central Region, North Central Regional Research Publication Number 221 (revised), Missouri Agricultural Experiment Station SB 1001, University of Missouri, Columbia, 53–58.

Combs, S.M., J. Denning, and K.D. Frank. 1998. Sulfate-sulfur, in J.R. Brown, Ed., Recommended Chemical Soil Test Procedures for the North Central Region, North Central Regional Research Publication 221 (revised), Missouri Agricultural Experiment Station SB 1001, University of Missouri, Columbia, 35–40.

Conyers, M.K. and B.G. Darey. 1988. Observations in some routine methods for soil pH determination. *Soil Sci.,* 145:29–36.

Cook, S.E. and R.G.V. Bramley. 2000. Coping with variability in agricultural production — implications for soil testing and fertilizer management. *Commun. Soil Sci. Plant Anal.,* 31:1331–1351.

Corey, R.B. 1990. Physical-chemical aspects of nutrient availability, in R.L. Westerman, Ed., *Soil Testing and Plant Analysis*, 3rd ed., SSSA Book Series Number 3, Soil Science Society of America, Madison, WI, 11–24.

Cottenie, A. 1980. Soil and Plant Testing as a Basis of Fertilizer Recommendations, FAO Soils Bulletin 38/2, Food and Agriculture Organization of the United Nations, Rome, Italy.

Cox, F.R. and E.J. Kamprath. 1972. Micronutrient soil test, in J.J. Mortvedt, P.N. Giordano, and W.L. Lindsay, Eds., *Micronutrients in Agriculture*, Soil Science Society of America, Madison, WI, 289–317.

Cox, F.R. and J.I. Wear, Eds. 1977. Diagnosis and Correlation of Zinc Problems in Corn and Rice Production, Southern Cooperative Series Bulletin 222, North Carolina State Experiment Station, Raleigh.

Crépin, J. and R.L. Johnson. 1993. Soil sampling for environmental assessment, in R.L. Carter, Ed., *Soil Sampling and Methods of Analysis*, CRC Press, Boca Raton, FL, 5–18.

Crock, J.G. 1996. Mercury, in D.L. Sparks, Ed., *Methods of Soil Analysis,* Part 3, *Chemical Methods*, SSSA Book Series Number 5, Soil Science Society of America, Madison, WI, 769–791.

Dahnke, W.C. and G.V. Johnson. 1990. Testing soils for available nitrogen, in R.L. Westerman, Ed., *Soil Testing and Plant Analysis*, 3rd ed., SSSA Book Series Number 3, Soil Science Society of America, Madison, WI, 127–139.

Dahnke, W.C. and R.A. Olson. 1990. Soil test correlation, calibration, and recommendations, in R.L. Westerman, Ed., *Soil Testing and Plant Analysis,* 3rd ed., SSSA Book Number 3, Soil Science Society of America, Madison, WI, 45–71.

David, M.B. 1988. Use of loss-on-ignition to assess soil organic carbon in forest soils. *Commun. Soil Sci. Plant Anal.,* 19:1593–1599.

Davidescu, D. and V. Davidescu. 1972. *Evaluation of Fertility by Plant and Soil Analysis*, Abacus Press, Kent, England.

Davies, B.E. 1974. Loss-on-ignition as an estimate of soil organic matter. *Soil Sci. Soc. Am. Proc.,* 38:150–151.

Day, P.R. 1965. Particle fractionation and particle-size analysis, in C.A. Black, Ed., *Methods of Soil Analysis,* Part 1, Agronomy Monograph Number 9, American Society of Agronomy, Madison, WI, 545–567.

De Temmerman, L.O., H. Hoenig, and P.O. Scokart. 1984. Determination of "normal" levels and upper limit values of trace elements in soils. *Z. Pflanzenernachr. Bodenk.,* 147:687–694.

Doerge, T.A. and E.H. Gardner. 1988. Comparison of four methods for interpreting the Shoemaker–McLean–Pratt (SMP) lime requirement test. *Soil Sci. Soc. Am. J.,* 52:1054–1059.

Doll, E.C. and R.E. Lucas. 1973. Testing soils for potassium, calcium, and magnesium, in L.M. Walsh and J.D. Beaton, Eds., *Soil Testing and Plant Analysis*, Soil Science Society of America, Madison, WI, 133–151.

Donohue, S.J., Ed. 1988. Reference Soil Test and Media Diagnostic Procedures for the Southern Region of the United States, Southern Cooperative Series Bulletin, VPI, Blacksburg, VA.

Dorich, R.A. and D.W. Nelson. 1984. Evaluation of manual cadmium reduction methods for determination of nitrate in potassium chloride extracts in soils. *Soil Sci. Soc. Am. J.,* 48:72–75.

Dudal, R. 1976. Inventory of the major soils of the world with special reference to mineral stress hazards, in M.J. Wright, Ed., *Plant Adaptation to Mineral Stress in Problem Soils*, Cornell University Press, Ithaca, NY, 3–14.

Eckert, D.L. 1987. Soil test interpretations: basic cation saturation ratios and sufficiency levels, in *Soil Testing: Sampling, Correlations, Calibration, and Interpretation*, SSSA Special Publication 21, Soil Science Society of America, Madison, WI, 53–64.

Eckert, D.L. 1988. Recommended pH and lime requirement tests, in W.C. Dahnke, Ed., Recommended Chemical Soil Test Procedures for the North Central Region, North Central Regional Publication Number 221 (revised), North Dakota Agricultural Experiment Station Bulletin Number 499 (revised), North Dakota State University, Fargo, 6–8.

Eckert, D.L. 1991. Recommended soil pH and lime requirement tests, in J.T. Sims and A. Wolf, Eds., Recommended Soil Testing Procedures for the Northeastern United States, Northeast Regional Publication, Agricultural Experiment Station Bulletin 493, University of Delaware, Newark, 35–40.

Eckert, D.L. and J.T. Sims. 1995. Recommended soil pH and lime requirement, in J.T. Sims and A. Wolf, Eds., Recommended Soil Testing Procedures for the Northeastern United States. 2nd ed., Northeast Regional Publication, Agricultural Experiment Station Bulletin 493, University of Delaware, Newark, 16–21.

Eliason, R. 1998. Laboratory factors of importance to soil extraction, in J.R. Brown, Ed., Recommended Chemical Soil Test Procedures for the North Central Region, North Central Regional Publication Number 221 (revised), Missouri Agricultural Experiment Station SB1001, University of Missouri, Columbia, 11–12.

Epstein, E. 1972. *Mineral Nutrition of Plants: Principles and Perspectives*, John Wiley & Sons, New York.

Evans, C.E. and E.J. Kamprath. 1970. Lime response as related to percent Al saturation, solution Al and organic matter content. *Soil Sci. Soc. Am. Proc.*, 34:893–896.

Fisher, J.E. 1984. Measurement of pH. *Am. Lab.*, 16:54–60.

Fitts, J.W. and W.L. Nelson. 1956. The determination of lime and fertilizer requirements of soils through chemical tests. *Adv. Agron.*, 8:241–282.

Fixen, P.E. and J.H. Grove. 1990. Testing soils for phosphorus, in R.L. Westerman, Ed., *Soil Testing and Plant Analysis*, 3rd ed., SSSA Book Series Number 3, Soil Science Society of America, Madison WI, 141–180.

Fixen, P.E., R.H. Gelderman, J.R. Gerwing, and B.G. Farber. 1987. Calibration and implementation of a soil Cl test. *J. Fertil. Issues*, 4:91–97.

Flannery, R.L. and D.K. Markus. 1972. Determination of phosphorus, potassium, calcium and magnesium simultaneously in North Carolina, ammonium acetate, and Bray P1 soil extracts by AutoAnalyzer, in L.M. Walsh, Ed., *Instrumental Methods for Analysis of Soils and Plant Tissue*, Soil Science Society of America, Madison, WI, 97–112.

Flannery, R.L. and D.K. Markus. 1980. Automated analysis of soil extracts for phosphorus, potassium, calcium, and magnesium. *J. Assoc. Off. Anal. Chem.*, 63:779–787.

Fotyma, M., T. Jadczyszyn, and G. Jozefaciuk. 1998. Hundredth molar calcium chloride extraction procedure. II. Calibration with conventional soil testing methods for pH. *Commun. Soil Sci. Plant Anal.*, 29:1625–1632.

Fox, R.H. 1980. Comparisons of several lime requirement methods for agricultural soils in Pennsylvania. *Commun. Soil Sci. Plant Anal.*, 11:57–69.

Foy, C.D. 1984. Physiological effects of hydrogen, aluminum, and manganese toxicities in acid soil, in F. Adams, Ed., *Soil Acidity and Liming*, 2nd ed., Agronomy Number 12, American Society of Agronomy, Madison, WI, 57–97.

Frank, K., D. Beegle, and J. Denning. 1998. Phosphorus, in J.R. Brown, Ed., Recommended Chemical Soil Test Procedures for the North Central Region, North Central Regional Research Publication Number 221 (revised), Missouri Agricultural Experiment Station SB 1001, University of Missouri, Columbia, 21–29.

Frankenberger, W.T., Jr., M.A. Tabatabi, D.C. Adriano, and H.E. Doner. 1996. Bromine, chlorine, and fluorine, in *Methods of Soils Analysis,* Part 3, *Chemical Methods*, SSSA Book Series Number 5, D.L. Sparks, Ed., Soil Science Society of America, Madison, WI, 833–867.

Gallardo, J.F. and J. Saavedra. 1987. Soil organic matter determination. *Commun. Soil Sci. Plant Anal.,* 18(6):699–707.

Gambrell, R.P. 1996. Manganese, in D.L. Sparks, Ed., *Methods of Soils Analysis,* Part 3, *Chemical Methods*, SSSA Book Series Number 5, Soil Science Society of America, Madison, WI, 665–683.

Gardner, G. 1996. Shrinking Fields: Cropland Loss in a World of Eight Billion, WorldWatch Paper 131, WorldWatch Institute, Washington, D.C.

Gartley, K. 1995. Recommended soluble salts tests, in J.T. Sims and A.M. Wolf, Eds., Recommended Soil Testing Procedures for the Northeastern United States, 2nd ed., Northeast Regional Publication Number 493 (revised), Agricultural Experiment Station, University of Delaware, Newark, 70–75.

Gascho, G.J., T.P. Gaines, and C.O. Plank. 1990. Comparison of extractants for testing coastal plain soil. *Commun. Soil Sci. Plant Anal.,* 21:1052–1077.

Gascho, G.J., M.B. Parker, and T.P. Gaines. 1996. Reevaluation of suspension solutions for soil pH. *Commun. Soil Sci. Plant Anal.,* 27:773–782.

Gavlak, R.G., D.A. Horneck, and R.O. Miller, Eds. 1994. Plant, Soil, and Water Reference Methods for the Western Region, Western Regional Extension Publication WREP 125, University of Alaska Cooperative Extension Service, Fairbanks.

Gee, G.W. and J.W. Bauder. 1986. Particle-size analysis, in A. Klute, Ed., *Methods of Soil Analysis,* Part 1, *Physical and Mineralogical Methods*, 2nd ed., American Society of Agronomy, Madison, WI, 383–412.

Gelderman, R.H. and D. Beegle. 1998. Nitrate-nitrogen, in J.R. Brown, Ed., Recommended Chemical Soil Test Procedures for the North Central Region, North Central Regional Research Publications Number 221 (revised), Missouri Agricultural Experiment Station SB 1001, University of Missouri, Columbia, 17–20.

Gelderman, R.H. and A.P. Mallarino. 1998. Soil sample preparation, in J.R. Brown, Ed., Recommended Chemical Soil Test Procedures for the North Central Region, North Central Regional Publication Number 221 (revised), Missouri Agricultural Experiment Station SB1001, University of Missouri, Columbia, 5–6.

Gelderman, R.H., J.L. Denning, and R.J. Goos. 1998. Chlorides, in J.R. Brown, Ed., Recommended Chemical Soil Test Procedures for the North Central Region, North Central Regional Research Publication Number 221 (revised), Missouri Agricultural Experiment Station SB 1001, University of Missouri, Columbia, 49–52.

Genrich, D.A. and J.M. Bremner. 1974. Isolation of soil particle-size fractions. *Soil Sci. Soc. Am. Proc.*, 38:222–225.

Geraldson, C.M. 1957. Soil soluble salts — determination of and association with plant growth. *Proc. Fla. State Hortic. Soc.*, 71:121–127.

Geraldson, C.M. 1970. Intensity and balance concept as an approach to optimal vegetable production. *Commun. Soil Sci. Plant Anal.*, 1:187–196.

Gilliam, G.P., R.C. Bruce, B.C. Davey, J.M. Kimble, P.L. Searle, and J.O. Skjemstad. 1983. A comparison of methods used for determination of cation exchange capacity. *Commun. Soil Sci. Plant Anal.*, 14:1005–1014.

Gilroy, T.E. 1969. Relationship between Extractable Soil Zinc and Zinc Concentration in Corn Leaves, M.S. thesis, University of Missouri, Columbia.

Glass, D.M. 1989. *Plant Nutrition: An Introduction to Current Concepts*, Jones and Bartlett Publishers, Boston, MA.

Glenn, R.C. 1983. Reliability of volumetric sampling as compared to weighed samples in quantitative soil test interpretations. *Commun. Soil Sci. Plant Anal.*, 14:199–207.

Golden, A. 1987. Reassessing the use of loss-on-ignition for estimating organic matter contents in noncalcareous soils. *Commun. Soil Sci. Plant Anal.*, 18:111–116.

Goulding, K.W.T. 1987. Potassium fixation and release. *Proc. Colloq. Int. Potash Inst.*, 20:137–154.

Gourley, C.J.P. 1999. Potassium, in K.I. Peverill, L.A. Sparrow, and D.J. Reuter, Eds., *Soil Analysis: An Interpretation Manual*, CSIRO Publishing, Collingwood, Australia, 229–245.

Graham, E.R. 1948. Determination of soil organic matter by means of a photoelectric colorimeter. *Soil Sci.*, 65:181–183.

Graham, E.R. 1959. An Explanation of Theory and Methods of Soil Testing, Missouri Agricultural Experiment Station Bulletin 734, Missouri Agricultural Experiment Station, Columbia.

Graham, E.R., S. Powell, and M. Carter. 1956. Soil Magnesium and the Growth and Composition of Plants, Missouri Agricultural Experiment Station Research Bulletin 607, Missouri Agricultural Experiment Station, Columbia.

Greweling, T. and M. Peech. 1965. Chemical Soil Tests, Cornell Agricultural Experiment Station Bulletin 960, Cornell Agricultural Experiment Station, Ithaca, NY.

Griffin, G., W. Jokela, and D. Ross. 1995. Recommended soil nitrate-N tests, in J.T. Sims and A.M. Wolf, Eds., Recommended Soil Testing Procedures for the Northeastern United States, 2nd ed., Northeast Regional Publication Number 493 (revised), University of Delaware, Newark, 22–29.

Gupta, U.C. 1967. A simplified method for determining hot water-soluble boron in podzol soils. *Soil Sci.,* 103:424–428.

Gupta, U.C. 1990. Boron, molybdenum, and selenium, in *Soil Sampling and Methods of Analysis*, M.R. Carter, Ed., Lewis Publishers, Boca Raton, FL, 91–99.

Haby, V.A., M.P. Russelle, and E.O. Skogley. 1990. Testing soils for potassium, calcium, and magnesium, in R.L. Westerman, Ed., *Soil Testing and Plant Analysis,* 3rd ed., SSSA Book Series Number 3, Soil Science Society of America, Madison, WI, 181–227.

Hajek, B.F., F. Adams, and J.T. Cope. 1972. Rapid determination of exchangeable bases, acidity and base saturation for soil characterization. *Soil Sci. Soc. Am. Proc.,* 36:436–438.

Halliday, D.J. and M.E. Trenkel, Eds. 1992. *IFA World Fertilizer Use Manual*, International Fertilizer Industry Association, Paris.

Haneklaus, S. and E. Schnug. 2000. Decision-making strategies for the variable-rate application of compound fertilizers. *Commun. Soil Sci. Plant Anal.,* 31:1863–2000.

Hanlon, E.A. and G.V. Johnson. 1984. Bray/Kurtz, Mehlich-III, AB/DTPA, and ammonium acetate extractions for P, K, and Mg in four Oklahoma soils. *Commun. Soil Sci. Plant Anal.,* 15:277–294.

Hanway, J.J., S.A. Barber, and R.H. Bray. 1962. North Central Regional Potassium Studies III. Field Studies with Corn, Iowa State Experiment Station Research Bulletin 503, Ames.

Harter, R.D. 1991. Micronutrient adsorption–desorption reactions in soils, in J.J. Mortvedt, Ed., *Micronutrients in Agriculture*, 2nd ed., SSSA Book Series Number 4, Soil Science Society of America, Madison, WI, 59–87.

Hatfield, A.L. 1972. Soil test reporting: a nutrient index system. *Commun. Soil Sci. Plant Anal.,* 3:425–436.

Hayes, M.H.B., R.S. Swift, R.E. Wardle, and J.K. Brown. 1975. Humic materials from an organic soil: a comparison of extractants and properties of extracts. *Geoderma,* 13:231–245.

Helmke, P.A. and D.L. Sparks. 1996. Lithium, sodium, potassium, rubidium, and calcium, in D.L. Sparks, Ed., *Methods of Soil Analysis,* Part 3, *Chemical Methods*, SSSA Book Series Number 5, Soil Science Society of America, Madison, WI, 551–574.

Helyar, K.R. and G.H. Price. 1999. Making recommendations based on soil tests, in K.I. Peverill, L.A. Sparrow, and D.J. Reuter, Eds., *Soil Analysis: An Interpretation Manual*, CSIRO Publishing, Collingwood, Australia, 331–357.

Hendershot, W.H., H. Lalande, and M. Duquette. 1993. Soil reaction and exchangeable acidity, in M.R. Carter, Ed., *Soil Sampling and Methods of Analysis*, Lewis Publishers, Boca Raton, FL, 141–145.

Hergert, G.W. 1998. A futuristic view of soil and plant analysis and nutrient recommendations. *Commun. Soil Sci. Plant Anal.,* 29:1441–1454.

Hesse, P.R. 1971. *Soil Chemical Analysis*, Chemical Publishing Co., New York.

Hipp, A.W. and G.W. Langdale. 1971. Use of solid-state chloride electrode for chloride determination in soil extractions. *Commun. Soil Sci. Plant Anal.,* 2:237–240.

Holmgren, G.G.S., S.W. Meyer, R.L. Chaney, and R.B. Daniels. 1993. Cadmium, lead, zinc, copper, and nickel in agricultural soils in the United States of America. *J. Environ. Qual.,* 22:335–348.

Hoskins, B. and D. Ross. 1995. Soil sample preparation and extraction, in J.T. Sims and A.M. Wolf, Eds., Recommended Soil Testing Procedures for the Northeastern United States, Northeast Regional Publication Number 493 (revised), Agricultural Experiment Station, University of Delaware, Newark, 3–10.

Houba, V.J.G. and I. Novozamsky. 1998. Influence of storage time and temperature of air-dried soils on pH and extractable nutrients using 0.01 *M* $CaCl_2$. *Fresenius J. Anal. Chem.,* 360:362–365.

Houba, V.J.G., I. Novozamsky, A.W.M. Huijbregts, and J.J. van der Lee. 1986. Comparison of soil extractions by 0.01 *M* $CaCl_2$, by EUF and by some conventional extraction procedures. *Plant Soil,* 96:433–437.

Houba, V.J.G., I. Novozamsky, J. Uittenbogaard, and J.J. van der Lee. 1987. Automatic determination of "total soluble nitrogen" in soil extracts. *Landwirtsch. Forsch.* 40:295–302.

Houba, V.J.G., I. Novozamsky, Th.M. Lexmond, and J.J. van der Lee. 1990. Applicability of 0.01 *M* $CaCl_2$ as a single extractant for the assessment of the nutrient status of soils and other diagnostic purposes. *Commun. Soil Sci. Plant Anal.,* 21:2281–2290.

Houba, V.J.G., W.J. Chardon, and K. Roelse. 1993. Influence of grinding of soil on apparent chemical composition. *Commun. Soil Sci. Plant Anal.,* 24:1591–1602.

Houba, V.J.G., E.J.M. Temminghoff, G.A. Gaikhorst, and W. van Vark. 2000. Soil analysis procedures using 0.01 *M* calcium chloride as extraction reagent. *Commun. Soil Sci. Plant Anal.,* 31:1299–1396.

Huang, P.M. and R. Fujii. 1990. Selenium and arsenic, in D.L. Sparks, Ed., *Methods of Soil Analysis,* Part 3, *Chemical Methods*, SSSA Book Series Number 5, Soil Science Society of America, Madison, WI, 793–831.

Huffman, S.A. and K.A. Barbarick. 1981. Soil nitrate analysis by cadmium reduction. *Commun. Soil Sci. Plant Anal.,* 122:231–264.

Ihnat, M. 1993. Reference materials for quality data, in M.R. Carter, Ed., *Soil Sampling and Methods of Analysis*, Lewis Publishers, Boca Raton, FL, 247–262.

Isaac, R.A. 1992. Determination of boron by hot water extraction, in S.J. Donohue, Ed., Reference Soil and Media Diagnostic Procedures for the Southern Region of the United States, Southern Cooperative Series Bulletin Number 374, Virginia Agricultural Experiment Station, Blacksburg, 23–24.

Jackson, M.L. 1958. *Soil Chemical Analysis*, Prentice-Hall, Englewood Cliffs, NJ.

James, D.W. and K.L. Wells. 1990. Soil sample collection and handling: techniques based on source and degree of field variability, in R.L. Westerman, Ed., *Soil Testing and Plant Analysis*, 3rd ed., SSSA Book Series Number 3, Soil Science Society of America, Madison, WI, 25–44.

Janzen, H.H. 1993. Soluble salts, in M.R. Carter, Ed., *Soil Sampling and Methods of Analysis*, CRC Press, Boca Raton, FL, 161–166.

John, M.K. 1973. A batch-handling technique for hot-water extraction of boron from soils. *Soil Sci. Soc. Am. Proc.,* 37:332–333.

Johnson, G.V. 1992a. Determination of zinc, manganese, copper, and iron by DTPA extraction, in S.J. Donohue, Ed., Reference Soil and Media Diagnostic Procedures for the Southern Region of the United States, Southern Cooperative Series Bulletin Number 374, Virginia Agricultural Experiment Station, Blacksburg, 16–18.

Johnson, G.V. 1992b. Determination of nitrate-nitrogen by specific ion electrode, in S.J. Donohue, Ed., Reference Soil and Media Diagnostic Procedures for the Southern Region of the United States, Southern Cooperative Series Bulletin Number 374, Virginia Agricultural Experiment Station, Blacksburg, 25–27.

Johnson, G.V. 1992c. Determination of sulfate–sulfur by monocalcium phosphate extraction, in S.J. Donohue, Ed., Reference Soil and Media Diagnostic Procedures for the Southern Region of the United States, Southern Cooperative Series Bulletin Number 374, Virginia Agricultural Experiment Station, Blacksburg, 13–15.

Johnson, G.V. and S.J. Donohue. 1992. Testing method for waste amended soil, in S.J. Donohue, Ed., Reference Soil and Media Diagnostic Procedures for the Southern Region of the United States, Southern Cooperative Series Bulletin Number 374, Virginia Agricultural Experiment Station, Blacksburg, 46–47.

Johnson, G.V. and P.E. Fixen. 1990. Testing soils for sulfur, boron, molybdenum, and chlorine, in R.L. Westerman, Ed., *Soil Testing and Plant Analysis*, 3rd ed., SSSA Book Series Number 3, Soil Science Society of America, Madison, WI, 265–273.

Jones, J.B., Jr. 1973. Soil testing in the United States. *Commun. Soil Sci. Plant Anal.,* 4:307–322.

Jones, J.B., Jr. 1983. Soil test works when used right. 2. *Solution,* 27:61–70.

Jones, J.B., Jr. 1985. Soil testing and plant analysis: guides to the fertilization of horticultural crops. *Hortic. Rev.,* 7:1–68.

Jones, J.B., Jr. 1990. Universal soil extractants: their composition and use. *Commun. Soil Sci. Plant Anal.,* 21:1091–1101.

Jones, J.B., Jr. 1998a. Soil test methods: past, present, and future. *Commun. Soil Sci. Plant Anal.,* 29:1543–1552.

Jones, J.B., Jr. 1998b. *The Plant Nutrition Manual*, CRC Press, Boca Raton, FL.

Junus, M.A. and F.R. Cox. 1987. A zinc soil test calibration based upon Mehlich 3 extractable zinc, pH, and cation exchange capacity. *Soil Sci. Soc. Am. J.*, 51:678–683.

Kabata-Pendias, A. 2000. *Trace Elements in Soil and Plants*, 3rd ed., CRC Press, Boca Raton, FL.

Kabata-Pendias, A. and D.C. Adriano. 1995. Trace elements, in J.E. Rechcigl, Ed., *Soil Amendments and Environmental Quality*, Lewis Publishers, Boca Raton, FL, 139–167.

Kahn, A. 1979. Distribution of DTPA-extractable Fe, Zn, and Cu in soil particle-size fractions. *Commun. Soil Sci. Plant Anal.*, 10:1211–1218.

Kahn, A. and P.N. Soltanpour. 1978. Effect of wetting and drying on DTPA-extractable Fe, Zn, Mn, and Cu in soils. *Commun. Soil Sci. Plant Anal.*, 9:193–202.

Kalra, Y.P. 1995. Determination of pH of soils by different methods: collaborative study. *J. Assoc. Off. Anal. Chem.*, 78:310–324.

Kalra, Y.P. 1996. Soil pH: first soil analysis methods validated by the AOAC International. *J. For. Res.*, 1:61–64.

Kammin, W.R., S. Cull, R. Knox, J. Ross, M. McIntosh, and D. Thomson. 1995. Labware cleaning protocols for the determination of low-level metals by ICP-MS. *Am. Environ. Lab.*, 11/95–12/95:1–3.

Kamprath, E.J. 1970. Exchangeable aluminum as a criterion for liming leached mineral soils. *Soil Sci. Soc. Am. Proc.*, 34:252–254.

Keeney, D.R. and D.W. Nelson. 1982. Nitrogen-inorganic forms, in A.L. Page, Ed., *Methods of Soil Analysis,* Part 2, 2nd ed., Agronomy Monograph Number 9, American Society of Agronomy, Madison, WI, 643–698.

Keren, R. 1996. Boron, in D.L. Sparks, Ed., *Methods of Soils Analysis,* Part 3, *Chemical Methods*, SSSA Book Series Number 5, Soil Science Society of America, Madison, WI, 603–626.

Khasawneh, F.E. et al., Eds., 1980. *The Role of Phosphorus in Agriculture*, American Society of Agronomy, Madison, WI.

Kidder, G. 1992. Determination of pH, soluble salts, nitrate, phosphorus, potassium, calcium, magnesium, sodium, and chloride in potting media (non-soil mixes) by saturation extraction, in S.J. Donohue, Ed., Reference Soil and Media Diagnostic Procedures for the Southern Region of the United States, Southern Cooperative Series Bulletin Number 374, Virginia Agricultural Experiment Station, Blacksburg, 28–32.

Kovar, J.L. 1994. Sampling soils under reduced row crop tillage, in W.O. Thom and W.E. Sabbe, Eds., Soil Sampling Procedures for the Southern Region of the United States, Southern Cooperative Series Bulletin Number 377, Kentucky Cooperative Extension Service, University of Kentucky, Lexington, 19–24.

Kuehny, J.S. and B. Morales. 1998. How salinity and alkalinity levels in media affect plant growth. *Greenhouse Prod. News,* 8(6):42–43.

Kuo, S. 1996. Phosphorus, in R.L. Sparks, Ed., *Methods of Soil Analysis,* Part 3, *Chemical Methods*, SSSA Book Series Number 5, Soil Science Society of America, Madison, WI, 869–919.

Lamond, R., V. Martin, T. Maxwell, R. Bowden, and S. Duncan. 1999. Chloride fertilization on winter wheat. *Better Crops,* 83(4):21–23.

Lamond, R., V. Martin, C. Olsen, and K. Rector. 2000. Chloride fertilization increases yields of corn and grain sorghum. *Better Crops,* 84(4):10–11.

Levesque, M. and M. Schnitzer. 1967. The extraction of soil organic matter by base and chelating resin. *Can. J. Soil Sci.,* 47:76–78.

Lewis, D.C. 1999. Sulfur, in K.I. Peverill, L.A. Sparrow, and D.J. Reuter, Eds., *Soil Analysis: An Interpretation Manual,* CSIRO Publishing, Collingwood, Australia, 221–228.

Liang, J. and R.E. Karamanos. 1993. DTPA-extractable Fe, Mn, Cu and Zn, in M.R. Carter, Ed., *Soil Sampling and Methods of Analysis*, Lewis Publishers, Boca Raton, FL, 87–90.

Liebhardt, W.C. 1981/1982. In pursuit of the better soil test. *The New Farm,* 3(3) and 4(1–5), Emmaus Press, Emmaus, PA.

Lindsay, W.L. 1979. *Chemical Equilibria in Soils*, Wiley-Interscience, New York.

Lindsay, W.L. 1991. Inorganic equilibria affecting micronutrients in soils, in J.J. Mortvedt, Ed., *Micronutrients in Agriculture,* SSSA Book Series Number 4, Soil Science Society of America, Madison, WI, 89–112.

Lindsay, W.L. and E.C. Moreno. 1960. Phosphate phase equilibria in soils. *Soil Sci. Soc. Am. Proc.,* 24:177–182.

Lindsay, W.L. and W.A. Norvell. 1978. Development of a DTPA micronutrient soil test for zinc, iron, manganese, and copper. *Soil Sci. Soc. Am. J.,* 42:421–428.

Loch, J., I. Jaszberenyi, and I. Vago. 1998. Hundredth molar calcium chloride extraction procedure. III. Calibration with conventional soil testing methods for magnesium. *Commun. Soil Sci. Plant Anal.,* 29:1633–1640.

Lockman, R.B. and C.J. Molley. 1984. Seasonal variations in soil test results. *Commun. Soil Sci. Plant Anal.,* 15:741–757.

Loeppert, R.L. and W.P. Inskeep. 1996. Iron, in D.L. Sparks, Ed., *Methods of Soils Analysis,* Part 3, *Chemcial Methods,* SSSA Book Series Number 5, Soil Science Society of America, Madison, WI.

Lowther, J.R., P.J. Smethurst, J.C. Carlyle, and E.K.S. Mabiar. 1990. Methods for determining organic carbon in Podzolic sands. *Commun. Soil Sci. Plant Anal.,* 21:457–470.

Lucas, R.E., P.E. Rieke, and E.C. Doll. 1972. Soil saturated extract method for determining plant nutrient levels in peats and other soil mixes. *4th Int. Peat Congress,* 3:221–230.

Ludwick, A.E., Ed. 1998.*Western Fertilizer Handbook*, 2nd ed., Interstate Publishers. Danville, IL.

Lunt, H.A., C.L.W. Swanson, and H.G.M. Jacobson. 1950. The Morgan Soil Testing System. Connecticut Agricultural Experiment Station (New Haven) Bulletin 541, Connecticut Agricultural Experiment Station, New Haven.

Luscombe, P.C., J.K. Syers, and P.E.H. Greg. 1979. Water extraction as a soil testing procedure for phosphate. *Commun. Soil Sci. Plant Anal.,* 10:1361–1370.

Magdoff, F.R., D. Ross, and J. Amadon. 1984. A soil test for nitrogen availability to corn. *Soil Sci. Soc. Am. J.,* 48:1301–1305.

Magdoff, F.R., W.E. Jokela, R.H. Fox, and G.F. Griffin. 1990. A soil test for nitrogen availability in the northeastern U.S. *Commun Soil Sci. Plant Anal.,* 21:1102–1115.

Makarim, A.K. and F.R. Cox. 1983. Evaluation of the need for copper with several soil extractants. *Agron. J.*, 75:493–496.

Mallarino, A.P. 1997. Interpretation of soil phosphorus tests for corn in soils with varying pH and calcium carbonate content. *J. Prod. Agric.,* 10:163–167.

Mallarino, A.P. and W.R. Luellen. 1998. Choosing the best P test. *Agric. Consultant,* 54:21–24.

Martens, D.C. and W.L. Lindsay. 1990. Testing soils for copper, iron, manganese, and zinc, in R.L. Westerman, Ed., *Soil Testing and Plant Analysis*, 3rd ed., SSSA Book Series Number 3, Soil Science Society of America, Madison, WI, 229–264.

Mascagni, H.J. and F.R. Cox. 1983. Calibration of a manganese availability index for soybeans soil test data. *Soil Sci. Soc. Am. J.,* 49:382–386.

Mayland, H.F. and S.R. Wilkinson. 1989. Soil factors affecting magnesium availability in plant-animal systems: a review. *J. Anim. Sci.,* 67:3437–3444.

Maynard, D.N. and G.J. Hochmuth. 1997. *Knott's Handbook for Vegetable Growers*, 4th ed., John Wiley & Sons, New York.

Maynard, D.G. and Y.P. Kalra. 1993. Nitrate and exchangeable ammonium-nitrogen, in R.L. Carter, Ed., *Soil Sampling and Methods of Analysis*, Lewis Publishers, Boca Raton, FL, 25–38.

McFairlane, J.D. 1999. Iron, in K.I. Peverill, L.A. Sparrow, and D.J. Reuter, Eds., *Soil Analysis: An Interpretation Manual*, CSIRO Publishing, Collingwood, Australia, 295–301.

McKeague, J.A., Ed. 1978. *Manual on Soil Sampling and Methods of Analysis*, 2nd ed., Canada Society of Soil Science, Ottawa, Ontario, Canada.

McLean, E.O. 1973. Testing soils for pH and lime requirement, in L.M. Walsh and J.D. Beaton, Eds., *Soil Testing and Plant Analysis*, rev. ed., Soil Science Society of America, Madison, WI, 78–95.

McLean, E.O. 1977. Contrasting concepts in soil test interpretation: sufficiency levels of available nutrients vs. basic cation saturation ratios, in T.R. Peck, Ed., *Soil Testing: Correlating and Interpreting the Analytical Results*, ASA Special Publication Number 29, American Society of Agronomy, Madison, WI, 39–54.

McLean, E.O. 1978. Principles underlying the practice of determining lime requirements of acid soils by the use of buffer methods. *Commun. Soil Sci. Plant Anal.,* 9:699–715.

McLean, E.O. 1982a. Chemical equilibrium with soil buffer systems as bases for future soil testing programs. *Commun. Soil Sci. Plant Anal.,* 13:411–433.

McLean, E.O. 1982b. Soil pH and lime requirement, in A.L. Page, Ed., *Methods of Soil Analysis,* Part 2, *Chemical and Microbiological Properties*, 2nd ed., American Society of Agronomy, Madison, WI, 199–224.

McLean, E.O. and S. Mostaghimi. 1983. Improved corrective fertilizer recommendations based on a two-step alternative usage of soil tests. III. The BrayP-1 test on soils with concretions. *Soil Sci. Soc. Am. J.,* 47:966–971.

McLean, E.O., S.W. Dumford, and F. Coronel. 1966. A comparison on several methods of determining lime requirements of soils. *Soil Sci. Soc. Am. Proc.,* 30:26–30.

McLean, E.O., J.F. Trierweiler, and D.J. Eckert. 1977. Improved SMP buffer method for determining lime requirement of acid soils. *Commun. Soil Sci. Plant Anal.,* 8:667–675.

McLean, E.O., J.L. Adams, and R.C. Hartwig. 1982. Improved corrective fertilizer recommendations based on a two-step alternative usage of soil tests. II. Recovery of soil-equilibrated K. *Soil Sci. Soc. Am. J.,* 46:1198–1201.

McLean, E.O., D.J. Eckert, G.Y. Reddy, and J.F. Trierweiler. 1978. Use of double-buffer and quicktest features for improving the SMP method for determining lime requirement of acid soils. *Soil Sci. Soc. Am. J.,* 42:311–316.

McLean, E.O., R.C. Hartwig, D.J. Eckert, and G.B. Triplett. 1983. Basic cation saturation ratios as a basis for fertilizing and liming agronomic soils. II. Field studies. *Agron. J.,* 75:635–639.

Mebius, L.J. 1960. A rapid method for the determination of organic carbon in soil. *Anal. Chem. Acta*, 22:120–124.

Mehlich, A. 1939. Use of trithanolamine acetate-barium hydroxide buffer for the determination of some base exchange properties and lime requirement of soil. *Soil Sci. Soc. Am. Proc.,* 3:162–166.

Mehlich, A. 1953a. Determination of P, Ca, Mg, K, Na and NH_4 (mimeo), North Carolina Soil Test Division, Raleigh.

Mehlich, A. 1953b. Rapid determination of cation and anion exchange properties and pHe of soils. *J. Assoc. Off. Agric. Chem.,* 36:445–457.

Mehlich, A. 1972. Uniformity of expressing soil test results on a volume basis. *Commun. Soil Sci. Plant Anal.*, 3:417–424.

Mehlich, A. 1973. Uniformity of soil test results as influenced by volume weight. *Commun. Soil Sci. Plant Anal.*, 4:475–486.

Mehlich, A. 1974. Uniformity of soil test results as influenced by extractants and soil properties, in *Proceedings 7th International Colloquium Plant Analysis and Fertilizer Problems*, J. Wehrmann, Ed., Hanover, Germany, 295–305.

Mehlich, A. 1976. New buffer pH method for rapid estimation of exchangeable acidity and lime requirement of soils. *Commun. Soil Sci. Plant Anal.,* 7:637–652.

Mehlich, A. 1978. New extractant for soil test evaluation of phosphorus, potassium, magnesium, calcium, sodium, manganese and zinc. *Commun. Soil Sci. Plant Anal.*, 9:477–492.

Mehlich, A. 1984a. Mehlich 3 soil test extractant: a modification of Mehlich 2 extractant. *Commun. Soil Sci. Plant Anal.*, 15:1409–1416.

Mehlich, A. 1984b. Photometric determination of humic matter in soils. A proposed method. *Commun. Soil Sci. Plant Anal.,* 15:1417–1422.

Mehlich, A., S.S. Bowling, and A.L. Hatfield. 1976. Buffer pH acidity in relation to nature of soil acidity and expression of lime requirement. *Commun. Soil Sci. Plant Anal.,* 7:253–263.

Meier, P., A. Lohrum, and J. Gareiss. 1989. *Practice and Theory of pH Measurement: An Outline of pH Measurement, Information and Practical Hints*, Ingold Messtechnik AG, Urdorf, Switzerland.

Melsted, S.W. and T.R. Peck. 1973. The principles of soil testing, in L.M. Walsh and J.D. Beaton, Eds., *Soil Testing and Plant Analysis,* rev. ed., Soil Science Society of America, Madison, WI, 13–22.

Merkle, F.G. and E.C. Dunkle. 1944. The soluble salt content of greenhouse soils as a diagnostic aid. *J. Am. Soc. Agron.,* 36:10–19.

Mikkelsen, R.L. and J.J. Camberato. 1995. Potassium, sulfur, lime, and micronutrient fertilizers, in J.E. Rechcigl, Ed., *Soil Amendments and Environmental Quality*, Lewis Publishers, Boca Raton, FL, 109–137.

Miller, E.R, X. Lei, and D.E. Ullrey. 1991. Trace elements in animal nutrition, in J.J. Mortvedt, Ed., *Micronutrients in Agriculture*, 2nd ed., SSSA Book Series Number 4, Soil Science Society of America, Madison, WI, 593–662.

Millham, P.J., A.S. Awad, R.E. Paul, and J.H. Bull. 1970. Analysis of plants, soils, and waters for nitrate by using an ion-specific electrode. *Analyst,* 95:751–759.

Mills, H.A. 1980. Nitrogen specific-ion electrodes for soil, plant, and water analysis. *J. Assoc. Off. Anal. Chem.*, 63:797–801.

Moody, P.W. and M.D.A. Bolland. 1999. Phosphorus, in K.I. Peverill, L.A. Sparrow, and D.J. Reuter, Eds., *Soil Analysis: An Interpretation Manual,* CSIRO Publishing, Collingwood, Australia, 187–220.

Moore, J.T. and R.H. Loeppert. 1987. Significance of potassium chloride pH of calcareous soils. *Soil Sci. Soc. Am. J.,* 51:908–912.

Moraghan, J.T. and H.J. Mascagni. 1991. Environmental and soil factors affecting micronutrient deficiencies and toxicities, in J.J. Mortvedt, Ed., *Micronutrients in Agriculture*, 2nd ed., SSSA Book Series Number 4, Soil Science Society of America, Madison, WI, 371–425.

Morgan, M.F. 1932. Microchemical Soil Tests, Connecticut Agricultural Experiment Station Bulletin 333.

Morgan, M.F. 1941. Chemical Diagnosis by the Universal Soil Testing System, Connecticut Agricultural Experiment Station Bulletin 450, New Haven.

Mortensen, J.L. 1965. Partial extraction of organic matter, in C.A. Black, Ed., *Methods of Soil Analysis*, Part 2, *Chemical and Microbiological Properties*, Agronomy Number 9, American Society of Agronomy, Madison, WI, 1401–1408.

Mortvedt, J.J., F.R. Cox, L.M. Shuman, and R.M. Welch, Eds. 1991. *Micronutrients in Agriculture*, 2nd ed., SSSA Book Series Number 4, Soil Science Society of America, Madison, WI.

Mosko, J.A. 1984. Automated determination of inorganic anions in water by ion chromatography. *Anal. Chem.*, 56:629–633.

Muchovej, R.M.C. and J.E. Rechcigl. 1995. Nitrogen fertilizers, in J.E. Rechcigl, Ed., *Soil Amendments and Environmental Quality*, Lewis Publishers, Boca Raton, FL, 1–64.

Mullenax, C.H., L.E. Baumann, E.A. Kihn, and W.E. Campbell. In press. Global synchrony in biospheric variations and influence on soil pH. *Commun. Soil Sci. Plant Anal.*

Mulvaney, R.L. 1996. Nitrogen — inorganic forms, in R.L. Sparks, Ed., *Methods of Soil Analysis,* Part 3, *Chemical Methods*, SSSA Book Series Number 5, Soil Science Society of America, Madison, WI, 1123–1184.

Murphy, J. and J.R. Riley. 1962. A modified single solution method for the determination of phosphate in natural waters. *Anal. Chem. Acta,* 27:31–36.

Murphy, W.M., D.T. Dugdale, and D.S. Ross. 1983. Problems in predicting legume fertility needs in steepland pasture improvement with soil tests. *Commun. Soil Sci. Plant Anal.,* 14:1193–1200.

Nelson, D.W. and L.E. Sommers. 1996. Total carbon, organic carbon, and organic matter, in D.L. Sparks, Ed., *Methods of Soil Analysis*, Part 3, *Chemical Methods,* SSSA Book Series Number 5, American Society of Agronomy, Madison, WI, 961–1010.

Nelson, J.L., L.C. Boawn, and F.G. Viets, Jr. 1959. A method for assessing the zinc status of soils using acid extractable zinc and "titratable alkalinity" values. *Soil Sci.,* 88:275–283.

Nelson, W.L., J.W. Fitts, L.D. Kardos, W.T. McGeorge, R.Q. Parks, and J. Fielding Reed. 1951. Soil Testing in the United States, 0-979953, National Soil and Fertilizer Research Committee, U.S. Government Printing Office, Washington, D.C.

Nelson, W.L., A. Mehlich, and E. Winters. 1953. The development, evaluation and use of soil tests for phosphorus availability, in W.H. Pierre and A.G. Norman, Eds., *Soil and Fertilizer Phosphorus*, Agronomy Number 4, American Society of Agronomy, Madison, WI, 153–188.

Novozamsky, I., R. van Eck, J.J. van der Lee, V.J.G. Houba, and E.J.M. Temminghoff. 1986. Determination of total sulphur and extractable sulfate in plant materials by inductively coupled plasma atomic emission spectrometry. *Commun. Soil Sci. Plant Anal.,* 17:1147–1157.

Novozamsky, I., Th.M. Lexmond, and V.J.G. Houba. 1993a. A single extraction procedure of soil for evaluation of uptake of some heavy metals by plants. *Int. J. Environ. Anal. Chem.,* 51:47–58.

Novozamsky, I., D. van Dijk, J.J. van der Lee, and V.J.G. Houba. 1993b. Automated determination of trace amounts of phosphate in soil extracts using malachite green. *Commun. Soil Sci. Plant Anal.,* 24:1065–1076.

Nowak, P. 1998. Agriculture and change: the promises and pitfalls of precision. *Commun. Soil Sci. Plant Anal.,* 29:1537–1541.

Odom, J.W. 1980. Kinetics of the hot-water soluble boron soil test. *Commun. Soil Sci. Plant Anal.,* 11:759–769.

Olsen, S.R. and L.A. Dean. 1965. Phosphorus, in C.A. Black, Ed., *Methods of Soil Analysis,* Agronomy Number 9, American Society of Agronomy, Madison, WI, 1044–1046.

Olsen, S.R., C.V. Cole, F.S. Watanabe, and L.A. Dean. 1954. Estimation of Available Phosphorus in Soils by Extraction with Sodium Bicarbonate, U.S.D.A. Circular Number 939, U.S. Government Printing Office, Washington, D.C.

Pais, I. and J.B. Jones, Jr. 1997. *The Handbook of Trace Elements*, St. Lucie Press, Boca Raton, FL.

Peck, T.R., Ed. 1977. *Soil Testing: Correlating and Interpreting the Analytical Results*, ASA Special Publication Number 29, American Society of Agronomy, Madison, WI.

Peck, T.R. 1980. Standard soil scoop, in W.C. Dahnke, Ed. Recommended Chemical Soil Test Procedures for the North Central Region, North Central Regional Publication 221 (revised), North Dakota Agricultural Experiment Station Bulletin 499 (revised), North Dakota State University, Fargo, 4–5.

Peck, T.R. 1983. Measurement of pH and electrode selection, in *Proceedings Ninth Soil-Plant Analyst's Workshop*, Council on Soil Testing and Plant Analysis, Athens, GA.

Peck, T.R. 1990. Soil testing: past, present, and future. *Commun. Soil Sci. Plant Anal.,* 21:1165–1186.

Peck, T.R. 1998. Standard soil scoop, in J.R. Brown, Ed., Recommended Chemical Soil Test Procedures for the North Central Region, North Central Regional Publication Number 221 (revised), Missouri Agricultural Experiment Station SB 1001, University of Missouri, Columbia, 7–9.

Peck, T.R. and R. Beck. 1998. Soil sampling and soil availability, past and present. *Commun. Soil Sci. Plant Anal.,* 29:1425–1428.

Peck, T.R. and S.W. Melsted. 1973. Field sampling for soil testing, in L.M. Walsh and J.D. Beaton, Eds., *Soil Testing and Plant Analysis,* rev. ed., Soil Science Society of America, Madison, WI, 67–76.

Peck, T.R. and P.N. Soltanpour. 1990. The principles of soil testing, in R.L. Westerman, Ed., *Soil Testing and Plant Analysis*, 3rd ed., SSSA Book Series Number 3, Soil Science Society of America, Madison, WI, 1–9.

Peck, T.R., J.T. Cope, Jr., and D.A. Whitney, Eds. 1977. *Soil Testing: Correlating and Interpreting the Analytical Results,* ASA Special Publication Number 29, American Society of Agronomy, Madison, WI.

Peech, M. 1965. Hydrogen-ion activity, in C.A. Black, Ed., *Methods of Soil Analysis*, Part 2. Agronomy Number 9, American Society of Agronomy, Madison, WI, 914–926.

Perkins, H.F. 1970. A rapid method of evaluating the zinc status of coastal plain soils. *Commun. Soil Sci. Plant Anal.,* 1:35–42.

Peterson, R.G. and L.D. Calvin. 1996. Sampling, in R.L. Sparks, Ed., *Methods of Soil Analysis,* Part 3, *Chemical Methods,* SSSA Book Series Number 5, Soil Science Society of America, Madison, WI, 1–17.

Peverill, K.I., L.A. Sparrow, and D.J. Reuter, Eds., 1999. *Soil Analysis: An Interpretation Manual*, CSIRO Publishing, Collingwood, Australia.

Plank, C.O. 1992a. Determination of soil pH on 0.01 *M* $CaCl_2$, in S.J. Donohue, Ed., Reference Soil and Media Diagnostic Procedures for the Southern Region of the United States, Southern Cooperative Series Bulletin Number 374, Virginia Agricultural Experiment Station, Blacksburg, 1–3.

Plank, C.O. 1992b. Determination of soil pH in 1 *N* KCl, in S.J. Donohue, Ed., Reference Soil and Media Diagnostic Procedures for the Southern Region of the United States, Southern Cooperative Series Bulletin Number 374, Virginia Agricultural Experiment Station, Blacksburg, 4–5.

Prianischnikov, D.N. 1913. *Landwirtsch. Vers. Stn.,* 79–80:667.

Prince, A.L., M. Zimmerson, and F.E. Bear. 1947. The magnesium-supplying powers of 20 New Jersey soils. *Soil Sci.,* 93:69–78.

Radojevic, M. and V.N. Bashkin. 1999. *Practical Environmental Analysis*, The Royal Society of Chemistry, Cambridge, U.K., 288–295.

Rappaport, B.D., J.D. Scott, D.C. Martens, R.B. Reneau, Jr., and T.W. Simpson. 1987. Availability and Distribution of Heavy Metals, Nitrogen, and Phosphorus from Sewage Sludge in the Plant–Soil–Water Continuum, VPI-VWRRC-Bulletin 154 5C, Virginia Water Resources Center, Blacksburg.

Rappaport, B.D., D.C. Martens, R.B. Reneau, Jr., and T.W. Simpson. 1988. Metal availability in sludge-amended soils with elevated levels. *J. Environ. Qual.,* 17:42–47.

Reed, S.T. and D.C. Martens. 1996. Copper and zinc, in D.L. Sparks, Ed., *Methods of Soil Analysis,* Part 3, *Chemical Methods*, SSSA Book Series Number 5, Soil Science Society of America, Madison, WI, 703–722.

Rehm, G.W. and A.C. Caldwell. 1968. Sulfur supplying power capacity of soils and the relationships to soil type. *Soil Sci.*, 105:355–361.

Reid, K., Ed. 1998a. Soil pH, liming and acidification, in *Soil Fertility Handbook*, Ministry of Agriculture, Food and Rural Affairs, Toronto, Ontario, Canada, 71–82.

Reid, K., Ed. 1998b. *Soil Fertility Handbook*, Ontario Ministry of Agriculture, Food and Rural Affairs, Queen's Printer for Ontario, Ontario, Canada.

Reisenauer, H.M., L.M. Walsh, and R.G. Hoeft. 1973. Testing soils for sulfur, boron, molybdenum, and chloride, in L.M. Walsh and J.D. Beaton, Eds., *Soil Testing and Plant Analysis*, rev. ed., Soil Science Society of America, Madison, WI, 173–200.

Rengel, Z., Ed. 1998. *Nutrient Use in Crop Production*, Haworth Press, Binghamton, NY.

Rhoades, J.D. 1996. Salinity: electrical conductivity and total dissolved solids, in R.L. Sparks, Ed. *Methods for Soil Analysis,* Part 3, *Chemical Methods*, SSSA Book Series Number 5, Soil Science Society of America, Madison, WI, 417–435.

Richards, L.A., Ed. 1969. Diagnosis and Improvement of Saline and Alkali Soils, U.S.D.A. Agricultural Handbook Number 60, U.S. Government Printing Office, Washington, D.C.

Risser, J.A. and D.L. Baker. 1990. Testing soils for toxic metals, in R.L. Westerman, Ed., *Soil Testing and Plant Analysis*, 3rd ed., SSSA Book Series Number 3, Soil Science Society of America, Madison, WI, 275–298.

Rodriguez, J.B., J.R. Self, and P.N. Soltanpour. 1994. Optical conditions for phosphorus analysis by the ascorbic acid-molybdenum blue method. *Soil Sci. Soc. Am. J.,* 58:866–870.

Römheld, V. and H. Marschner. 1991. Function of micronutrients in plants, in J.J. Morvedt, Ed., *Micronutrients in Agriculture*, 2nd ed., SSSA Book Series Number 4, Soil Science Society of America, Madison, WI, 297–328.

Ruzicka, J. and E.H. Hansen. 1988. *Flow Injection Analysis*, 2nd ed., John Wiley & Sons, New York.

Sabbe, W.W. and D.B. Marx. 1987. Soil sampling: spatial and temporal variability, in J.R. Brown, Ed., *Soil Testing: Sampling, Correlation, Calibration, and Interpretation*, SSSA Special Publication 21, Soil Science Society of America, Madison, WI.

Sartoretto, P. 1991. *The pH Factor*, W-A-Clearly Chemical, Dayton, NJ.

Schmisek, M.E., L.J. Cihacek, and L.J. Swenson. 1998. Relationships between the Mehlich-III soil test extraction procedure and standard soil test methods in North Dakota. *Commun. Soil Sci. Plant Anal.,* 29:1719–1729.

Schmitt, M.A. and G.W. Randall. 1994. Developing a soil test nitrogen test for improved recommendations for corn. *J. Prod. Agric.,* 7:328–334.

Schnug, E., K. Panten, and S. Haneklaus. 1998. Sampling and nutrient recommendations — the future. *Commun. Soil Sci. Plant Anal.,* 29:1455–1462.

Schofield, R.K. and A.W. Taylor. 1955. The measurement of soil pH. *Soil Sci. Soc. Am. Proc.,* 19:164–167.

Schollenberger, C.J. and R.H. Simon. 1945. Determination of exchange capacity and exchangeable bases in soil-ammonium acetate method. *Soil Sci.,* 59:13–24.

Schulte, E.E. and K. Eik. 1988. Recommended sulfate–sulfur test, in W.C. Dahnke, Ed., Recommended Chemical Soil Test Procedures for the North Central Region, North Central Regional Publication Number 221 (revised), North Dakota Agricultural Experiment Station Bulletin Number 499 (revised), North Dakota State University, Fargo, 17–19.

Schulte, E.E., C. Kaufmann, and J.B. Peter. 1991. The influence of sample size and heating time on soil weight loss-on-ignition. *Commun. Soil Sci. Plant Anal.,* 22:159–168.

Schwertmann, U. and M.L. Jackson. 1964. Influence of hydroxyaluminum ions on pK titration curves of hydronium-aluminum clays. *Soil Sci. Soc. Am. Proc.,* 28:179–182.

Sen Tran, T. and R.R. Simard. 1993. Mehlich III-extractable elements, in M.R. Carter, Ed., *Soil Sampling and Methods of Analysis*, CRC Press, Boca Raton, FL, 43–49.

Sen Tran, T. and W. van Lierop. 1993. Lime requirement, in R.L. Carter, Ed., *Soil Sampling and Methods of Analysis*, Lewis Publishers, Boca Raton, FL, 109–113.

Shaw, R.J. 1999. Soil salinity — electrical conductivity and chloride, in K.I. Peverill, L.A. Sparrow, and D.J. Reuter, Eds., *Soil Analysis: An Interpretation Manual*, CSIRO Publishing, Collingwood, Australia, 129–145.

Sheldrick, B.H. and C. Wang. 1993. Particle size distribution, in M.R. Carter, Ed., *Soil Sampling and Methods of Analysis*, Lewis Publishers, Boca Raton, FL, 499–511.

Shoemaker, H.E., E.O. McLean, and P.F. Pratt. 1962. Buffer methods for determination of lime requirement of soils with appreciable amount of exchangeable aluminum. *Soil Sci. Soc. Am. Proc.*, 25:274–277.

Shuman, L.M. 1980. Effects on soil temperature, moisture, and air-drying on extractable manganese, iron, copper, and zinc. *Soil Sci.,* 130:336–343.

Shuman, L.M. 1991. Chemical forms of micronutrients in soils, in J.J. Mortvedt, Ed., *Micronutrients in Agriculture*, 2nd ed., SSSA Book Series Number 4, Soil Science Society of America, Madison, WI, 113–144.

Shuman, L.M. 1998. Micronutrient fertilizers, in Z. Rengel, Ed., *Nutrient Use in Crop Production*, Haworth Press, Binghamton, NY, 165–195.

Shuman, L.M., V.A. Bandel, S.J. Donohue, R.A. Isaac, R.M. Lippert, J.T. Sims, and M.R. Tucker. 1992. Comparison of Mehlich-1 and Mehlich-3 extractable soil boron with hot water extractable boron. *Commun. Soil Sci. Plant Anal.,* 23:1–14.

Sillanpää, M. 1972. Trace Elements in Soils and Agriculture, Soils Bulletin Number 17, FAO, Rome, Italy.

Sillanpää, M. 1990. Micronutrient Assessment at the Country Level: An International Study, Soil Bulletin 63, FAO, Rome, Italy.

Sillanpää, M. and H. Jansson. 1992. Status of Cadmium, Lead, Cobalt, and Selenium in Soils and Plants of Thirty Countries, Soil Bulletin 65, FAO, Rome, Italy.

Simard, R.R. 1993. Ammonium acetate–extractable elements, in *Soil Sampling and Methods of Analysis*, R.L. Carter, Ed., CRC Press, Boca Raton, FL, 39–42.

Simonis, A.D. and H.B. Setatou. 1996. Assessment of available phosphorus and potassium in soils by the calcium chloride extraction method. *Commun. Soil Sci. Plant Anal.,* 27:685–694.

Sims, J.T. 1989. Comparison of Mehlich 1 and Mehlich 3 extractants for P, K, Ca, Mg, Mn, Cu, and Zn in Atlantic Coastal Plain soils. *Commun. Soil Sci. Plant Anal.,* 20:1707–1726.

Sims, J.T. 1995. Recommended soil tests for the micronutrients: manganese, zinc, copper, and boron, in J.T. Sims and A. Wolf, Eds., Recommended Soil Testing Procedures for the Northeastern United States, 2nd ed., Northeast Regional Publication, Agricultural Experiment Station Bulletin 493, University of Delaware, Newark, 40–45.

Sims, J.T. 1996. Lime requirement, in D.L. Sparks, Ed., *Methods of Soil Analysis,* Part 3, *Chemical Methods*, SSSA Book Series Number 5, Soil Science Society of America, Madison, WI, 491–515.

Sims, J.T. and D. Eckert. 1995. Recommended soil pH and lime requirement tests, in J.T. Sims and A. Wolf, Eds., Recommended Soil Testing Procedures for the Northeastern United States, 2nd ed., Northeast Regional Publication, Agricultural Experiment Station Bulletin 493, University of Delaware, Newark, 35–40.

Sims, J.T. and G.V. Johnson. 1991. Micronutrient soil tests, in J.J. Mortvedt, Ed., *Micronutrients in Agriculture*, 2nd ed., SSSA Book Series Number 4, Soil Science Science of America, Madison, WI, 427–476.

Singh, R., D.K. Bhumbia, and R.F. Keefer. 1995. Recommended soil sulfate-S tests, in J.T. Sims and A.M. Wolf, Eds., Recommended Soil Testing Procedures for the Northeastern United States, Northeast Regional Publication Number 493 (revised), University of Delaware, Newark, 46–50.

Slattery, W.J., M.K. Conyers, and R.L. Aitken. 1999. Soil pH, aluminium, manganese, and lime requirement, in K.I. Peverill, L.A. Sparrow, and D.J. Reuter, Eds., *Soil Analysis: An Interpretation Manual,* CSIRO Publishing, Collingwood, Australia, 103–128.

Small, H., T.S. Stevens, and W.A. Bauman. 1975. Novel ion exchange chromatography method using conductance detection. *Anal. Chem.*, 47:1801–1809.

Smith, F.W., B.G. Ellis, and J. Grava. 1957. Use of acid-fluoride solutions for extraction of available phosphorus in calcareous soils and in soils to which rock phosphate has been added. *Soil Sci. Soc. Am. Proc.,* 21:400–404.

Soltanpour, P.N. 1991. Determination of nutrient element availability and elemental toxicity by the AB–DTPA soil test and ICPS. *Adv. Soil Sci.,* 16:165–190.

Soltanpour, P.N. and A.P. Schwab. 1977. A new soil test for simultaneous extraction of macro- and micro-nutrients in alkaline soils. *Commun. Soil Sci. Plant Anal.,* 8:195–207.

Soltanpour, P.N. and S.M. Workman. 1979. Modification of NH_4HCO_3–DTPA soil test to omit carbon black. *Commun. Soil Sci. Plant Anal.,* 10:1411–1420.

Soltanpour, P.N., A. Khan, and A.P. Schwab. 1979. Effect of grinding variables on the NH_4HCO_3–DTPA soil test values for Fe, Zn, Mn, Cu, P, and K. *Commun. Soil Sci. Plant Anal.,* 10:903–909.

Soltanpour, P.N., C.W. Johnson, S.M. Workman, J.B. Jones, Jr., and R.O. Miller. 1996. Inductively coupled plasma emission spectrometry and inductively coupled plasma-mass spectroscopy, in D.L. Sparks, Ed., *Methods of Soil Analysis,* Part 3, *Chemical Methods*, SSSA Book Series Number 5, Soil Science Society of America, Madison, WI, 91–139.

Soltanpour, P.N., C.W. Johnson, S.M. Workman, J.B. Jones, Jr., and R.O. Miller. 1998. Advances in ICP emission and ICP mass spectrometry. *Adv. Agron.,* 64:28–113.

Sonneveld, C. and J. van den Ende. 1971. Soil analysis by a 1:2 volume extract. *Plant Soil,* 35:505–516.

Sonneveld, C., J. van den Ende, and P.A. van Dijk. 1974. Analysis of growing media by means of a 1:½ volume extract. *Commun. Soil Sci. Plant Anal.,* 5:183–202.

Soon, Y.K. and A. Abbound. 1993. Cadmium, chromium, lead, and nickel, in M.R. Carter, Ed., *Soil Sampling and Methods of Analysis*, Lewis Publishers, Boca Raton, FL, 101–108.

Sparks, D.L. 1987. Potassium soil dynamics in soils. *Adv. Soil Sci.,* 6:1–63.

Spurway, C.H. and K. Lawton. 1949. Soil Testing: A Practical System of Soil Diagnosis, Michigan Agricultural Experiment Station Bulletin 132.

Steinhardt, G.C. 1979. Particle-size distribution, in R.W. Fairbridge and C.W. Finkel, Jr., Eds., *The Encyclopedia of Soil Science,* Part 1 — *Physics, Chemistry, Biology, Fertility, and Technology,* Dowden, Hutchinson & Ross, Stroudsburg, PA, 350–358.

Stevenson, F.J. 1991. Organic matter–micronutrient reactions in soil, in J.J. Mortvedt, Ed., *Micronutrients in Agriculture*, 2nd ed., SSSA Book Series Number 4, Soil Science Science of America, Madison, WI, 145–186.

Storer, D.A. 1984. A simple high sample volume ashing procedure for determination of soil organic matter. *Commun. Soil Sci. Plant Anal.,* 15:759–772.

Strek, H.J. 1984. Improved Herbicide Rate Recommendations Using Soil and Herbicide Property Measurements, Ph.D. dissertation, Crop Science Department, North Carolina State University, Raleigh.

Strek, H.J. and J.B. Weber. 1983. Update of soil testing and herbicide rate recommendations, *Proc. S. Weed Sci. Soc.,* 36:398–403.

Strong, W.M. and M.G. Mason. 1999. Nitrogen, in K.I. Peverill, L.A. Sparrow, and D.J. Reuter, Eds., *Soil Analysis: An Interpretation Manual,* CSIRO Publishing, Collingwood, Australia, 171–185.

Suarez, D.L. 1996. Beryllium, magnesium, calcium, strontium, and barium, in D.L. Sparks, Ed., *Methods of Soil Analysis,* Part 3, *Chemical Methods*, SSSA Book Series Number 5, Soil Science Society of America, Madison, WI, 575–601.

Summer, M.E. 1992a. Determination of exchangeable acidity using $BaCl_2$–TEA buffer, in S.J. Donohue, Ed., Reference Soil and Media Diagnostic Procedures for the Southern Region of the United States, Southern Cooperative Series Bulletin Number 374, Virginia Agricultural Experiment Station, VPI & SU, Blacksburg, 39–40.

Summer, M.E. 1992b. Determination of exchangeable acidity and exchangeable Al using 1 *N* KCl, in S.J. Donohue, Ed., Reference Soil and Media Diagnostic Procedures for the Southern Region of the United States, Southern Cooperative Series Bulletin Number 374, Virginia Agricultural Experiment Station, VPI & SU, Blacksburg, 41–42.

Sumner, M.E. 1994. Measurement of soil pH: problems and solutions. *Commun. Soil Sci. Plant Anal.,* 25:859–879.

Sumner, M.E. and W.P. Miller. 1996. Cation exchange capacity and exchange capacity, in D.L. Sparks, Ed., *Methods of Soil Analysis*, Part 3, *Chemical Methods,* SSSA Book Series Number 5, American Society of Agronomy, Madison, WI, 1201–1229.

Swaine, D.J. 1969. The Trace Element Content of Soils, Commonwealth Bureau of Soil Science, Technical Communication Number 26, Commonwealth Agricultural Bureaux, Farnham Royal, Bucks, England.

Swift, R.S. 1996. Organic matter characterization, in D.L. Sparks, Ed., *Methods of Soil Analysis*, Part 3, *Chemical Methods,* SSSA Book Series Number 5, American Society of Agronomy, Madison, WI, 1011–1069.

Tan, K.H. 1998. *Principles of Soil Chemistry*, 3rd ed., Marcel Dekker, New York, 357–394.

Tel, D.A. and C. Heseltine. 1990a. The analyses of KCl soil extracts for nitrate, nitrite, and ammonium using a TRAACS 800 Analyzer. *Commun. Soil Sci. Plant Anal.,* 21:1681–1688.

Tel, D.A. and C. Heseltine. 1990b. Chloride analyses of soil leachate using the TRAACS 800 analyzer. *Commun. Soil Sci. Plant Anal.,* 21:1689–1693.

Thom, W.O. and W.E. Sabbe. 1994. Soil Sampling Procedures for the Southern Region of the United States, Southern Cooperative Series Bulletin Number 377, Kentucky Cooperative Extension Service, University of Kentucky, Lexington.

Thomas, G.W. 1996. Soil pH and soil acidity, in D.L. Sparks, Ed., *Methods of Soil Analysis,* Part 3, *Chemical Methods*, SSSA Book Series Number 5, Soil Science Society of America, Madison, WI, 475–490.

Thomas, G.W. and W.L. Hargrove. 1984. The chemistry of soil acidity, in F. Adams, Ed., *Soil Acidity and Liming*, 2nd ed., American Society of Agronomy, Madison, WI, 3–56.

Thomas, G.W. and D.E. Peaslee. 1973. Testing soils for phosphorus, in L.M. Walsh and J.D. Beaton, Eds., *Soil Testing and Plant Analysis*, 2nd ed., Soil Science Society of America, Madison, WI, 115–132.

Tucker, M.R. 1984. Volumetric soil measures for routine soil testing. *Commun. Soil Sci. Plant Anal.,* 15:833–840.

Tucker, M.R. 1992a. Determination of phosphorus by Mehlich 3 extraction, in S.J. Donohue, Ed., Reference Soil and Media Diagnostic Procedures for the Southern Region of the United States, Southern Cooperative Series Bulletin Number 374, Virginia Agricultural Experiment Station, Blacksburg, 6–8.

Tucker, M.R. 1992b. Determination of potassium, calcium, magnesium, and sodium by Mehlich 3 extraction, in S.J. Donohue, Ed., Reference Soil and Media Diagnostic Procedures for the Southern Region of the United States, Southern Cooperative Series Bulletin Number 374, Virginia Agricultural Experiment Station, Blacksburg, 9–12.

Tucker, M.R. 1992c. Determination of zinc, manganese, and copper by Mehlich 3 extraction, in S.J. Donohue, Ed., Reference Soil and Media Diagnostic Procedures for the Southern Region of the United States, Southern Cooperative Series Bulletin Number 374, Virginia Agricultural Experiment Station, Blacksburg, 19–22.

Tucker, M.R. 1992d. Colorimetric determination of humic matter with 0.2 *N* NaOH extraction, in S.J. Donohue, Ed., Reference Soil and Media Diagnostic Procedures for the Southern Region of the United States, Southern Cooperative Series Bulletin Number 374, Virginia Agricultural Experiment Station, Blacksburg, 43–45.

Tucker, M.R. and L.T. Kurtz. 1955. A comparison of several chemical methods for extracting zinc from soils. *Soil Sci. Soc. Am. Proc.,* 19:477–481.

Tucker, M.R., J.K. Messick, and C.C. Canter. 1997. Crop fertilization based on North Carolina Soil Tests, Circular No. 1 (revised), North Carolina Department of Agriculture and Consumer Affairs, Raleigh.

Uren, N.C. 1999. Manganese, in K.I. Peverill, L.A. Sparrow, and D.J. Reuter, Eds., *Soil Analysis: An Interpretation Manual*, CSIRO Publishing, Collingwood, Australia, 287–294.

van Campen, D.R. 1991. Trace elements in human nutrition, in J.J. Mortvedt, Ed., *Micronutrients in Agriculture*, 2nd ed., SSSA Book Series Number 4, Soil Science Society of America, Madison, WI, 663–701.

van Erp, P.J. and M.L. van Beusichem. 1998. Soil and plant testing programs as a tool for optimizing fertilizer strategies, in Z. Rengel, Ed., *Nutrient Use in Crop Production*, Haworth Press, Binghamton, NY, 53–80.

van Erp, P.J., V.J.G. Houba, and M.L. Van Beusichem. 1998. One hundredth molar calcium chloride extraction procedure. I. A review of chemical, analytical, and plant nutrition aspects. *Commun. Soil Sci. Plant Anal.,* 29:1603–1623.

van Lierop, W. 1981. Laboratory determination of field bulk density for improving fertilizer recommendations of organic soils. *Can. J. Soil Sci.,* 61:475–482.

van Lierop, W. 1989. Effect of assumptions on accuracy of analytical results and liming recommendations when testing a volume or weight of soil. *Commun. Soil Sci. Plant Anal.,* 20:121–137.

van Lierop, W. 1990. Soil pH and lime requirement determinations, in R.L. Westerman, Ed., *Soil Testing and Plant Analysis*, SSSA Book Series Number 3, Soil Science Society of America, Madison, WI, 73–126.

van Raij, B. 1994. New diagnostic techniques, universal soil extractants. *Commun. Soil Sci. Plant Anal.,* 25:799–816.

Verloo, M.G. 1980. Peat as a natural complexing agent for trace elements. *Acta Hortic.,* 99:51–56.

Viets, F.G., Jr. and L.C. Boawn. 1965. Zinc, in C.A. Black, Ed., *Methods of Soil Analysis*, Part 2, Agronomy Number 9, American Society of Agronomy, Madison, WI, 1090–1101.

Viets, F.G., Jr. and W.L. Lindsay. 1973. Testing soils for zinc, copper, manganese, and iron, in L.M. Walsh and J.D. Beaton, Eds., *Soil Testing and Plant Analysis*, rev. ed., Soil Science Society of America, Madison, WI, 153–172.

Voss, R. 1998. Fertility recommendations: past and present. *Commun. Soil Sci. Plant Anal.,* 29:1429–1440.

Walkley, A. 1947. A critical examination of a rapid method for determination of organic carbon in soils: effect of variation in digestion conditions and of inorganic soil constituents. *Soil Sci.,* 63:251–257.

Warncke, D.D. 1986. Analyzing greenhouse growth media by the saturation extraction method. *HortScience,* 21:223–225.

Warncke, D.D. 1990. Testing artificial growth media and interpreting the results, in R.L. Westerman, Ed., *Soil Testing and Plant Analysis,* 3rd ed., SSSA Book Series Number 3, Soil Science Society of America, Madison, WI, 337–357.

Warncke, D.D. 1995. Recommended test procedures for greenhouse growth media, in J.T. Sims and A.M. Wolf, Eds., Recommended Soil Testing Procedures for the Northeastern United States, 2nd ed., Northeast Regional Publication Bulletin 493, Agricultural Experiment Station, University of Delaware, Newark, 76–82.

Warncke, D.D. 1998. Greenhouse root media, in J.R. Brown, Ed., Recommended Chemical Soil Test Procedures for the North Central Region, North Central Regional Research Publication Number 221 (revised), Missouri Agricultural Experiment Station SB 1001, Columbia, 61–64.

Warncke, D.D. and J.R. Brown. 1998. Potassium and other basic cations, in J.R. Brown, Ed., Recommended Chemical Soil Test Procedures for the North Central Region, North Central Regional Publication Number 221 (revised), Missouri Agricultural Experiment Station SB 1001, University of Missouri, Columbia, 31–33.

Warncke, D.D. and D.M. Krauskopf. 1983. Greenhouse Growth Media: Testing and Nutritional Guidelines, Michigan State University Cooperative Extension Bulletin E-1736.

Watanabe, F.S. and S.R. Olsen. 1965. Test of an ascorbic acid method for determining phosphorus in water and $NaHCO_3$ extracts from soil. *Soil Sci. Soc. Am. Proc.,* 29:677–678.

Watson, M.E. 1998. Boron test, in J.R. Brown, Ed., Recommended Chemical Soil Test Procedures for the North Central Region, North Central Regional Publication Number 221 (revised), Missouri Agricultural Experiment Station SB 1001, University of Missouri, Columbia, 45–48.

Watson, M.E. and J.R. Brown. 1998. pH and lime requirement, in J.R. Brown, Ed., Recommended Chemical Soil Test Procedures for the North Central Region, North Central Regional Research Publication Number 22 (revised), Missouri Agricultural Experiment Station SB 1001, University of Missouri, Columbia, 13–16.

Watson, M.E. and R.A. Isaac. 1990. Analytical instruments for soil and plant analysis, in R.L. Westerman, Ed., *Soil Testing and Plant Analysis,* 3rd ed., SSSA Book Series Number 3, Soil Science Society of America, Madison, WI, 691–740.

Wear, J.I. 1968. Boron, in C.A. Black, Ed., *Methods of Soil Analysis*, Part 2, Agronomy Number 9, American Society of Agronomy, Madison, WI, 1059–1063.

Wear, J.I. and C.E. Evans. 1968. Relationships of zinc uptake of corn and sorghum to soil zinc measured by three extractants. *Soil Sci. Soc. Am. Proc.,* 32:543–546.

Weber, J.B. and C.J. Peter. 1982. Absorption, bioactivity, and evaluation of soil tests for Alachlor, Acetochlor and Metolochlor. *Weed Sci.,* 30:14–20.

Welsh, R.M., W.H. Allaway, W.A. House, and J. Kubota. 1991. Geographic distribution of trace element problems, in J.J. Mortvedt, Ed., *Micronutrients in Agriculture*, 2nd ed., SSSA Book Series Number 4, Soil Science Society of America, Madison, WI, 31–57.

Westerman, R.L., Ed. 1990. *Soil Testing and Plant Analysis*, 3rd ed., SSSA Book Series Number 3, Soil Science Society of America, Madison, WI.

Westervelt, J.D. and H.F. Reetz. 2000. *GIS in Site-Specific Agriculture*, Interstate Publisher, Danville, IL.

Whipker, B.E., T. Kirk, and P.A. Hammer. 1994. Industry root media analysis results: useful in determining greenhouse nutrition problems and educational opportunities. *Commun Soil Sci. Plant Anal.*, 25:1455–1469.

Whitney, D.A. 1998a. Micronutrients: zinc, iron, manganese, and copper, in J.R. Brown, Ed. Recommended Chemical Soil Test Procedures for the North Central Region, North Central Regional Research Publication Number 221 (revised), Missouri Agricultural Experiment Station SB 1001, University of Missouri, Columbia, 41–44.

Whitney, D.A. 1998b. Soil salinity, in J.R. Brown, Ed., Recommended Chemical Soil Test Procedures for the North Central Region, North Central Regional Research Publication Number 221 (revised), Missouri Agricultural Experiment Station SB 1001, University of Missouri, Columbia, 59–60.

Withers, P.J. and A.N. Sharpley. 1995. Phosphorus fertilizers, in J.E. Rechcigl, Ed., *Soil Amendments and Environmental Quality*, Lewis Publishers, Boca Raton, FL, 65–107.

Wolf, A.M. and D.E. Baker. 1989. Comparison of soil test phosphorus by Olsen, Bray P1, Mehlich I and Mehlich III methods. *Commun. Soil Sci. Plant Anal.,* 16:467–484.

Wolf, A.M. and D. Beegle. 1995. Recommended soil tests for the macronutrients: phosphorus, potassium, calcium, and magnesium, in J.T. Sims and A.M. Wolf, Eds., Recommended Soil Testing Procedures for the Northeastern United States, Northeast Regional Publication Number 493 (revised), Agricultural Experiment Station, University of Delaware, Newark, 30–38.

Wolf, B. 1971. The determination of boron in soil extracts, plant materials, composts, manures, water, and nutrient solution. *Commun. Soil Sci. Plant Anal.,* 2:363–374.

Wolf, B. 1974. Improvements in the Azomethine-H method for the determination of boron. *Commun. Soil Sci. Plant Anal.,* 5:39–44

Wolf, B. 1982. An improved universal extracting solution and its use for diagnosing soil fertility. *Commun. Soil Sci. Plant Anal.,* 13:1005–1033.

Wolnik, K.A., F.L. Fricke, S.G. Capar, G.L. Braude, M.W. Meyer, R.D. Satzger, and E. Bonnin. 1983. Elements in major raw agricultural crops in the United States. I. Cadmium and lead in lettuce, peanuts, potatoes, soybeans, sweet corn, and wheat. *J. Agric. Food Chem.,* 31:1240–1244.

Wolnik, K.A., F.L. Fricke, S.G. Capar, M.W. Meyer, R.D. Satzger, E. Bonnin, and C.M. Gaston. 1985. Elements in major raw agricultural crops in the United States. III. Cadmium, lead, and other elements in carrots, field corn, onions, rice, spinach, and tomatoes. *J. Agric. Food Chem.,* 33:807–811.

Woodruff, C.M. 1961. Brom cresol purple as an indicator of soil pH. *Soil Sci.,* 91:272.

Woodruff, C.M. 1967. Crop response to lime in the midwestern United States, in R.W. Pearson and F. Adams, Eds., *Soil Acidity and Liming*, American Society of Agronomy, Madison, WI, 207–227.

Woodruff, J. 1994. Soil sampling for soil testing purposes: conventional row crop tillage, in W.O. Thom and W.E. Sabbe, Eds., Soil Sampling Procedures for the Southern Region of the United States, Southern Cooperative Series Bulletin Number 377, Kentucky Cooperative Extension Service, University of Kentucky, Lexington, 15–18.

Wright, N.A. 1998. Soil fertility variograms from "True Point Sampling™" on 20.0, 0.9, and 0.1 meter grids in two fields. *Commun. Soil Sci. Plant Anal.,* 29:1649–1666.

Wright, R.J. and T.I. Stuczynski. 1996. Atomic absorption and flame emission spectrometry, in D.L. Sparks, Ed., *Methods of Soil Analysis,* Part 3, *Chemical Methods,* SSSA Book Series Number 5, Soil Science Society of America, Madison, WI, 65–90.

Yuan, T.L. 1959. Determination of exchangeable hydrogen in soils by a titration method. *Soil Sci.,* 88:164–167.

Yuan, T.L. 1974. A double buffer method for the determination of lime requirement of acid soils. *Soil Sci. Soc. Am. Proc.*, 38:437–440.

Yuan, T.L. 1975. Lime requirement determination of sandy soils by different rapid methods. *Soil Crop Sci. Soc. Fla. Proc.,* 25:274–277.

Chapter 3

Plant Analysis

A. Purpose

Plant analysis, sometimes referred to as leaf analysis, is the technique for determining the elemental content of tissue of a particular plant part. Plant analysis can play a major role when diagnosing mineral nutrition problems, whether for research purposes or for solving practical field problems. The principles of the plant analysis technique have been reviewed and discussed by Thomas (1945), Ulrich (1952), Smith (1962), Munson and Nelson (1990), Munson (1998), and Reuter and Robinson (1997). Jones (1984) has produced a video and written a chapter on the plant analysis technique in his *Plant Nutrition Manual* (Jones, 1998a).

Goodall and Gregory (1947) were among the first to relate the nutrient element composition of a plant to its nutritional status, grouping work done prior to 1947 into four categories: (1) investigations of nutritional disorder made manifest by definite symptoms, (2) interpretation of the results of field trials, (3) development of rapid testing methods for use in advisory work, and (4) use of plant analysis as a method of nutritional survey. These categories are still applicable today in terms of research as well as plant analysis utilization in crop production decision making.

Normally, a plant analysis refers to a laboratory analysis of collected plant tissue. By contrast, tissue testing is an analysis of extracted cellular sap normally carried out in the field, making use of specially prepared papers and reagents. Tissue testing is discussed in Chapter 4.

A plant analysis is carried out as a series of steps:

- Sampling
- Sample preparation
- Laboratory analysis
- Interpretation

as is illustrated in Figure 3.1.

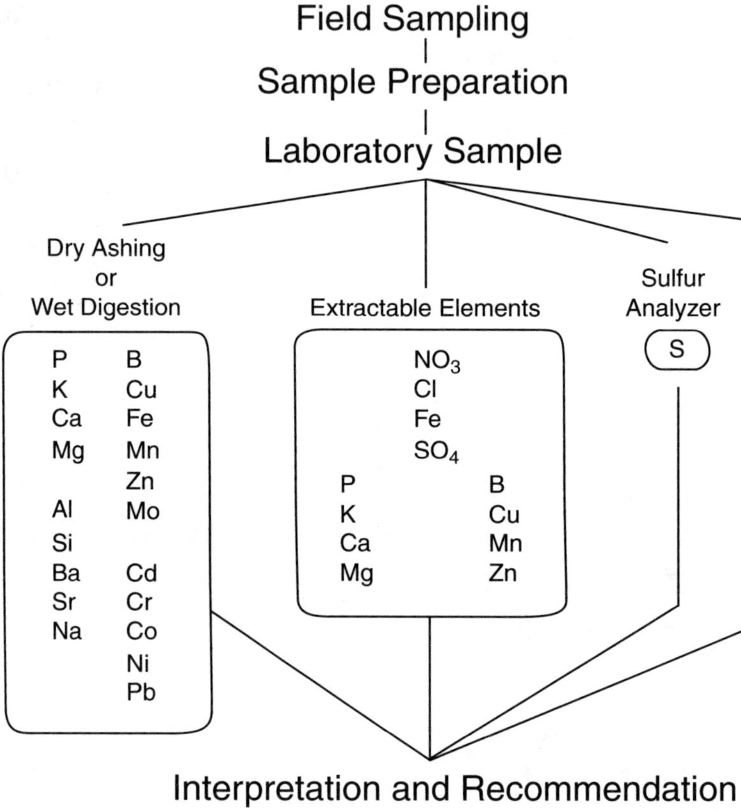

Figure 3.1
Sequence of procedures for conducting a plant analysis.

B. Sampling

1. Techniques

As a general rule, taking as a sample mature leaves exposed to full sunlight just below the growing tip on main branches or stems just prior to or at the time the plant begins its reproductive stage of growth is the preferred technique. In some situations, sampling may be necessary at earlier periods in the growth cycle of the plant, collecting leaf tissue of the same maturity.

The important components for proper plant tissue collection are as follows:

- Definite plant part taken at a specific location on the plant
- Stage of plant growth or specific time of sampling
- Number of parts taken per plant
- Number of plants selected for sampling

If the prescribed sampling directions are followed, the sampler should achieve reasonable statistical reliability.

Sampling instructions are quite specific in terms of plant part and stage of growth, since a comparison of an assay result with established *critical* or *standard values*, or *sufficiency ranges*, is based on a clearly identified plant part taken at a specified time.

When no specific sampling instructions are given or when they are unknown, the general rule of thumb is to select *upper mature leaves*.

Avoidance criteria are also crucial; plants to be avoided are ones that:

- Have suffered long-term climatic or nutritional stress;
- Have been damaged mechanically or by insects;
- Are infested with disease;
- Are covered with dust or soil or foliar-applied spray materials unless these extraneous substances can be removed effectively (decontamination procedures are discussed later in this chapter);
- Are border row plants or shaded leaves within the plant canopy;
- Contain dead plant tissue.

Since plant species, age, plant part, and time sampled (Bell, 2000) are variables that affect the interpretation of a plant analysis result, careful sampling is important. Steyn (1959) found that there were errors in the sampling of citrus and pineapple for a plant analysis related to time of sampling, number of plants to sample, and the amount of tissue taken from each plant, which could contribute to a misinterpretation. Environmental factors equally influence the elemental status of plants (Smith and Loneragan, 1997). In addition, most of the essential elements are not equally distributed in the plant or within its parts (Sayre, 1952; 1958; Jones, 1970; Oertli, 1994).

Procedures for collecting a plant tissue sample for either field or laboratory determination of its elemental content have been widely published. Chapman (1966), Kenworthy (1969), Jones et al. (1971), Reuter et al. (1986), Reuter and Robinson (1997), Mills and Jones (1996), and Jones (1998) have described plant tissue sampling techniques that have been generally accepted. The following is a partial list of recommended sampling procedures taken from these sources:

Suggested Sampling Procedures for Field and Vegetable Crops, Fruits and Nuts, and Ornamental Plants

Crop	Stage of growth	Plant part to sample	Number of plants to sample
		Field crops	
Corn	Seedling stage (<12 in.)	All the aboveground portion	20–30
	Prior to tasseling below the whorl	The entire leaf fully developed	15–25
	From tasseling and shooting to silking	The entire leaf at the ear node (or immediately above or below it)	15–25
Soybean or other beans	Seeding stage (<12 in.)	All the aboveground portion	20–30
	Prior to or during flowering[a]	Two or three fully developed leaves at the top of the plant	20–30
Small grains (including rice)	Seedling stage (<12 in.)	All the aboveground portion	50–100
	Prior to heading[b]	The fourth uppermost leaves	50–100
Hay, pasture, or forage grasses	Prior to seed head emergence or at the optimum stage for best-quality forage	The fourth uppermost leaf blades	40–50
Alfalfa	Prior to or at 1/10 bloom stage	Mature leaf blades taken about one third of the way down the plant	40–50
Clover and other legumes	Prior to bloom	Mature leaf blades taken about one thrid of the way down the plant	40–50
Sugar beets	Midseason	Fully expanded and mature leaves midway between the younger center leaves and the oldest leaf whorl on the outside	40–50
Tobacco	Before bloom	Uppermost fully developed leaf	8–12
Sorghum-milo	Prior to or at heading	Second leaf from top of plant	15–25
Peanuts	Prior to or at bloom stage	Mature leaves from both the main stem and either cotyledon lateral branch	40–50
Cotton	Prior to or at first bloom or when squares appear	Youngest fully mature leaves on main stem	30–40
		Vegetable crops	
Potato	Prior to or during early bloom	Third to sixth leaf from growing tip	20–30

Suggested Sampling Procedures for Field and Vegetable Crops, Fruits and Nuts, and Ornamental Plants (continued)

Crop	Stage of growth	Plant part to sample	Number of plants to sample
Head crops (cabbage, etc.)	Prior to heading	First mature leaves from center of the whorl	10–20
Tomato (field)	Prior to or during early fruit set	Third or fourth leaf from growing tip	20–25
Tomato (greenhouse)	Prior to or during fruit set	Young plants: leaves adjacent to second and third clusters	20–25
		Older plants: leaves from fourth to sixth clusters	20–25
Bean	Seedling stage (<12 in.)	All the aboveground portion	20–30
	Prior to or during initial flowering	Two or three fully developed leaves at the top of the plant	
Root crops (carrots, onions, beets, etc.)	Prior to root or bulb enlargement	Center mature leaves	20–30
Celery	Midgrowth (12 to 15 in. tall)	Petiole of youngest mature leaf	15–30
Leaf crops (lettuce, spinach, etc.)	Midgrowth flowering	Youngest mature leaf from the top of the plant	35–60
Peas	Prior to or during initial flowering	Leaves from the third node down	30–60
Sweet corn	Prior to tasseling	The entire fully mature leaf below the whorl	
	At tasseling	The entire leaf at the ear node	20–30
Melons (watermelon, cucumber, muskmelon)	Early stages of growth prior to fruit set	Mature leaves near the base portion of plant on main stem	20–30
		Fruits and nuts	
Apple, apricot, almond, prune, peach, pear, cherry	Midseason	Leaves near base of current year's growth or from spurs	5–100
Strawberry	Midseason	Youngest fully expanded mature leaves	50–75
Pecan	6 to 8 weeks after bloom	Middle pair of leaflets from midportion of terminal growth	3–5

(continued)

Suggested Sampling Procedures for Field and Vegetable Crops, Fruits and Nuts, and Ornamental Plants (continued)

Crop	Stage of growth	Plant part to sample	Number of plants to sample
Walnut	6 to 8 weeks after bloom	Middle pair of leaflets from mature shoots	30–35
Lemon, lime	Midseason	Mature leaves from last flush or growth on nonfruiting terminals	20–30
Orange	Midseason	Spring cycle leaves, 4 to 7 months old from nonbearing terminals	20–30
Grapes	End of bloom period	Petioles from leaves adjacent to fruit clusters	60–100
Raspberry	Midseason	Youngest mature leaves on lateral or "primo" canes	20–40
		Ornamentals and flowers	
Ornamental trees, shrubs	Current year's growth	Fully developed leaves	30–100
Turf	During normal growing period	Leaf blades; clip by hand to avoid contamination with soil or other material	¼ L
Roses	During flowering	Upper leaves on the flowering production stem	20–30
Chrysanthemums	Prior to or at flowering	Upper leaves on flowering stem	20–30
Carnations	Unpinched plants	Fourth or fifth leaf pairs from base of plant	20–30
	Pinched plants	Fifth and sixth leaf pairs from top of primary laterals	20–30
Poinsettias	Prior to or at flowering	Most recently mature fully expanded leaves	15–20

[a] Sampling after pods begin to set is not recommended.

[b] Sampling after heading is not recommended.

If the sampling procedure used does not conform to that recommended, interpretation of the plant analysis result may be difficult, if not impossible. Since there is a substantially large potential for error that can occur as a result of improper sampling technique, only thoroughly trained and experienced technicians should be responsible for collecting plant tissue samples.

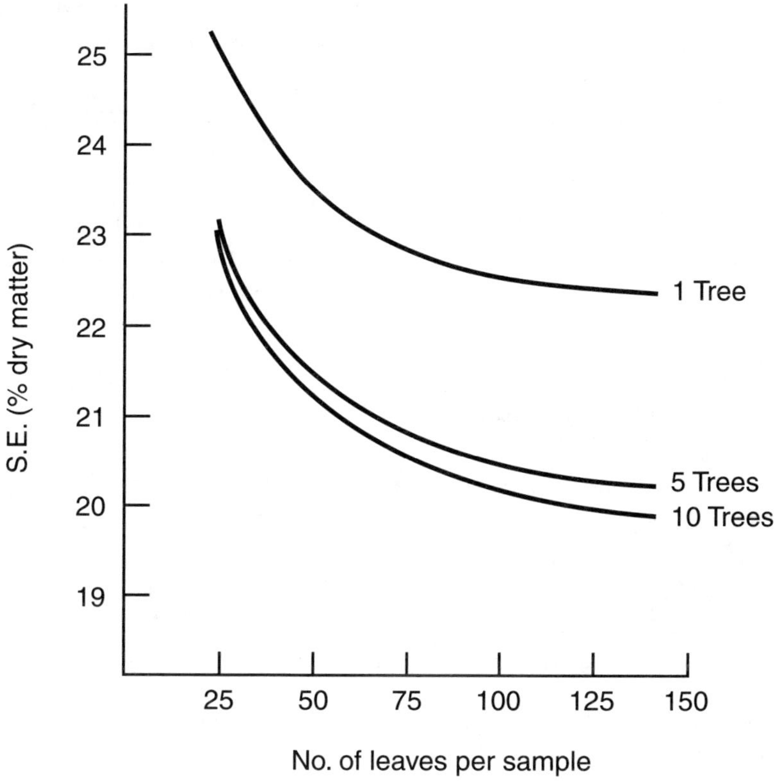

Figure 3.2
Variance associated with the number of leaves per sample.

2. Number of Plants to Sample

The number of plants to sample in a particular situation depends on the general condition of the plants, soil homogeneity, and the purpose for which the analysis result will be used. To ensure adequate representation, sampling as many plants as practical, collecting samples during a particular time of day, and collecting under calm climatic conditions are recommended.

Precision requirements will dictate the number of plant parts to be collected and what number of plants to be sampled to make a composite sample, as well as just how many composite samples will be necessary to ensure sufficient replication. Various studies indicate that the number of individual leaves and/or plants required is correlated with the desired variance to be obtained, as is illustrated in Figure 3.2.

In this example, the variance was more significantly affected by the number of trees sampled rather than by the number of leaves collected per tree. The combination of the number of plants selected for sampling and the number of samples taken per plant determines the variance associated with the result of the final analysis.

In addition, sampling requirements are considerably more complex when plants are under stress because of the high variability in plant characteristics that result from such conditions. Therefore, to achieve statistical verification, a more intensive sampling routine may be required.

Normally, the mean value of several composite sample assays gives a more accurate estimate than does a single assay result based on a single composite sample consisting of the same total number of individual samples.

3. Lack of Homogeneity

Colonna (1970) found a lack of homogeneity within a coffee plantation that influenced fertilizer procedures. How a plant tissue sample is selected is important because the distribution of the essential elements within the plant, and even within any one of its parts, is not homogeneous because of a number of factors (Sayre, 1952; 1958; Jones, 1970; Oertli, 1994). For example, as plant tissues mature, there are changes due to the following:

- The movement of mobile elements from the older tissue to newly developing tissues;
- An accumulation of nonmobile elements;
- A reduction in dry matter content.

One sign of increasing maturity in leaves is an increasing concentration (accumulation) of Ca and Mg, and a decreasing concentration (reduction) of N, P, and K.

Another factor contributing to the variation is the relative proportion of leaf blade to midrib and the size of the leaf, anatomical factors that can affect the concentration of elements found in the whole leaf. For example, the leaf midrib will normally contain a higher concentration of K than the blade. Similarly, the relative proportion of leaf blade to margin affects the B and Mn contents of the whole leaf since these two elements accumulate to fairly high concentrations in the leaf margins (Jones, 1970).

A sampling procedure that enhances the distribution effect of elements within the leaf will affect the analysis result. A sampling procedure, for example, in which only the leaf tips are sampled or only blade punches are collected, will produce an analytical result different from the result of an assay of the entire intact leaf. The effect on elemental concentration of dividing a corn leaf into four equal sections is shown in the following table.

Element Content of a Whole Corn Ear Leaf and Four Equal-Length Sections

Element	Whole leaf	Equal quarter-length sections of the ear leaf			
		Tip	Upper middle	Lower middle	Base
	%				
Nitrogen (N)	2.93	3.20	3.65	2.75	1.95
Phosphorus (P)	0.22	0.22	0.23	0.25	0.18
Potassium (K)	1.22	1.26	1.19	1.44	1.23
Calcium (Ca)	0.48	0.75	0.58	0.48	0.35
Magnesium (Mg)	0.39	0.40	0.41	0.45	0.40
	mg/kg				
Boron (B)	11	25	14	8	6
Copper (Cu)	9	12	10	10	8
Iron (Fe)	96	110	102	75	57
Manganese (Mn)	73	125	79	62	49
Zinc (Zn)	22	30	22	22	18

4. Petioles

Petioles are not a part of the leaf blade and should not be included in a leaf sample. However, for some crops, such as grape, sugar beet (Ulrich and Hills, 1990), and cotton (Maples et al., 1990; Constable et al., 1991; Davis, 1995), the petiole is the plant part to be assayed rather than the leaf blade. Petioles, as conductive tissue, are normally higher in elements, such as K, P, and NO_3–N, than is the attached leaf blade.

5. Compound Leaves

Compound leaves pose a problem for sampling since compound leaves are a mixture of petioles, conductive tissue, and leaf blade. In general, the recommended procedure is to collect a select leaf or leaves, for such plants as tomato and potato, the terminal leaf at the end of the compound leaf, and the middle pair of leaves for nut trees, such as pecan and walnut.

6. Comparative Sampling

Sampling two different populations of plants for comparative purposes, a highly desirable diagnostic procedure, poses a difficult sampling problem, particularly when the type of stress has resulted in substantial differences in plant growth. For example, when two or more sets of plants exhibit varying signs of a possible nutrient element insufficiency, collecting tissue for comparative purposes can be difficult in part because of the effect of the nutrient element stress on plant growth and development. Therefore, it is important, whenever possible, to obtain plant tissue samples when the symptoms of stress first appear, not waiting until there is a substantial difference in plant character.

Finally, great care must be taken to ensure that representative samples are collected for such comparisons, and that the interpretation of the analysis result takes into consideration the condition of the plants when they were sampled, whether normal in physical appearance or not, because of some type of stress.

7. Inappropriate Plant Tissue

Although it is possible to assay just about any plant part, or even the whole plant itself, the biological significance of such an analysis result is dependent on the availability of interpretative data for the plant part collected, stage of plant growth, etc. For example, the assay of fruits and grain, or an analysis of the whole plant or one of its parts at maturity or at harvest, does not usually provide reliable information on the nutritional status of the plant during its earlier growth period.

When conducting a plant analysis, the primary objective should be to obtain that plant part for which assay results can be compared with known interpretative values.

C. Sample Preparation

1. Introduction

Sample preparation procedures have been described by Steyn (1959), Grier (1966), and Campbell and Plank (1998), suggesting errors that can occur in the various preparation stages, transporting, decontaminating, drying, and particle size reduction, prior to elemental analysis.

2. Initial Handling

Fresh plant tissue is perishable and, therefore, must be kept cool and in a drying atmosphere prior to delivery to the laboratory. It is best to transport plant tissue in clean paper or cloth bags, not in airtight containers or plastic bags. If possible, the tissue should be air-dried prior to shipment, particularly if the tissue is succulent and/or if the time in transit will be greater than 24 h. Keeping the tissue at a reduced temperature (40 to 50°C; 104 to 122°F) will prevent decay. Any deterioration will result is reduced dry weight, which in turn will affect the analysis result (Lockman, 1970).

Maintaining the integrity of the collected sample is also crucial to ensure accurate assay results. When collecting plant tissue, care should be taken to ensure that the sample is not altered chemically or contaminated by extraneous materials as a result of contact with sampling tools and containers.

3. Decontamination (Washing)

Plant tissue that is covered with dust, soil particles, or coated with foliar-applied materials that contain elements of interest in the plant analysis determination will require decontamination prior to drying. Only fresh, fully turgid plant tissue can be subjected to a decontamination procedure. Decontamination by washing must be done either in the field as tissue is being collected or, if the tissue is to be decontaminated later in the laboratory, by keeping collected plant tissue in a fully turgid state in a cool and moist atmosphere.

Normally, decontamination (washing) is not recommended unless absolutely necessary. Mechanical wiping or brushing may be sufficient to remove large soil particles. Washing the fresh plant tissue in a 0.1 to 0.3% P-free detergent solution followed by a rinse in pure water can effectively remove most extraneous materials (Wallace et al., 1980). Sonneveld and van Dijk (1982) recommend dipping tissue for 15 s in an ample volume of solution containing a concentration of Teepol at 0.1% and hydrochloric acid (HCl) at 0.1 *M*, then followed by rinsing in pure water.

The element that is most affected by washing is Fe (Sonneveld and van Dijk, 1982; Wallace et al., 1982; Jones and Wallace, 1992) as well as Al, Mn, and Si (Jones and Case, 1990). In studies conducted by Bahn et al. (1959), Jones (1963), Baker et al. (1964), Labanauskas (1968), Ashby (1969), Smith and Storey (1976), Wallace et al. (1982), and Sonneveld and van Dijk (1982), they found that washing resulted in significant reductions for only the elements Fe, Mn, and Zn, and moderate reductions for Cu. For example, washing citrus leaves resulted in a reduction in Fe from 186 to 61, Mn from 182 to 94, Zn from 123 to 68, and Cu 5.6 to 5.1 mg/kg (Labanauskas, 1968).

Contaminants on rough or pubescent plant tissue surfaces are difficult to remove, posing a serious problem, as washing may not be able to remove the contaminants effectively. Unfortunately, there is no satisfactory alternative decontamination procedure.

In some instances, the decontamination procedure itself may significantly alter the elemental composition of the tissue. The procedure may add elements to the tissue, or it may leach elements, such as K, Cl, B, and NO_3, during the washing process.

Following an analysis, contamination can be detected by noting if the concentrations of Al, Si, and Fe in the assayed tissue are all equally high or if their concentrations among a series of plant analysis results track each other. Cherney and Robinson (1982) used the determination of Ti as a means of judging contamination. Plant tissue that is dry-ashed as the means of organic matter destruction will generally give lower values for Al, Si, and Fe (probably Ti also), than when the tissue is wet-acid-digested (see Section 6.b in this chapter).

4. Moisture Removal (Oven-Drying)

It should be remembered that the elemental concentration used for interpretation is based on that in the dry weight of tissue; therefore, any condition that affects the dry weight of collected plant tissue will affect its elemental composition (Lockman, 1970). If collected plant tissue begins to decay, significant reductions in dry weight can occur; in addition, some elements, particularly N and S, may be lost by volatilization.

Fresh plant tissue is best dried in a dust-free, forced draft oven at a temperature of 80°C (176°F), which is a temperature sufficient to remove moisture without causing appreciable thermal decomposition (Isaac and Jones, 1972). Drying temperatures lower than 80°C (176°F) may not be sufficient to remove all the moisture (Mills and Jones, 1996), and temperatures above 80°C (176°F) can result in thermal decomposition (Jones et al., 1991). Once dried (which may take more than 24 h), the dried tissue should be stored in a moisture-free atmosphere prior to further processing.

Those plant tissues high in soluble sugars are not easily oven-dried and, therefore, moisture removal is best done by either freeze-drying or vacuum-oven-drying (Horwitz, 1980).

Plant tissue may be quickly dried in a microwave oven, although the procedure is somewhat tedious and is not suited for drying large quantities of materials (Carlier and van Hee, 1971; Shuman and Ruazi, 1981; Jones et al., 1991).

5. Particle Size Reduction (Grinding)

To reduce the dried plant tissue to a particle size suitable for laboratory analysis and, at the same time, to ensure a greater degree of uniformity in sample composition, the tissue is mechanically ground or crushed. Particle size reduction can be done by cutting action using a Wiley or hammer mill, by abrasion in a cyclonic mill, or by crushing in a ball mill. In most mills, particles of the contact surfaces will be added to the sample, such as Cu and Zn additions from brass fittings, and even Fe when fittings, cutting, and crushing surfaces are made of steel. Therefore, to avoid Fe contamination, tissue samples are best reduced either by hand cutting or by crushing in an agate ball mill. Munter et al. (1984) assayed unground filter paper and then ground the filter paper in an Fe mill and a stainless steel mill and found Fe to be 8.7, 41.0, and 10.5 mg/kg, respectively. Grinding devices with Al, plastic (adding Na), and rubber (adding Zn) fittings are potential sources for contamination. Hood et al. (1944) found that most mechanical mills contaminate the sample to some extent with one or more elements.

A Wiley mill fitted with a 20-mesh screen is commonly used for grinding tissue samples. The finer the screen (40 or 60 mesh), the more homogeneous the sample will be. However, addition of elements from the cutting surfaces will also be greater because of the longer contact time between the tissue sample and the cutting mill itself (Hood et al., 1944). When grinding dry plant material, segregation of the finer particles occurs, which must be controlled by eliminating static electricity buildup (Nelson and Boodley, 1965; Smith et al., 1968). Adherence of fine plant particles to the grinding mill components can be partially overcome by attaching a vacuum system to the grinding mill (Graham, 1972) or by using pulsing air (Ulrich, 1984).

Some types of samples are not easily ground because of the presence of pubescence (such as on apple leaves), coarse fiber tissues, or because the tissues are highly deliquescent. Therefore, special care is required when reducing these types of tissues in particle size.

After milling, most tissues are sensitive to thermal decomposition at temperatures >80°C (>176°F) (Steyn, 1961) and, therefore, this drying temperature should not be exceeded prior to weighing and analysis. For long-term storage, milled tissue should be kept in an airtight container and stored in a cool (<10°C; <50°F), dark environment (Steyn, 1961).

6. Organic Matter Destruction

Organic matter destruction can be accomplished either by high temperature thermal oxidation or by wet-acid digestion; the former method is frequently referred to as dry ashing, and the latter as wet acid digestion or wet ashing. There continues to be considerable controversy between advocates of each method in terms of ease of execution, elemental losses and/or additions during the ashing process, and suitability of use with some types of tissues. Major works on this subject have been authored by Gorsuch (1959; 1970; 1976), Bock (1978), and Tolg (1974). The number of papers written about both methods of organic matter destruction is considerable. Huang and Schulte (1985) compared a number of organic destruction procedures prior to assay by ICP-AES.

a. High-temperature thermal oxidation (dry ashing)

Dry ashing is done in a muffle furnace at a maximum temperature of 500°C (932°F). An ashing temperature <500°C (<932°F) can result in incomplete organic matter destruction, whereas temperatures >500°C (>932°F) can result in elemental losses. Both conditions result in low elemental (Al, B, Cu, Fe, K, and Mn) recoveries (Isaac and Jones, 1972; Labanauskas and Handy, 1973; Miller, 1998b). The extent of losses occurring at temperatures >500°C (>932°F) may be significantly less for tissues relatively high (>1.0%) in Ca. Quartz crucibles are the best ashing vessels, although acid-washed, well-glazed, porcelain crucibles and Pyrex beakers have been satisfactorily used. Munter et al. (1984) found small additions of B, Cu, Fe, and Mn to corn leaf tissue when porcelain crucibles were used as the ashing vessels as compared with those made of quartz.

After placing the ashing vessel in a cool muffle furnace, the furnace is slowly brought to 500°C (932°F) in about 1 to 1.5 h. The minimum ashing time at 500°C (932°F) is 4 h, although an overnight ash of 6 to 8 h is preferred.

An ashing aid has been suggested, particularly when ashing high carbonaceous-type tissues, i.e., high in sugar and soluble carbohydrates. The most commonly used ashing aid is magnesium nitrate [$Mg(NO_3)_2 \cdot 6H_2O$].

Dry Ashing Procedure

1. Weigh 0.5 g dried (80°C; 176°F), 0.84-mm (20-mesh-screened) plant tissue into a 30-mL, high-form porcelain and/or quartz crucible.

Note: *If an ashing aid is needed, add either 5 mL HNO_3, or 5 mL 7% $Mg(NO_3)_2 \cdot 6H_2O$. Dry on a hot plate and then continue with Step 2.*

2. Place crucible in a rack, and the rack in a cool muffle furnace.
3. Set furnace temperature to reach set temperature (500°C; 932°F) in about 2 h.
4. After 4 to 8 h of muffling at 500°C (932°F), remove the crucible rack from the furnace and let cool.
5. Add 10 mL dilute acid mixture [300 mL hydrochloric acid (HCl) and 100 mL nitric acid (HNO_3) in 1000 mL water] to dissolve the ash.

Note: *Dilute HNO_3 alone is frequently used to minimize the corrosive character of HCl when in contact with metal in the elemental analysis procedure.*

Crucible and contents may be heated to assist in solubilization of ash by weighing the crucible, heating to acid fumes, cooling, and adding pure water to original weight.

6. Allow suspended material to settle to the bottom of the crucible. Clear solution is ready for elemental analysis.

b. Wet-acid digestion (wet ashing)

Numerous wet-acid digestion procedures have been proposed, but they all make use of some combination of three acids — nitric (HNO_3), sulfuric (H_2SO_4), and perchloric acids ($HClO_4$) — with or without 30% hydrogen peroxide (H_2O_2) as described by Tolg (1974). Zasoski and Burau (1977) and Miller (1998c) provide details on digestion in a mixture of HNO_3 and $HClO_4$, and a digestion mixture of H_2SO_4 and $HClO_4$ for the determination of N, P, K, Ca, and Mg is given by Cresser and Parsons (1979). The use of $HClO_4$ requires special precautions as described by Horwitz (1980). The digestion can be carried out in beakers on a hot plate (Tucker, 1974; Halvin and Soltanpour, 1980; Adler and Wilcox, 1985; Zarcinas et al., 1987), in an enclosed container under pressure (Vigler et al., 1980; Okamoto and Fuwa, 1984; Sung et al., 1984; Knapp, 1985; Anderson and Henderson, 1986; Sah and Miller, 1992), or in open or closed vessels placed in a microwave oven (White and Douthit, 1985; Kingston and Jassie, 1988; Kalra et al., 1989; Stripp and Bogan, 1989; Kalra and Maynard, 1998; Miller, 1998c).

The inclusion of H_2SO_4 in the digestion mixture is limited to tissues low (<1.0%) in Ca, since calcium sulfate ($CaSO_4$) can be formed during the digestion step, trapping elements within the formed precipitate and if S is one of the determined elements. Parkinson and Allen (1975) and Wolf (1982) have successfully used a mixture of H_2SO_4 and 30% H_2O_2 as the digestion mixture for the determination of N as well as other major elements and micronutrients in plant tissue.

Digestion in a Mixture of HNO_3 and $HClO_4$

1. Weigh 0.5 g dried (80°C; 176°F), 0.84-mm (20-mesh screened) plant tissue into a beaker or digestion tube.
2. Add 2.5 mL concentrated HNO_3. Cover the beaker with a watch glass or place a funnel into the mouth of the digestion tube. Let stand overnight.
3. Place covered beaker on a hot plate or digestion tube into a port of a digestion block and digest at 80°C (176°F) for 1 h. Remove beaker or digestion tube from hot plate or block, and let cool.
4. Add 2.5 mL $HClO_4$, replace watch glass or funnel, and heat at 180 to 200°C (356 to 392°F) for 2 to 3 h, or until digest is clear.

5. Remove watch glass or funnel, lower heat to 100°C (212°F) until fumes of $HClO_4$ dissipate. If digest is not colorless at this point, repeat Step 4.
6. Remove from hot plate or digestion block and let cool.
7. Add pure water to digest to bring to 10 mL or other appropriate volume. Digest is ready for elemental assay.

Acid Digestion in a Mixture of HNO_3 and 30% H_2O_2

1. Weigh 0.5 g dried (80°C; 176°F), 0.84-mm (20-mesh screened) plant tissue into a beaker or digestion tube.
2. Add 5.0 mL concentrated HNO_3. Cover with watch glass or place funnel into mouth of the digestion tube. Let stand overnight.
3. Place covered beaker on a hot plate or digestion tube into a port of a digestion block and digest at 125°C (257°F) for 1 h. Remove beaker or digestion tube from plate or block and let cool.
4. Add 3 mL 30% H_2O_2 to the beaker or digestion tube and digest at 125°C (257°F). Repeat additions of 30% H_2O_2 until digest is clear. Add HNO_3 as needed to prevent digest from going to dryness.
5. When the digest is clear, remove the watch glass or funnel and reduce temperature of hot plate or block to 80°C (176°F). Take nearly to dryness. Residue should be colorless. If not, repeat Step 4.
6. Add 1:10 HNO_3 or HCl to bring to final volume of 10 mL. Clear solution is ready for elemental assay.

Acid Digestion in a Mixture of H_2SO_4 and 30% H_2O_2

1. Weigh 0.5 g dried (80°C; 176°F), 0.84-mm (20-mesh screened) plant tissue into a beaker or digestion tube.
2. Add 3.5 mL concentrated H_2SO_4 and let stand for 30 min.
3. Add 3.5 mL 30% H_2O_2.
4. Cover the beaker or place a funnel into the mouth of the digestion tube. Place beaker on a hot plate or digestion tube into a port of the digestion block. Heat at 350°C (662°F) for 30 min.
5. Remove beaker or digestion tube from hot plate or digestion block and let cool.
6. Add 2-mL aliquots of 30% H_2O_2 and repeat digestion step until cool digest is clear.
7. Once the digest is clear, dilute to 20 mL with pure water. Digest is ready for elemental assay.

If B is a determined element, it is recommended that organic matter destruction be performed by dry ashing (Wikner, 1986), since B can be lost during the wet digestion procedure. However, the extent of loss will vary with tissue type and Ca tissue content (Feldman, 1961). van der Lee et al. (1987) found that wet ashing in a mixture of hydrochloric (HCl) and hydrofluoric acid (HF) gave B results similar to those obtained when using dry ashing as the organic destruction procedure. Gestring and Soltanpour (1981) obtained certified B concentrations by wet-digesting plant tissue in nalgene bottles with concentrated nitric acid (HNO_3).

The method of organic matter destruction can also affect the determination of Fe and Al, elements that are influenced by the presence of soil or dust on the surface of the tissue. For example, Jones and Wallace (1992) compared assay results for unwashed leaves and found Fe and Al contents determined following dry ashing vs. wet-acid digestion to be 640 and 1221 mg Fe/kg, and 1160 and 1553 mg Al/kg, respectively; and for washed leaves dry ashed vs. wet-acid digestion, 246 and 401 mg Fe/kg, and 318 and 330 mg Al/kg, respectively. It has been the author's experience that Fe and Al levels in plant tissue, washed or unwashed, are almost always less when organic matter destruction is by the dry ash vs. wet digestion techniques, differences that are not consistently seen for other elements.

Since there are conflicting published results as well as considerable disagreement about how best to destroy plant tissue organic matter, the analyst should carefully verify the results obtained by the ashing procedure used by including reference standards of verified elemental concentration, such as those available from the National Institute of Standards and Technology (NIST) (Alvarez, 1980) and other sources (Ihnat, 1998), as check samples.

D. Elemental Analysis for the Mineral Elements

The analyst today has a wide range of analytical procedures from which to choose. Some factors that affect the choice are described by McLaughlin et al. (1979). Hislop (1980) has developed analytical criteria for selection of an analysis technique based on accuracy, precision, limit of detection, elemental coverage, single or multielement, and determined chemical form.

Those in search of a suitable analytical procedure may find the review articles by Morrison (1979), Stika and Morrison (1981), Watson and Isaac (1990), and Watson (1998) and the book edited by Walsh (1971) helpful as they compare the relative sensitivity and precision of various analytical procedures. In those instances in which the micronutrient concentration in plant tissue is high (>100 mg/kg), detection limit considerations are of less significance than those factors that affect precision. Munter et al. (1984) discuss the sources of variation from plant tissue preparation to analysis for the micronutrients, finding that the major source of variation exists in the preparation steps. For most instrumental procedures, coefficients of variability are usually less than 5%, although they increase as the detection limit of the analytical procedure is approached, as well as at high concentrations (Horwitz, 1982), which must be considered in the interpretation of a plant analysis result.

Horwitz (1982), based on his many years with the Association of Official Analytical Chemists (AOAC), has discussed the practical limits of acceptable variability for methods of analysis, focusing on the important aspects of reliability, reproducibility, repeatability, systematic error of bias, specificity, and limit of reliable measurement. The impact of these aspects on any analytical procedure varies considerably in terms of sample size, determinations made, concentration of the determined element, and the characteristics of the analytical instrument used. An additional criterion is the ruggedness factor, which sets the limits for each step in the analytical procedure, which, if exceeded, will invalidate the obtained result. Unfortunately, most plant analysis procedures have not been so described. Examples would include limit critieria for sample preparation procedures, such as moisture removal techniques, ashing and digestion temperatures, length of time, etc.

Advances in analytical chemistry in the past two decades have significantly improved the ease and speed for the determination of elements found in plant tissue ash or digests. For most essential plant nutrient elements, the more traditional wet chemistry procedures have been replaced by various instrumental procedures that employ either emission or absorption spectrophotometry.

The classical spectrophotometric (colorimetric) procedures described by Piper (1942), Jackson (1958), Johnson and Ulrich (1959), and Chapman and Pratt (1982) are still in use today, although they are frequently automated by employing an autoanalyzer or a flow-injection analyzer (Isaac and Jones, 1970; Steckel and Flannery, 1971; Watson, 1998). Jackson (1958), Johnson and Ulrich (1959), and Chapman and Pratt (1982) describe various spectrophotometric and instrumental methods of plant tissue analysis for the micronutrients. Losche (1982) describes a microanalytical procedure for determining B in plant tissue using the Azomethine-H reagent (Gaines and Mitchell, 1979). Most spectrophotometric procedures require careful sample preparation and are frequently subject to both matrix and interelement interference (see Chapter 5, Section B).

Flame emission spectrophotometry for the determination of K and Na has had a long history of use (Horneck and Hanson, 1998). Similarly, flame atomic absorption spectrophotometry is routinely used to determine Ca, Mg, Cu, Fe, Mn, and Zn in plant tissue digests (Hanlon, 1998). For very low-concentration determinations, such as the determination of Mo, flameless atomic absorption spectrophotometry must be used (Gupta, 1998). All these procedures are affected to varying degrees by the matrix and the concentration ranges for each element in the digest. In addition, sample digests usually require considerable manipulation to bring the elemental concentration within the usable analytical range of these analytical instruments. Although these instrumental techniques are in common use today for elemental determination in prepared plant tissue digests, they are slow and cumbersome when compared with multielement instrumental procedures, such as spark and inductively coupled plasma emission spectrometry (frequently referred to by the acronyms ICP, ICP-AES, or ICAP). Over the last several decades, there has been a succession of excitation sources coupled with emission spectrometers. AC and DC arcs (Mitchell, 1964) were followed by the AC spark (Jones, 1976), the DC plasma (DeBolt, 1980), and finally ICP (Jones, 1977; Dalquist and Knoll, 1978; Munter and Grande, 1981; Keller, 1992; Isaac and Johnson, 1998; Soltanpour et al., 1998). The advantages of plasma excitation are considerable in terms of minimal matrix and spectral interference, excellent sensitivity (frequently <0.1 mg/kg), and dynamic reading range (usually 3 to 5 decades).

Another instrumental method for the determination of S, Fe, Mn, and Zn in plant tissue is X-ray fluorescence (Alexander, 1965; Kubota and Lazar, 1971). This method is matrix sensitive and is little used today for routine assay of plant tissue. These various methods of analysis are discussed in more detail in Chapter 5.

E. Total Nitrogen (N) Determination

1. Introduction

The N in plant tissue can be determined by two analytical procedures: (1) Kjeldahl digestion (Bradstreet, 1965; Bremner and Mulvaney, 1982; Jones, 1991) and (2) the Dumas technique (Bremner and Mulvaney, 1982; McGeehan and Naylor, 1988; Schmitter and Rihns, 1989; Horneck and Miller, 1998). Simmone et al. (1994) compare the various methods of plant N determination and find some significant differences.

The Kjeldahl digestion procedure dates to the late 1800s; the first published procedure appeared in 1883. Reviews of those first 100 years have been written by Morries (1983) and Scarf (1988). The method requires two steps: first, high-temperature (330 to 350°C; 626 to 662°F) digestion in concentrated sulfuric acid (H_2SO_4) in the presence of a catalyst (such as Cu, Hg, Se, Ti, or mixtures of two or more), which converts organic N to inorganic ammonium (NH_4) and, second, determination of formed NH_4. Because Hg is an element of concern today as an environmental pollutant, either Cu or Se is the more commonly used catalyst today.

Hundreds of papers have been written about the Kjeldahl procedure as numerous modifications have been proposed to speed the analysis and to improve precision, accuracy, and N recovery. Nelson and Sommers (1980) have reviewed and evaluated many of the modifications proposed for the Kjeldahl procedure, Jones (1991) published a treatise on the Kjeldahl method, and Kane (1987) conducted a collaborative study on the use of mercuric oxide (HgO) and copper sulfate ($CuSO_4$)–titanium oxide (TiO_2) as catalysts. Automated Kjeldahl procedures have been developed; one method has been recently described by Wright and Wilkinson (1993). Simonne et al. (1993) found that sample size, catalyst, and digestion conditions have an influence on the Kjeldahl N determination, factors that must be considered when choosing the conditions under which the assay is conducted.

Depending on sample size, the Kjeldahl digestion procedure has been classed as either macro (1.0 g or greater), or semimicro (1.0 to 0.5 g), or micro (less than 0.5 g), with the size of the digestion/distillation apparatus scaled accordingly. Precision declines with decreasing sample size as a result of nonhomogeneous particles and the increasing need for a finely ground sample to ensure adequate homogeneity. The trend from macro- to micro-Kjeldahl digestion is an attempt to reduce the laboratory space and equipment required, as well as to reduce reagent use.

The trend today is toward the use of a digestion block, which can be either constructed by the analyst (Gallaher et al., 1975) or obtained commercially. The digestion is carried out in a digestion tube set in the heated block with the tube size dictated by sample size. The formed NH_4 is determined by titration being transferred by steam distillation into a trapping solution.

Nitrogen in plant tissue as either nitrate (NO_3) or nitrite (NO_2) is not recovered in Kjeldahl digestion unless converted to NH_4 by pretreatment of the sample with either reduced iron under acidic conditions or pretreatment in a moisture-free environment with salicylic acid. The Kjeldahl procedure, with and without NO_3 or NO_2 recovery, and NH_4 determination by alkaline distillation are given in the AOAC Manual (Horwitz, 2000).

Some have recommended the addition of either perchloric acid ($HClO_4$) or 30% hydrogen peroxide (H_2O_2) to speed the digestion. Recent findings indicate that H_2O_2 additions will reduce N recovery by about 15%.

The addition of either potassium sulfate (K_2SO_4) or sodium sulfate (Na_2SO_4) to the digestion mixture will increase the temperature of the digestion from 330°C (626°F) with pure H_2SO_4 to 370°C (698°F), which in turn speeds the digestion time and increases N recovery. The amount of sulfate to acid is 0.3 to 0.5 g/mL H_2SO_4. Potassium sulfate is preferred. If during the digestion step solidification occurs, some N will be lost by volatilization, a condition that should be avoided.

Adequate digestion time is important for complete conversion of organic N to NH_4 to occur. After the digestion mixture clears, an additional time period of two to three times the length of time it took for the mixture to clear is required to obtain complete N conversion. At clearing, about 92 to 93% of the organic N has been converted, and the additional boiling time is needed to obtain the remaining 7 to 8%.

2. Kjeldahl Methods

Standard Kjeldahl Digestion Procedure

For determination in Kjeldahl flasks or digestion tubes:

1. Weigh 500 mg of dried (80°C; 176°F) ground (40-mesh screened) plant tissue into a Kjeldahl flask or digestion tube.
2. Add 5.0 g of digestion mixture (100:1:1000 $CuSO_4{\cdot}5H_2O/Se/K_2SO_4$ or 1:60:1670 ($CuSO_4/TiO_2/K_2SO_4$).
3. Add 10 mL concentrated H_2SO_4.
4. Place Kjeldahl flask on digestion rack or place funnel in the neck of the digestion tube, and place the tube into a port of the digestion block. Heat to rolling boil in Kjeldahl flask or at 360 to 410°C (680 to 770°F) for digestion block procedure.
5. Continue to heat for 60 min after clearing.
6. Discontinue heating and let Kjeldahl flask or digestion tube and contents cool.
7. Dilute to appropriate volume with pure water and determine NH_4 content in digest.

Kjeldahl Nitrogen Digestion of Plant Tissue to Include Nitrate

Based on duPreez and Bale (1989).

1. Weigh 0.10 g plant tissue into a digestion tube.
2. Add 4 mL concentrated H_2SO_4 and 1.2 g of catalyst mixture; mix by grinding 100 g potassium sulfate (K_2SO_4), 10 g copper sulfate ($CuSO_4$), and 1 g Se together.
3. Add 0.01 g phenyl acetate ($C_6H_5{\cdot}O{\cdot}CO{\cdot}CH_3$).
4. Add 10 glass beads.
5. Place the digestion tube into a block digestor port and allow it to boil until clear blue-green is achieved.
6. Heat at the boiling temperature for an additional hour.
7. Remove the digestion tube from the digestion block and let it cool.
8. Dilute to an appropriate volume with pure water.
9. The digest is ready for NH_4 determination.

Excluding Nitrate from the Kjeldahl Digestion

Based on Bowman et al. (1988).

1. Weigh 100 mg sample into a digestion tube.
2. Add 1 boiling chip.
3. Add 1.0 mL 1 *M* potassium dihydrogen phosphate (KH_2PO_4, pH 6.0) and 3 mL 30% H_2O_2.
4. Place digestion tube in a digestion block port and heat the mixture at 100°C (212°F) for 15 min.
5. Remove digestion tube from the digestion block and let it cool.
6. Slowly add down the side of the digestion tube 4.5 mL concentrated sulfuric acid (H_2SO_4).
7. After the reaction ceases, add 1.5 g potassium sulfate (K_2SO_4) and 0.33 mL 8% copper sulfate ($CuSO_4{\cdot}5H_2O$) solution.
8. Place the digestion tube in a port of a preheated digestion block and digest at 360°C (680°F) for 60 min.
9. Remove digestion tube from the digestion block and let it cool.
10. Dilute to appropriate volume for NH_4 determination.

3. Determination of Ammonium in Kjeldahl Digest

Alkaline distillation (Bremner and Keeney, 1965) and determination of NH_4 by acidimetric titration (Munsinger and McKinney, 1982), or UV-VIS spectrophotometry (Isaac and Johnson, 1976; Smith, 1980; Baethegen and Alley, 1989), or NH_4 specific-ion electrode (Bremner and Tabatabia, 1972; Eastin, 1976; Gallaher et al., 1976; Powers et al., 1981) are the commonly used procedures. If Hg is the catalyst selected, then thiosulfate (S_2O_3) must be added to the digest in the distillation step to break the Hg–NH_4 complexes that are formed.

4. Non-Kjeldahl Methods

Currently, there are three non-Kjeldahl methods for N determination in plant tissue that are being investigated. One method calls for direct distillation, placing a plant tissue sample into alkali and steam distilling for a specified time (Arneh et al., 1983). The amount of amino-N converted to NH_4 is measured and compared with a determined Kjeldahl value to form a calibration curve. Precise control of the distillation is required to obtain reliable results.

Near infrared reflectance (NIR) is another procedure that looks promising for determining plant N (Dorsheimer and Isaac, 1982). A beam of infrared radiation is focused on a finely ground, dried plant tissue sample and the reflected radiation measured. The technique is fast and nondestructive; however, the instrument is quite expensive and the calibration procedure requires the use of standards of the same plant species as those to be analyzed. At the present time, the results obtained by NIR would be considered only "quite" good when compared with what is obtainable by Kjeldahl digestion (Isaac and Johnson, 1983).

A combustion method originally described by Dumas (Ebeling, 1968) and later modified by Sweeney (1989) quantitatively determines the amount of N in all forms (NH_4, NO_3, protein, and heteocyclic N) in plant tissue using an induction furnace and a thermal conductive detector (McGeehan and Naylor, 1988; Hansen, 1989; Horneck and Miller, 1998). Plant tissue samples are ignited in an induction furnace at approximately 900°C (1652°F) in a helium and oxygen environment, and the N forms released are converted to N_2 gas whose concentration is determined by thermal conductivity. The method has a detection limit of 0.10% N (dry sample basis) and is generally reproducible to within ±5%.

Kjeldahl-determined N in plant tissue is not total N since some forms of N in the plant tissue are not recovered. Simmone et al. (1993) found that total N determination by the Dumas procedure usually gives slightly higher N values than that obtained using Kjeldahl digestion.

F. Total Sulfur (S) Determination

1. Introduction

Numerous methods have been used for the determination of total S in plant material, with different procedures for converting the various forms of plant S into one form before its final quantification (Beaton et al., 1968; Blanchar, 1986; Abraham and DeMan, 1987; Kowalenko and van Laerhoven, 1998). In some cases, conversion and quantification have been combined into one instrumental procedure (e.g., commercially available S analyzers based on dry ashing) or determined without physical conversion by direct atomic methods, for example, X-ray (Kubota and Lazar, 1971; Murdock and Murdock, 1977) or neutron activation (Helmke, 1996). In those methods with separate conversion and quantification, quantification is usually either by barium precipitation (turbidity) or assayed as sulfide (Kowalenko and van Low, 1972) after reduction by a hydriodic acid reagent. More recently, quantification can be done by either ICP-AES (Pritchard and Lee, 1984; Hogan and Maynard, 1984; Topper and Kotuby-Amacher, 1990; Zhao et al., 1994; Kovács et al., 1996; Kowalenko and van Laerhoven, 1998) or ion chromatography (Sterrett et al., 1987; Artiola and Ali, 1990; Peverill, 1993).

Conversion by wet digestion in a mixture of nitric (HNO_3) and perchloric ($HClO_4$) acids can result in volatile losses of S by heating in an acidic condition (Randall and Spencer, 1980; Hafez et al., 1991; Zhao et al., 1994) unless there is careful temperature control. Wet-acid digestion (see pages 205–207) is an acceptable procedure for those situations where extreme accuracy for S determination may be compromised for purposes of multiple element analyses (Wolf, 1982; Pritchard and Lee, 1984).

Dry ashing (see page 204) is considered simple and fast for the preparation of large numbers of samples (Tabatabai et al., 1988). The procedure should be performed in the presence of an alkali to prevent the volatile loss of S; various types of alkali are used (Beaton et al., 1968). Although magnesium nitrate [$Mg(NO_3)_2$] has been used frequently (Johnson and Nishita, 1952; Wolf, 1982; Guthrie and Lowe, 1984; Cunniff, 1995; Jones, 1996) as the alkali, a procedure employing sodium bicarbonate ($NaHCO_3$) and silver oxide (Ag_2O) as the alkali has been used based on recent work that shows its adaptability with different methods of SO_4 quantification (Lea and Wells, 1980; Tabatabai et al., 1988; Artiola and Ali, 1990; Perrott et al., 1991).

The SO_4–S content in the digests and/or solubilized ash can be determined by either the barium sulfate turbidity method or by one of several UV-VIS spectrophotometric procedures (Beaton et al., 1968). The barium sulfate turbidity method has been adapted for use with an autoanalyzer (Wall et al., 1980) and the HACH kit (Jones, 1996). In a properly prepared plant tissue digest that retains S as the SO_4 anion, the turbidity procedure described in Chapter 2, Section P.4.a of this guide, can be used to determine S.

Automated combustion using a LECO Sulfur Analyzer (Jones and Isaac, 1972; Kirsten, 1979; Hern, 1984; Jackson et al., 1985; Kirsten and Nordenmark, 1987; Matrai, 1989; David et al., 1989) is another method for determining total plant S. A prepared plant tissue sample is placed into the induction furnace of the analyzer and heated to 1350°C (2462°F) in a steam of oxygen (O_2). Plant S is oxidized to sulfur dioxide (SO_2) gas, which is either trapped in an indicator solution and the amount of SO_2 evolved determined by a back-titration (Jones and Isaac, 1972) or the SO_2 is passed through an infrared analyzer (Hern, 1984). The titration technique requires ashing [at 500°C (932°F) for 2 h] the plant tissue mixed with magnesium oxide (MgO) to remove interfering chloride and N prior to S analysis.

In a properly prepared plant tissue digest and/or solubilized ash that retains S, S can be determined by ICP emission spectrometry, along with other elements, using a spectrometer than can detect emission lines in the ultraviolet range, with the S spectral line at 182.04 nm.

Kowalenko (1998) describes a method of determining SO_4–S extracted from plant tissue and then the S determined by hydriodic acid reduction.

2. Interpretation

The ratio of SO_4–S to total S has been suggested as the best means of determining the S status of a plant, although Scaife and Burns (1986) express caution on this method of interpretation. Kowalenko (1998) suggests that the SO_4–S determination alone is the best interpreter.

G. Methods for Expressing Elemental Content

The elemental concentration in plant tissue is expressed on a dry weight basis as a percentage for the macronutrients N, P, K, Ca, Mg, and S and for the micronutrients B, Cl, Cu, Fe, Mo, Mn, and Zn, parts per million (ppm). Using International (SI) Units, the macronutrients are expressed as grams per kilogram (g/kg) or the cations may also be expressed as centimoles per kilogram (cmol/kg). The micronutrients are expressed as either milligrams per kilogram (mg/kg) or micrograms per gram (μg/g).

For the macronutrients expressed as percent, the concentration is to the nearest 0.01%. For the micronutrients expressed as parts per million, concentrations greater that 10 ppm are expressed as whole numbers and concentrations greater than 1 ppm but less than 10 ppm are expressed as 0.01 ppm.

The following are methods of expressing elemental concentrations in plant tissue on a dry weight basis in various units (numbers selected only for illustrative purposes only):

Element	Percent	g/kg	cmol(p+)/kg	cmol/kg
Nitrogen (N)	3.15	31.5	225	225
Phosphorus (P)	0.32	3.2	—	—
Potassium (K)	1.95	19.50	50	50
Calcium (Ca)	2.00	20.00	25	50
Magnesium (Mg)	0.48	4.80	10	20
Sulfur (S)	0.32	3.20	10	20
	ppm	mg/kg	cmol(p+)/kg	mmol/kg
Boron (B)	20	20	—	1.85
Copper (Cu)	12	12	0.09	1.85
Iron (Fe)	111	111	0.66	1.98
Manganese (Mn)	55	55	0.50	1.00
Zinc (Zn)	33	33	0.25	0.50

Conversion factors for calculating milliequivalents and microequivalents are given in the following table:

Calculation of Milliequivalents (m.e.) and Microequivalents (p.e.)/100 g, from Percent (%) and Parts per Million (ppm), respectively

Element	Converting from	Valence	Equivalent weight	Factor
Nitrogen (N)	% to m.e.	3	4.6693	214.6
Phosphorus (P)	% to m.e.	5	6.1960	161.39
Potassium (K)	% to m.e.	1	39.096	25.578
Calcium (Ca)	% to m.e.	2	20.040	49.900
Magnesium (Mg)	% to m.e.	2	12.160	82.237
Boron (B)	ppm to p.e.	3	3.6067	27.726
Copper (Cu)	ppm to p.e.	2	31.770	3.1476
Iron (Fe)	ppm to p.e.	3	18.617	5.3726
Manganese (Mn)	ppm to p.e.	2	27.465	3.6410
Zinc (Zn)	ppm to p.e.	2	32.690	3.0590
Sulfur (S)	% to m.e.	2	16.033	62.377
Sodium (Na)	% to m.e.	1	22.991	43.496
Chlorine (Cl)	% to m.e.	1	35.457	28.175

Note: Milliequivalents can be converted to percentage by multiplying by equivalent weight/1000 and microequivalents can be converted to parts per million (ppm) by multiplying by equivalent weight/100. Factor x% = m.e./100 g, and factor x ppm = p.e./100 g.

H. Interpretation of Results

Difficulties have been encountered in the use and interpretation of a plant analysis result, although the quantitative association between absorbed nutrient element and growth has been the subject of many studies. Questions raised at the 1959 Plant Analysis and Fertilizer Problems Colloquium (Reuther, 1961) regarding the limitations of the plant analysis technique are still applicable today. These questions concern the reliability of interpretative data, utilization of ratio and balance concepts, hybrid influences, and changing physiological processes occurring at varying elemental concentrations. Bell (2000) evaluated the difficulties in interpreting a plant analysis result, primarily focusing on the micronutrients B and Zn and on time of evaluation and assay. Smith and Loneragan (1997) have pointed out environmental factors that induce intermittent symptoms that can influence the interpretation of a plant analysis result. In addition, reliable interpretative data are lacking

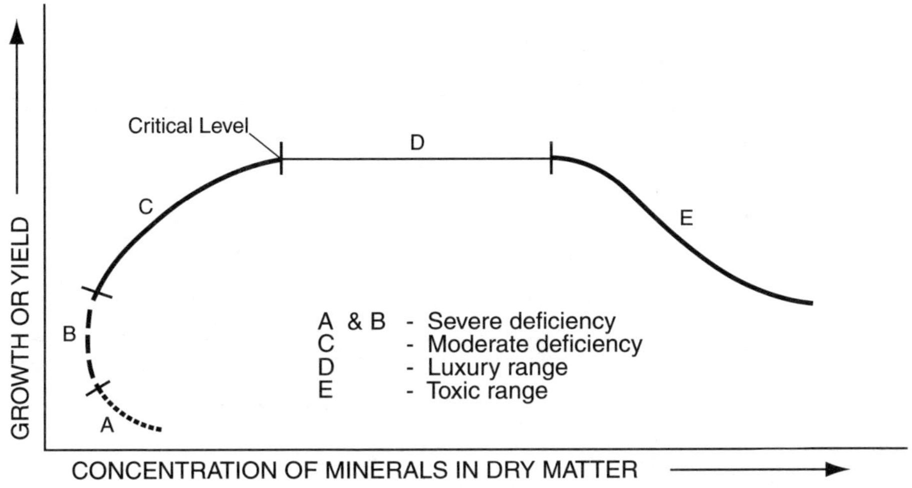

Figure 3.3
General relationship between plant growth or yield and elemental content of the plant. (From Smith, V.R., *Annu. Rev. Plant Physiol.*, 13, 81, 1962. With permission.)

for Cl, for all nutrient elements for use with most ornamental plants, for all plants during their early growth stages, and for identification of those concentrations considered excessive and/or toxic. It is also questionable whether the determination of the Fe concentration in a particular tissue can be used to establish the degree of Fe sufficiency (Chaney, 1984; Jones and Wallace, 1992).

Initially, single concentration values, such as critical (Macy, 1936; Ulrich, 1952; Smith, 1962) or standard (Kenworthy, 1961) concentrations, were sought. A critical value is that concentration below which deficiency occurs or above which toxicity occurs. Bates (1971) has reviewed those factors that can affect the use of critical values for interpreting a plant analysis result. As a single value, it is difficult to use when interpreting a plant analysis result if the assay concentration is considerably higher or lower than the critical value. Several elements have critical values that have universal application at both the deficiency as well as the excess level. For example, below 15 mg Zn/kg, 20 mg Mn/kg, and 50 mg Fe/kg are contents considered deficient for many plants, particularly those sensitive to these elements. Copper may also be added to this list, with <3 mg/kg the probable deficiency critical value, but there may be a number of plant species exceptions. At the excess level, P at the 1.00% or greater level would adversely affect plant performance.

But today those who interpret plant analysis results for diagnostic purposes prefer working with the full range in concentration, from deficiency to excess. Such interpretative data are obtained from response curves such as those described by Prevot and Ollagnier (1961) and Smith (1962), and shown in Figure 3.3. Others have drawn similar response curves with varying

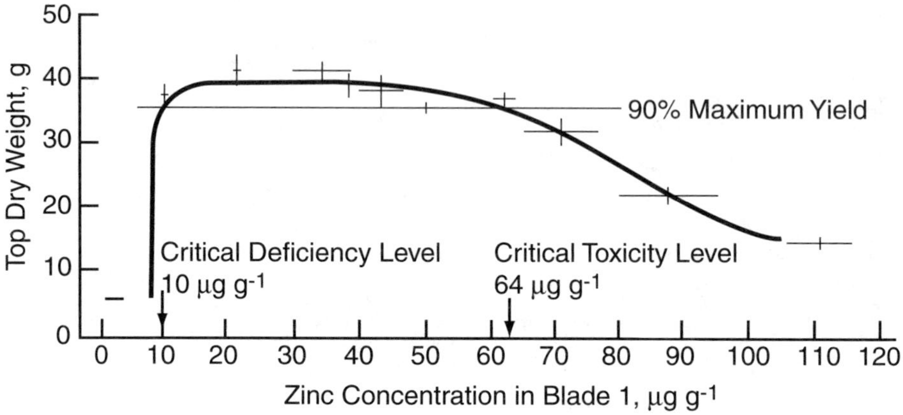

Figure 3.4
Relationship between zinc content of blade 1 of grain sorghum and top dry weight. (From Ohki, K., *Agron. J.*, 76, 253, 1984. With permission.)

slopes within the deficiency range, such as those obtained by Ohki (1984) and shown in Figure 3.4. The slope and general configuration shown in Figure 3.3 are typical for describing the association between yield or plant response and macronutrient concentration in the leaf or plant, whereas Figure 3.4 better typifies the association between yield and the micronutrient concentration.

The C shape of the left-hand portion of Figure 3.3 has been termed the "Steenbjerg Effect," and is the result of a combination of either elemental concentration or dilution, effects that have been discussed in some detail by Jarrell and Beverly (1981). Therefore, a misinterpretation of a plant analysis result can occur if the interpreter is not familiar with the interactive relationship between element concentration and dry matter accumulation.

The steep left-hand slope shown in Figure 3.4 poses a significant sampling and analytical problem since a very small change in concentration results in a significant change in plant growth and/or yield. This is particularly true for the micronutrients Mn and Zn, where a concentration change of only 1 or 2 mg/kg in the leaf tissue can define the difference between deficiency and sufficiency (Viets et al., 1954; Ohki, 1975; 1981).

In an ever-increasing number of instances, identifying at what concentration a major element or micronutrient becomes excessive or toxic is becoming as important as the determination of the concentration considered deficient.

Diagnosing a plant analysis result based either on critical or standard values or on sufficiency ranges requires that the plant part and time of sampling be identical for the diagnosed tissue as that for the source of the interpretative values. Because nutrient element concentrations in the plant can vary depending on plant part, stage of growth, genotype, and geographic location, these traditional techniques of plant analysis interpretation have their limitations.

Sufficiency range values, which are the most frequently used for the interpretation of a plant analysis result and sufficiency range data, can be found in the publications by Chapman (1966), Jones (1967), Shear and Faust (1980), Martin-Prevel et al. (1987), Halliday and Trenkel (1992), Mills and Jones (1996), and Reuter and Robinson (1997). Typical sufficiency ranges for a variety of different crop species are given in the following table:

Sufficiency Ranges for Selected Crops

Element	Corn[a]	Soybean[b]	Tomato[c]	Apple[d]	Pecan[e]
			%		
Nitrogen (N)	2.70–4.00	4.00–5.50	2.50–5.00	1.90–2.60	1.75–3.50
Phosphorus (P)	0.25–0.50	0.25–0.50	0.35–0.50	0.09–0.40	0.10–0.30
Potassium (Ca)	1.70–3.00	1.70–2.50	2.50–5.00	1.20–2.00	0.65–2.50
Calcium (Ca)	0.21–1.00	0.35–2.00	1.50–3.00	0.80–1.60	0.75–1.75
Magnesium (Mg)	0.20–1.00	0.25–1.00	0.50–1.00	0.25–0.45	0.25–0.30
			mg/kg		
Boron (B)	5–25	20–55	25–100	25–50	30–75
Copper (Cu)	6–20	10–30	5–20	6–25	10–20
Iron (Fe)	20–250	50–350	60–300	50–300	75–200
Manganese (Mn)	20–200	20–100	40–150	25–200	50–400
Zinc (Zn)	25–100	20–50	25–75	20–100	20–120

[a] Ear leaf taken at initial silk.

[b] Mature leaves from new growth.

[c] End leaflet from recently mature leaf.

[d] Mature leaves from new growth.

[e] Leaflet pairs from new growth.

Single concentration values, such as critical (Macy, 1936; Ulrich, 1952; Smith, 1962) or standard (Kenworthy, 1961) concentrations, are still in common use today to evaluate a plant analysis result. An interesting concept of what one might refer to as a "standard value" is that suggested by Market (1994), who has developed what he has defined as the "Reference Plant" composition values, values that can be compared to any plant analysis to establish a means of comparison. His Reference Plant values are as follows:

Element	%	Element	mg/kg
Nitrogen (N)	2.5	Boron (B)	40
Phosphorus (P)	0.2	Copper (Cu)	10
Potassium (K)	1.9	Iron (Fe)	150
Calcium (Ca)	1.0	Manganese (Mn)	200
Magnesium (Mg)	0.2	Molybdenum (Mo)	0.5
Sulfur (S)	0.3	Zinc (Zn)	50
Chlorine (Cl)	0.2	Sodium (Na)	150
Silicon (Si)	0.1	Aluminum (Al)	80

Heavy Metals	mg/kg
Arsenic (As)	0.1
Cadmium (Cd)	0.05
Chromium (Cr)	1.5
Lead (Pb)	1.0
Nickel (Ni)	1.5

It should be remembered that these levels are based on general plant composition, and are not necessarily applicable to any particular plant. If one has a plant analysis result and is looking for comparative values, these Reference Plant values may be helpful.

A different concept of plant analysis interpretation is embodied in the Diagnosis and Recommendation Integrated System (DRIS) proposed by Beaufils (1971; 1973). The DRIS technique of interpretation is based on a comparison of calculated elemental ratio indices with established norms. The DRIS approach was designed to (1) provide a valid diagnosis irrespective of plant age or tissue origin; (2) rank nutrients in their limiting order; and (3) stress the importance of nutrient balance. Beaufils (1973) used the survey approach by examining the world's published literature to obtain a plot of elemental leaf concentration vs. yield, a distribution that is normally skewed. To normalize the distribution curve, the yield component is divided into low- and high-yield groups. The DRIS method has been applied primarily for interpretation based on the major elements, because the database for the major elements is considerably larger than that for the micronutrients. Therefore, the reliability of a micronutrient DRIS index would be less than that for a major element. It is doubtful that the DRIS method of plant analysis interpretation will ever be exclusively used in lieu of the more traditional critical value or sufficiency range techniques. Those interested in investigating the DRIS technique should refer to the book by Beverly (1991).

The use of plant analyses for aiding in determining elemental sufficiency and/or the fertilization requirements of crops has been discussed by Ulrich and Hills (1990) for sugar beet, Bowen (1990) for sugarcane, Sabbe and Zelinski (1990) for cotton, Westfall et al. (1990) for small grains, Jones et al. (1990) for corn and grain sorghum, Geraldson and Tyler (1990) and Jones (1985) for vegetable crops, Righetti et al. (1990) for orchard crops, Kelling and Matocha (1990) for forage crops, Miner and Tucker (1990) for tobacco, and Weetman and Wells (1990) for forest species.

Clements (1960) uses a data-logging technique to monitor sugarcane crops, a procedure that can be very valuable for maintaining nutrient element sufficiency for crops with long growing periods as well as for following the yearly nutrient element status of annual crops, a procedure that is highly recommended by the author (Jones, 1986). Processing procedures for presenting plant analysis data, information sheet and analytical data entry/acquisition methods, calculations, and tracking procedures are discussed by Karamanos (1998). Similar data processing procedures are included in the review by Donohue and Gettier (1990), and various types of plant analysis report forms may be found in the book by Mills and Jones (1996).

I. Extractable Elements

1. Introduction

Extraction procedures for evaluating the Ca (Gallaher and Jones, 1976), Fe (Machold and Stephen, 1969; Katyal and Sharma, 1980; Chaney, 1984), N and NO_3 (Mills, 1980; Baker and Smith, 1996; Miller, 1998e), and S and SO_4 (Spencer et al., 1978; Scaife and Burns, 1986; Miller, 1998f), Cl (Miller, 1998c; Liu, 1998a), PO_4 (Miller, 1998f), and K (Miller, 1998f) status of plants have been proposed. Ulrich et al. (1959) used 2% acetic acid (CH_3COOH) to extract P and K from sugar beet (*Beta vulgaris* L.) petioles, and Miller (1998f) described the use of the same extraction reagent for the determination of extractable Cl, PO_4, K, and SO_4. Sahrawat (1980; 1987), Hunt (1982), and Miyazawa et al. (1984) used dilute hydrochloric acid (HCl) as an extraction reagent to determine Ca, Mg, K, Mn, Cu, Zn, Fe, and P content in dried tissue, attempting to bypass the traditional methods of organic matter destruction (see Section C.6 of this chapter). In most studies, K determined by extraction (by water, dilute HCl, 2% CH_3COOH) is comparable (equal) to that obtained by traditional organic matter destruction methods (see Section C.6 of this chapter). Baker and Greweling (1967) have developed an extraction procedure that provides results comparable with those for dry-ashed samples for the elements Ca, Mg, K, Mn, Cu, and Zn. Nicholas (1956) has also obtained good correlations between results obtained by extraction vs. total for the elements Ca, K, Mn, and P. The determination of B by extraction using a mixture of hydrochloric (HCl)–hydrofluoric (HF) acids has been suggested by van der Lee et al. (1987). Nitrate–N and PO_4 are the two most commonly determined elements since there are substantial interpretative data.

2. Extraction Procedures for Nitrate (NO_3) in Plant Tissue

The most frequent chemical form obtained by extraction is NO_3–N. A widely used method of extraction first described by Baker and Smith (1969) and then modified by Heanes (1982) is described below.

Baker and Smith Extraction Method

1. Weigh 400 mg oven-dried ground (80°C; 176°F) 20-mesh-screened tissue into shaking bottle.
2. Add 40 mL 0.025 *M* aluminum sulfate [$Al_2(SO_4)_3$] [weigh 8.55 g $Al_2(SO_4)_3$ into a 1000-mL volumetric flask and bring to the mark with water] containing 10 µg/mL NO_3–N and 1 mL/L preservative.
3. Shake for 15 min.
4. Filter and save filtrate for NO_3–N assay.

Heanes Extraction Method

1. Weigh 400 mg oven-dried ground (80°C; 176°F) 20-mesh-screened tissue into shaking bottle.
2. Add 50 mg of oxidized activated charcoal (AC).
3. Add 40 mL 0.025 *M* aluminum sulfate [$Al_2(SO_4)_3$] [weigh 8.55 g $Al_2(SO_4)_3$ into a 1000-mL volumetric flask and bring to the mark with water].
4. Shake for 30 min.
5. Filter into shaking bottle containing 500 mg of oxidized AC.
6. Shake filtrate plus AC for 30 min.
7. Filter and save filtrate for NO_3–N assay.

Mills (1980) suggests water as a satisfactory extraction reagent. A review of the techniques and procedures for extraction and determination of NO_3–N has been written by Keeney and Nelson (1982). The method of NO_3 determination varied from use of an ion-specific electrode (Milham et al., 1970; Carlson and Keeney, 1971; Carlson et al., 1990; Baker and Thompson, 1992; Miller, 1998) to ion chromatography (Barak and Chen, 1987). Procedures of NO_3 determination are also given in Chapter 2, Sections O.4 and O.5.

3. Extraction of Phosphorus (PO_4) in Plant Tissue Using 2% Acetic Acid

Two percent acetic acid (CH_3COOH) has been used by Ulrich et al. (1959) to extract PO_4 (as well as K) from sugar beet petioles, and Miller (1998) described its use for the determination of extractable PO_4 (as well as Cl, K, and SO_4) in plant tissue.

Reagent

Extraction Reagent

Pipette 20 mL *glacial* acetic acid (CH_3COOH) into a 1000-mL volumetric flask and bring to volume using water.

Extraction Procedure

Weigh 0.20 g oven-dried plant tissue into a 125-mL extraction vessel.
Add 50 mL Extraction Reagent and shake for 30 min.
Filter, refilter if filtrate is cloudy, and retain for analysis.

All of these extraction procedures are laboratory-conducted tests using oven-dried and milled tissue, and should not be confused with procedures called tissue tests, which are tests conducted in the field on extracted sap from fresh tissue. Tissue testing will be discussed in Chapter 4.

4. Interpretation for Extractable Nitrogen (NO_3) and Phosphorus (PO_4)

The following table provides interpretation values for NO_3–N and PO_4–P, which probably have wider application than only for "Western Crops," as specified in the title.

Nitrogen (NO_3) and Phosphorus (PO_4) Interpretation Guide for Western Crops

Plant	Time of sampling	Plant part		Nutrient level[a] (mg/kg) Deficient	Sufficient
Asparagus	Midgrowth of fern	4-in. tip section of new fern branch	N	100	500
			P	800	1,600
Bean, bush snap	Midgrowth	Petiole of fourth leaf from growing tip	N	2,000	4,000
			P	1,000	3,000

(continued)

Nitrogen (NO_3) and Phosphorus (PO_4) Interpretation Guide for Western Crops (continued)

Plant	Time of sampling	Plant part		Nutrient level[a] (mg/kg) Deficient	Sufficient
	Early bloom	Petiole of fourth leaf from growing tip	N	1,000	2,000
			P	800	2,000
Broccoli	Midgrowth	Midrib of young, mature leaf	N	7,000	10,000
			P	2,500	5,000
	First buds	Midrib of young. mature leaf	N	5,000	9,000
			P	2,000	4,000
Brussels sprout	Midgrowth	Midrib of young, mature leaf	N	5,000	9,000
			P	2,000	3,500
	Late growth	Midrib of young, mature leaf	N	2,000	4,000
			P	1,000	3,000
Cabbage	At heading	Midrib of wrapper leaf	N	5,000	9,000
			P	2,500	3,500
Cantaloupe	Early growth (short runners)	Petiole of sixth leaf from growing tip	N	8,000	12,000
			P	2,000	4,000
	Early fruit	Petiole of sixth leaf from growing tip	N	5,000	9,000
			P	1,500	2,500
	First mature fruit	Petiole of sixth leaf from growing tip	N	2,000	4,000
			P	1,000	2,000
Carrot	Midgrowth	Petiole of young, mature leaf	N	5,000	10,000
			P	2,000	4,000
Cauliflower	Buttoning	Midrib of young, mature leaf	N	5,000	9,000
			P	2,500	3,500
Celery	Midgrowth	Petiole of newest fully elongated leaf	N	5,000	9,000
			P	2,000	4,000
	Near maturity	Petiole of newest fully elongated leaf	N	4,000	6,000
			P	2,000	4,000
Cucumber	Early fruit-set (picking)	Petiole of sixth leaf from growing tip	N	5,000	9,000
			P	1,500	2,500
Lettuce	At heading	Midrib of wrapper leaf	N	4,000	8,000
			P	2,000	4,000
	At harvest	Midrib of wrapper leaf	N	3,000	6,000
			P	1,500	2,500
Pepper, chili	Early growth	Petiole of young, mature leaf	N	5,000	7,000
			P	2,000	3,000

Nitrogen (NO_3) and Phosphorus (PO_4) Interpretation Guide for Western Crops (continued)

Plant	Time of sampling	Plant part		Nutrient level[a] (mg/kg) Deficient	Sufficient
	Early fruit-set	Petiole of young, mature leaf	N	1,000	2,000
			P	1,500	2,500
Pepper, sweet	Early growth	Petiole of young, mature leaf	N	8,000	12,000
			P	2,000	4,000
	Early fruit-set	Petiole of young, mature leaf	N	3,000	5,000
			P	1,500	2,500
Potatoes	Early season	Petiole of fourth leaf from growing tip	N	8,000	12,000
			P	1,200	2,000
	Midseason	Petiole of fourth leaf from growing tip	N	6,000	9,000
			P	800	1,600
	Late season	Petiole of fourth leaf from growing tip	N	3,000	6,000
			P	500	1,000
Rose clover	Flowering	Leaves	N	—	—
			P	1,200	1,500
Spinach	Midgrowth	Petiole of young, mature leaf	N	4,000	8,000
			P	2,000	4,000
Subclover	Third flower	Fully expanded leaves	N	—	—
			P	800	1,000
Sweet corn	Tasseling	Midrib of first leaf above primary ear	N	500	1,500
			P	500	1,000
Sweet potato	Midgrowth	Petiole of sixth leaf from growing tip	N	1,500	3,500
			P	1,000	2,000
Tomato	Early bloom (canning)	Petiole of fourth leaf from growing tip	N	8,000	12,000
			P	2,000	3,000
	Fruit 1-in. (canning)	Petiole of fourth leaf from growing tip	N	6,000	10,000
			P	2,000	3,000
	First color (canning)	Petiole of fourth leaf from growing tip	N	2,000	4,000
			P	2,000	3,000
Watermelon	Early fruit (canning)	Petiole of sixth leaf from growing tip	N	5,000	9,000
			P	1,500	2,500

[a] Unless otherwise noted, values are N = NO_3–N ppm; P = acetic acid-soluble PO_4–P ppm.

Source: Ludwick, A.E., *Western Fertilizer Handbook,* Horticulture ed., Interstate Publishers, Danville, IL, 1990. With permission.

5. Extractable Ammonium (NH_4) in Plant Tissue

To assess the relative proportion of NH_4 absorption and accumulation of NH_4 in the plant, an analysis for NH_4 is required. Water-extractable NH_4 has been used to show NH_4 accumulation when plants are subjected to only NH_4–N nutrition (Carlson et al., 1990; Liu and Shelp, 1992).

Extraction of Ammonium (NH_4) from Plant Tissue

1. Grind oven-dried plant material to a fine powder.
2. Weigh 250 mg of the fine powder plant material into a 90-mL glass vial.
3. Add 25 mL water and cap the vial.
4. Place the vial on a reciprocating shaker and shake at 250 rpm for 30 min.
5. Remove from the shaker and allow to stand for 15 min.
6. Filter through Whatman 42 filter paper into a plastic vial for NH_4 analysis.

6. Extractable Sulfate (SO_4) in Plant Tissue

The sulfate–sulfur (SO_4–S) content of plants has been used as an indicator of their S nutrient status (Beaton et al., 1968). Sulfate is assumed to be transitory in the plant since it is reduced quickly for incorporation into plant components, and will accumulate only when it occurs in excess of plant requirement. The accumulation of SO_4 in plants, however, is influenced by many factors (e.g., age of plant, type of plant, status of other nutrient elements), which can sometimes make interpretation of measurements complex.

Many different extraction solutions have been used, but it has not been determined if the solution extracts all the SO_4 present in the plant or if reduced ions of organic S are converted to sulfate during extraction and subsequent quantification. A number of methods have been used to quantify the SO_4 in the extract, all of which have important implications for interpretation.

The most common methods of quantification have been the hydriodic acid reduction method and gravimetry, UV-VIS spectrophotometry, or turbidimetry involving precipitation with Ba. Barium-based methods are assumed to be specific to inorganic SO_4, but are not particularly sensitive and are subject to interference from many organic and inorganic compounds. The hydriodic acid method is quite sensitive and free from interference, but includes organic as well as inorganic SO_4 and is operationally slow. More recently, ion chromatography (Stevens, 1985) and ICP-AES (Novozamsky et al., 1986) have been found capable of S determinations. Ion chromatography measures the inorganic SO_4 form of S quite specifically, but requires specialized instrumentation. Since the ICP spectrometer measures total S (organic and inorganic, oxidized and reduced), SO_4 must be separated from other forms before quantification to make it S form specific. Novozamsky et al. (1986) used Ba precipitation for S species separation, making the method similar to other Ba-based methods. To date, there are little data by these newer methods on which to evaluate the interpretation of determinations.

The application of the hydriodic acid reduction method directly on plant material has the advantage of eliminating the extraction step, in addition to its relative sensitivity and freedom from interference. There are a number of studies with this direct measurement that have shown its effectiveness for determining the S nutritional status of plants in a variety of situations (Scott et al., 1983; 1984; Janzen and Bettany, 1984; Millard et al., 1985; Scaife and Burns, 1986; Pinkerton and Randall, 1995). Although the method includes both organic and inorganic SO_4, it is quite specific to SO_4 (Tabatabai, 1982). Adoption of the bismuth sulfide (BiS) instead of methylene blue colorimetric determination of the hydrogen sulfide (H_2S) produced by the hydriodic acid reagent (Kowalenko and van Lowe, 1972) and redesigning of the digestion–distillation apparatus (Kowalenko, 1985) have reduced the time for an analysis to less than 10 min from the 60 to 120 min required by the original method proposed by Johnson and Nishita (1952).

Sulfate can also be determined by ion chromatography (Busman et al., 1983).

7. Extractable Chloride (Cl) in Plant Tissue

Although Cl is classified as an essential plant micronutrient (not more than 100 mg Cl/kg for biochemical functions), plants can normally accumulate much higher concentrations in the range of 2000 to 20,000 mg Cl/kg. The majority of the Cl present in the plant is in the ionic form; therefore, Cl in the plant is quantitatively extracted with water or diluted acid or diluted salts (Gaines et al., 1984). The Cl in the filtrate can be analyzed using the colorimetric method on the TRAACS 800™ AutoAnalyzer (Tel and Heseltine, 1990). In this method, the sample is mixed with the color reagent and dialyzed into the color reagent again. The procedure is based on the release of thiocyanate ions from mercuric thiocyanate by Cl^- ions in the sample. The liberated thiocyanate reacts with ferric iron to form a red color complex of ferric thiocyanate. The color of the resulting solution is stable and directly proportional to the original Cl concentration.

Chloride can also be determined by ion chromatography (Busman et al., 1983).

Extraction of Chloride (Cl) from Plant Tissue

1. Grind oven-dried plant material to a fine powder.
2. Weigh 250 mg of the fine powder into a 90-mL glass vial.
3. Add 25 mL water and cap the vial.
4. Place the vial on a reciprocating shaker and shake at 250 rpm for 30 min.
5. Remove from the shaker and allow to stand for 15 min.
6. Filter through Whatman 42 filter into a plastic vial for Cl analysis.

8. Extraction of Chloride (Cl), Nitrate (NO_3), Orthophosphate (PO_4), Potassium (K), and Sulfate (SO_4) in Plant Tissue Using 2% Acetic Acid

The method semiquantifies the concentration of Cl, NO_3–N, PO_4–P, K, and SO_4–S in plant tissue by extraction with a 2% acetic acid (CH_3COOH) solution (Miller, 1998f). Dilute (CH_3COOH) does not quantitatively extract these anions from the tissue.

Extraction of Cl, NO_3, PO_4, K, and SO_4 from Plant Tissue

1. Weigh 200.0 ± 1.0 mg oven-dried plant tissue and place in a 125-mL extraction vessel. Include a method blank.
2. Add 50.0 ± 0.2 mL 2% Acetic Acid Extraction Solution [pipette 20 mL *glacial* acetic acid (CH_3COOH) into a 1000-mL volumetric flask and dilute to volume with water] and place on reciprocating mechanical shaker for 30 min.
3. Filter (and refilter if filtrate is cloudy) and retain for analysis.

Nitrate is determined by UV-VIS spectrophotometry at 520 nm by the Griess–Ilasvay method (cadmium reduction); K is determined by atomic emission or absorption spectrophotometry; PO_4–P in the extract is determined by UV-VIS spectrophotometry at 660 nm by reacting with paramolybdate; and Cl is determined by coulometric titration or ion-selective electrode (Watson and Isaac, 1990). The method has been used primarily to determine NO_3–N, K, PO_4–P, SO_4–S, and Cl for assessing plant nutrition and Cl status (Johnson and Ulrich, 1959; Chapman and Pratt, 1961). The method can also be used to determine extractable NH_4–N. Generally, the method detection limit is approximately 10 mg/kg (sample dry basis) and is generally reproducible to within ±10.0%.

9. Extractable Iron (Fe) in Plant Tissue

Total Fe has been found to be highly questionable in the interpretation of the Fe status of a plant (Chaney, 1984); therefore, a measure of "active" Fe [some have suggested that this form of Fe is the ferrous (Fe^{2+}) ion] in plant tissue would provide a better evaluation. How to determine "active" Fe has been investigated using a variety of extraction procedures (Chaney, 1984), with 1 *M* acetic acid (CH_3COOH) as the only extraction reagent giving consistent results. Mehrotta et al. (1985) found that 1 *M* CH_3COOH–extractable Fe in chlorotic corn leaves was <20 mg Fe/kg, and in normal green leaves >30 mg Fe/kg.

References

Abraham, V. and J.M. DeMan. 1987. Determination of total sulfur in canola oil. *J. Am. Oil Chem. Soc.,* 64:384–387.

Adler, P.R. and G.E. Wilcox. 1985. Rapid perchloric acid methods for analysis of major elements in plant tissue. *Commun. Soil Sci. Plant Anal.,* 16:1153–1163.

Alexander, G.V. 1965. An X-ray fluorescence method for the determination of calcium, potassium, chlorine, sulfur, and phosphorus in biological tissues. *Anal. Chem.,* 37:1671–1674.

Alvarez, R. 1980. NBS plant tissue standard reference materials. *J. Assoc. Off. Anal. Chem.,* 63:806–808.

Anderson, D.L. and L.J. Henderson. 1986. Sealed chamber digestion for plant nutrient analysis. *Agron. J.,* 78:937–939.

Arneth, W., B. Herold, and E. Schwab. 1983. Use of digest distillation technique for rapid determination of protein content in meat and meat products. *Tecator J. Technol. Chem. Anal.,* 6(1):11.

Artiola, J.F. and A.S. Ali. 1990. Determination of total sulfur in soil and plant samples using sodium bicarbonate/silver oxide, dry ashing and ion chromatography. *Commun. Soil Sci. Plant Anal.*, 21:941–949.

Ashby, D.L. 1969. Washing techniques for the removal of nutrient element deposits from the surface of apple, cherry, and peach leaves. *J. Am. Soc. Hortic. Sci.,* 94:266–268.

Baethegen, W.E. and M.M. Alley. 1989. A manual colorimetric procedure for measuring ammonium nitrogen in soil and plant Kjeldahl digests. *Commun. Soil Sci. Plant Anal.,* 20:961–969.

Bahn, K.C., A. Wallace, and D.R. Hunt. 1959. Some mineral losses from leaves by leaching. *Proc. Am. Soc. Hortic. Sci.,* 73:289–293.

Baker, A.S. and R. Smith. 1969. Extracting solution for potentiometric determination of nitrate in plant tissue. *J. Agric. Food Chem.,* 17:1284–1287.

Baker, D.E., G.W. Gorsline, C.G. Smith, W.I. Thomas, W.E. Grube, and J.L. Ragland. 1964. Technique for rapid analyses of corn leaves for eleven elements. *Agron. J.,* 56:133–136.

Baker, J.H. and T. Greweling. 1967. Extraction procedure for quantitative determination of six elements in plant tissue. *J. Agric. Food Chem.,* 15:340–344.

Baker, W.H. and T.L. Thompson. 1992. Determination of nitrate–nitrogen in plant samples by selective ion electrode, in C.O. Plank, Ed., *Plant Analysis Reference Procedures for the Southern Region of the United States*, Southern Cooperative Series Bulletin 368, University of Georgia, Athens, 25–31.

Barak, P. and Y. Chen. 1987. Three-minute analysis of chloride, nitrate, and sulfate by single column anion chromatography. *Soil Sci. Soc. Am. J.,* 51:257–258.

Bates, T.E. 1971. Factors affecting critical nutrient concentrations in plants and their evaluation: a review. *Soil Sci.,* 112:116–130.

Beaton, J.D., G.R. Burns, and J. Platou. 1968. *Determination of Sulfur in Soils and Plant Material*, Technical Bulletin 14, The Sulphur Institute, Washington, D.C., 17–19.

Beaufils, E.R. 1971. Physiological diagnosis — a guide for improving maize production based on principles developed for rubber trees. *Fertil. Soc. S. Afr. J.,* 1:1–31.

Beaufils, E.R. 1973. Diagnosis and Recommendation Integrated System (DRIS), Soil Science Bulletin Number 1, University of Natal, South Africa.

Bell, R.W. 2000. Temporary nutrient deficiency — a difficult case for diagnosis and prognosis by plant analysis. *Commun. Soil Sci. Plant Anal.,* 31:1847–1861.

Beverly, R.B. 1991. *A Practical Guide to the Diagnosis and Recommendation Integrated System (DRIS)*, MicroMacro Publishing, Athens, GA.

Blanchar, R.W. 1986. Measurements of sulfur in soils and plants, in M.A. Tabatabai, Ed., *Sulfur in Agriculture*, American Society of Agronomy, Madison, WI.

Bock, R.A. 1978. *Handbook of Decomposition Methods in Analytical Chemistry*, International Textbook Co., Glasgow, Scotland.

Bowen, J.E. 1990. Plant tissue analysis sugarcane, in R.L. Westerman, Ed., *Soil Testing and Plant Analysis*, 3rd ed., SSSA Book Series No. 3, Soil Science Society of America, Madison, WI, 449–467.

Bowman, D.C., J.L. Paul, and R.M. Carlson. 1988. A method to exclude nitrate from Kjeldahl digestion of plant tissues. *Commun. Soil Sci. Plant Anal.,* 19:205–213.

Bradstreet, R.B. 1965. *The Kjeldahl Method for Organic Nitrogen*, Academic Press, New York.

Bremner, J.M. and D.R. Keeney. 1965. Steam distillation methods for the determination of ammonium, nitrate, and nitrite. *Anal. Chem.,* 32:485–495.

Bremner, J.M. and C.S. Mulvaney. 1982. Total nitrogen, in A.L. Page, R.H. Miller, and D.R. Kenney, Eds., *Method of Soil Analysis*, Part 2, Agronomy Monograph No. 9, American Society of Agronomy, Madison, WI, 595–624.

Bremner, J.M. and M.A. Tabatabai. 1972. Use of an ammonia electrode for determination of ammonium in Kjeldahl analysis of soils. *Commun. Soil Sci. Plant Anal.,* 3:159–165.

Busman, L.M., R.P. Dick, and M.A. Tabatabai. 1983. Determination of total sulfur and chlorine in plant materials by ion chromatography. *Soil Sci. Soc. Am. J.,* 47:1167–1170.

Campbell, C.R. and C.O. Plank. 1998. Preparation of plant tissue for laboratory analysis, in Y.P. Kalra, Ed., *Handbook of Reference Methods for Plant Analysis,* CRC Press, Boca Raton, FL, 37–49.

Carlier, L.A. and L.P. van Hee. 1971. Microwave drying of lucerne and grass samples. *J. Sci. Food Agric.,* 22:306–307.

Carlson, R.M. 1978. Automated separation and conductiometric determination of ammonia and dissolved carbon dioxide. *Anal. Chem.,* 48:1528–1531.

Carlson, R.M. and D.R. Keeney. 1971. Specific ion electrodes for soil, plant, and water analysis, in L.M. Walsh, Ed., *Instrumental Methods for Analysis of Soils and Plant Tissue,* Soil Science Society of America, Madison, WI, 39–63.

Carlson, R.M., R.I. Cabrera, I.L. Paul, J. Quick, and R.Y. Evans. 1990. Rapid direct measurement of ammonium and nitrate in soil and plant tissue extracts. *Commun. Soil Sci. Plant Anal.,* 21:1519–1529.

Chaney, R.L. 1984. Diagnostic practices to identify iron deficiency in higher plants. *J. Plant Nutr.,* 7:47–67.

Chapman, H.D., Ed. 1966. *Diagnostic Criteria for Plants and Soils,* Division of Agricultural Science, University of California, Riverside.

Chapman, H.D. and P.F. Pratt. 1982. *Methods of Analysis for Soils, Plants and Waters,* Priced Publication 4034, Division of Agricultural Science, University of California, Berkeley.

Cherney, J.H. and D.L. Robinson. 1982. A comparison of plant digestion methods for identifying contamination of plant tissue by Ti analysis. *Agron. J.,* 75:145–147.

Clements, H.F. 1960. Crop logging of sugar cane in Hawaii, in W. Reuther, Ed., *Plant Analysis and Fertilizer Problems,* American Institute of Biological Sciences, Washington, D.C., 131–147.

Colonna, J.P. 1970. The mineral diet of excelsior coffee plants. Natural variability of the mineral foliar composition on a homogeneous plantation. *Cah. O.R.S.T.O.M. Ser. Biol.,* 13:67–80.

Constable, G.A., I.J. Rochester, J.H. Betts, and D.F. Herridge. 1991. Prediction of nitrogen fertilizer requirement in cotton using petiole and sap nitrate. *Commun. Soil Sci. Plant Anal.,* 22:1315–1324.

Cresser, M.S. and J.W. Parsons. 1979. Sulfuric-perchloric acid digestion of plant material for the determination of nitrogen, phosphorus, potassium, calcium, and magnesium. *Anal. Chem. Acta,* 109:431–436.

Cunniff, P., Ed. 1995. *Official Methods of Analysis of AOAC International*, Association of Official Analytical Chemists International, Arlington, VA, 22.

Dalquist, R.L. and J.W. Knoll. 1978. Inductively-coupled plasma-atomic emission spectroscopy: analysis of biological materials and soil for major, trace, and ultra-trace elements. *Appl. Spectrosc.,* 32:1–30.

David, M.B., M.J. Mitchell, and D. Aldcorn. 1989. Analysis of sulfur in soil, plant and sediment materials: sample handling and use of an automated analyzer. *Soil Biol. Biochem.*, 21:119–123.

Davis, J.G. 1995. Impact of time of day and time since irrigation on cotton leaf blade and petiole nutrient concentration. *Commun. Soil Sci. Plant Anal.,* 26:2351–2360.

DeBolt, D.C. 1980. Multielement emission spectroscopic analysis of plant tissue using DC agron plasma source. *J. Assoc. Off. Anal. Chem.*, 63:803–805.

Dionex Corporation. 1987. *Dionex Ion Chromatography Cookbook*, Literature Part Number 032861, Sunnyvale, CA.

Donohue, S.J. and S.W. Gettier. 1990. Data processing in soil testing and plant analysis, in R.L. Westerman, Ed., *Soil Testing and Plant Analysis*, 3rd ed., SSSA Book Series 3, Soil Science Society of America, Madison, WI, 741–755.

Dorscheimer, W.T. and R.A. Isaac. 1982. Application of NIR analysis. *Am. Lab.,* 14:58–63.

duPreez, D.R. and G.C. Bale, 1989. A simple method for the quantitative recovery of nitrate–N during the Kjeldahl analysis of dry soil and plant samples. *Commun. Soil Sci. Plant Anal.,* 20:345–357.

Eastin, E.F. 1976. Use of ammonia electrode for total nitrogen determination in plants. *Commun. Soil Sci. Plant Anal.,* 7:477–481.

Ebeling, M.E. 1968. The Dumas method for nitrogen in feeds. *J. Assoc. Off. Anal. Chem.*, 51:766–770.

Feldman, C. 1961. Evaporation of boron from acid solutions and residues. *Anal. Chem.,* 33:1916–1920.

Gaines, T.P. and G.A. Mitchell. 1979. Boron determination in plant tissues by the Azomethine-H method. *Commun. Soil Sci. Plant Anal.,* 10:1099–1108.

Gallaher, R.N. and J.B. Jones, Jr. 1976. Total, extractable, and oxalate calcium and other elements in normal and mouse-ear pecan tree tissues. *J. Am. Soc. Hortic. Sci.,* 101:692–696.

Gallaher, R.N., C.O. Wilson, and J.G. Futral. 1975. An aluminum block digester for plant and soil analysis. *Soil Sci. Soc. Am. Proc.,* 39:803–806.

Geraldson, C.M. and K.B. Tyler. 1990. Plant analysis as an aid in fertilizing vegetable crops, in R.L. Westerman, Ed., *Soil Testing and Plant Analysis*, 3rd ed., SSSA Book Series 3, Soil Science Society of America, Madison, WI, 549–562.

Gestring, W.D. and P.N. Soltanpour. 1981. Evaluation of wet and dry digestion methods for boron determination in plant samples by ICP-AES. *Commun. Soil Sci. Plant Anal.,* 12:743–753.

Goodall, D.W. and F.G. Gregory. 1947. Chemical Composition of Plants as an Index of Their Nutritional Status, Imperial Bureau Horticultural Plant Crops (GB) Technical Communication No. 17, Ministry of Agriculture, London, England.

Gorsuch, T.T. 1959. Radiochemical investigation on the recovery for analysis of trace elements in organic and biological materials. *Analyst,* 84:135–173.

Gorsuch, T.T. 1970. *Destruction of Organic Matter*, International Series of Monographs in Analytical Chemistry, Vol. 39, Pergamon Press, New York.

Gorsuch, T.T. 1976. Dissolution of organic matter, in P.D. Lafluer, Ed., *Accuracy in Trace Analysis. Sampling, Sample Handling Analysis*, Vol. 1, Special Publication 422, National Bureau of Standards, Washington, D.C., 491–508.

Graham, R.D. 1972. Vacuum attachment for grinding mills. *Commun. Soil Sci. Plant Anal.,* 3:167–173.

Grier, J.D. 1966. Preparation of plant material for plant analysis. *J. Assoc. Off. Anal. Chem.,* 49:292–298.

Gupta, U.C. 1979. Boron nutrition of crops. *Adv. Agron.,* 31:273–307.

Gupta, U.C. 1998. Determination of boron, molybdenum, and selenium in plant tissue, in Y.P. Kalra, Ed., *Handbook of Reference Methods for Plant Analysis*, CRC Press, Boca Raton, FL, 171–182.

Guthrie, T.F. and L.E. Lowe. 1984. A comparison of methods for total sulphur analysis of tree foliage. *Can. J. For. Res.,* 14:470–473.

Hafez, A.A., S.S. Goyal, and D.W. Rains. 1991. Quantitative determination of total sulfur in plant tissues using acid digestion and ion chromatography. *Agron. J.,* 83:148–IS3.

Halliday, D.J. and M.E. Trenkel, Eds. 1992. *IFA World Fertilizer Use Manual*, International Fertilizer Industry Association, Paris, France.

Halvin, J.L. and P.N. Soltanpour. 1980. A nitric acid plant tissue digest method for use with inductively coupled plasma spectrometry. *Commun. Soil Sci. Plant Anal.,* 11:969–980.

Hanlon, E.A. 1998. Elemental determination by atomic spectrophotometry, in Y.P. Kalra, Ed., *Handbook of Reference Methods for Plant Analysis*, CRC Press, Boca Raton, FL, 157–164.

Hansen, B. 1989. Determination of nitrogen as elementary N: an alternative to Kjeldahl. *Acta Agric. Scand.*, 39:113–118.

Heanes, D.L. 1982. Determination of nitrate-N in plants by an improved extraction procedure adapted for ultraviolet spectrophotometry. *Commun. Soil Sci. Plant Anal.,* 13:805–818.

Helmke, P.A. 1996. Neutron activation analysis, in D.L. Sparks, Ed., *Methods of Soil Analysis,* Part 3, *Chemical Methods*, SSSA Book Series Number 5, Soil Science Society of America, Madison, WI, 141–159.

Hern, J.L. 1984. Determination of total sulfur in plant materials using an automated sulfur analyzer. *Commun. Soil Sci. Plant Anal.*, 15:99–107.

Hislop, J.S. 1980. Choice of the analytical method, in P. Bratter and P. Schramel, Eds., *Trace Element Analytical Chemistry in Medicine and Biology*, DeGruyter, Berlin, Germany, 747–767.

Hogan, G.D. and D.G. Maynard. 1984. Sulphur analysis of environmental materials by vacuum inductively coupled plasma atomic emission spectrometry (ICP-AES), in *Proceedings of Sulphur-84*, Sulphur Development Institute of Canada (SUDIC), Calgary, Alberta, Canada, 676–683.

Hood, S.L., R.Q. Parks, and C. Hurwitz. 1944. Mineral contamination resulting from grinding plant samples. *Ind. Eng. Chem. Anal. Ed.,* 16:202–205.

Horneck, D.A. and D. Hanson. 1989. Elemental determination by atomic absorption spectrophotometry, in Y.P. Kalra, Ed., *Handbook of Reference Methods for Plant Analysis*, CRC Press, Boca Raton, FL, 157–164.

Horneck, D.A. and R.O. Miller. 1998. Determination of total nitrogen in plant tissue, in Y.P. Kalra, Ed., *Handbook of Reference Methods for Plant Analysis*, CRC Press, Boca Raton, FL, 75–83.

Horwitz, W., Ed. 1980. *Official Methods of Analysis of the Association of Official Analytical Chemists*, 13th ed., Association of Official Analytical Chemists, Arlington, VA.

Horwitz, W. 1982. Evaluation of analytical methods used for regulation of foods and drugs. *Anal. Chem.,* 54:67A–76A.

Horwitz, W., Ed. 2000. *Official Methods of Analysis of the Association of Official Analytical Chemists*, 17th ed., Association of Official Analytical Chemists, Arlington, VA.

Huang, C.L. and E.E. Schulte. 1985. Digestion of plant tissue for analysis by ICP emission spectroscopy. *Commun. Soil Sci. Plant Anal.,* 16:945–958.

Hunt, J. 1982. Dilute hydrochloric acid extraction of plant material for routine cation analysis. *Commun. Soil Sci. Plant Anal.,* 13:49–55.

Ihnat, M. 1998. Plant and related reference materials for data quality control of elemental content, in Y.P. Kalra, Ed., *Handbook of Reference Methods for Plant Analysis*, CRC Press, Boca Raton, FL, 235–284.

Isaac, R.A. and W.C. Johnson, Jr. 1976. Determination of total nitrogen in plant tissue. *J. Assoc. Off. Anal. Chem.,* 59:98–100.

Isaac, R.A. and W.C. Johnson, Jr. 1983. Determination of protein nitrogen in plant tissue using near infrared spectroscopy. *J. Assoc. Off. Anal. Chem.*, 66:506–509.

Isaac, R.A. and W.C. Johnson, Jr. 1998. Elemental determination by inductively coupled plasma atomic emission spectrometry, in Y.P. Kalra, Ed., *Handbook of Reference Methods for Plant Analysis*, CRC Press, Boca Raton, FL, 165–170.

Isaac, R.A. and J.B. Jones, Jr. 1970. Auto-analyzers systems for the analysis of soil and plant tissue extracts, in *Advances in Automated Analysis, Technicon Congress Proceedings,* Technicon Corp., Tarrytown, NY, 57–64.

Isaac, R.A. and J.B. Jones, Jr. 1972. Effects of various drying temperatures on the determination of five plant tissues. *Commun. Soil Sci. Plant Anal.,* 3:261–269.

Jackson, L.L., E.L. Engleman, and J.L. Peard. 1985. Determination of total sulfur in lichens and plants by combustion-infrared analysis. *Environ. Sci. Tech.,* 19:437–441.

Jackson, M.L. 1958. *Soil Chemical Analysis*, Prentice-Hall, Englewood Cliffs, NJ.

Janzen, H.H. and J.R. Bettany. 1984. Sulfur nutrition of rapeseed. I. Influences of fertilizer nitrogen and sulfur rates. *Soil Sci. Soc. Am. J.,* 48:100–107.

Jarrell, W.M. and R.B. Beverly. 1981. The dilution effect in plant nutrition studies. *Adv. Agron.,* 34:197–224.

Johnson, C.M. and N. Nishita. 1952. Microestimation of sulfur in plant materials, soils and irrigation waters. *Anal. Chem.,* 24:736–742.

Johnson, C.M. and A. Ulrich. 1959. Analytical Method for Use in Plant Analysis, Bulletin 766, University of California Experiment Station, Berkeley, 26–78.

Jones, J.B., Jr. 1963. Effect of drying on ion accumulation in corn leaf margins. *Agron. J.,* 55:579–580.

Jones, J.B., Jr. 1967. Interpretation of plant analysis for several agronomic crops, in G.W. Hardy, Ed., *Soil Testing and Plant Analysis, Plant Analysis*, Part 2, SSSA Special Publication No. 2, Soil Science Society of America, Madison, WI, 49–58.

Jones, J.B., Jr. 1970. Distribution of 15 elements in corn leaves. *Commun. Soil Sci. Plant Anal.,* 1:27–34.

Jones, J.B., Jr. 1976. Elemental analysis of biological substances by direct reading spark emission spectroscopy. *Am. Lab.,* 8:15–18.

Jones, J.B., Jr. 1977. Elemental analysis of soil extracts and plant tissue ash by atomic emission spectroscopy. *Commun. Soil Sci. Plant Anal.,* 8:345–365.

Jones, J.B., Jr. 1984. *Tissue Testing* (VHS-video), On Sat One, Anderson, SC.

Jones, J.B., Jr. 1985. Soil testing and plant analysis: guides to the fertilization of horticultural crops. *Hortic. Rev.,* 7:1–68.

Jones, J.B., Jr. 1986. Crop essential elements status assessment in the field, in W.G. Gensler, Ed., *Advanced Agricultural Instrumentation — Design Allure*, NATO ASI Series E, Applied Science 11, Martinus Nijhoff Publishers, Dordrecht, the Netherlands, 311–348.

Jones, J.B., Jr. 1991. *Kjeldahl Method for Nitrogen (N) Detemination*, MicroMacro Publishing, Athens, GA.

Jones, J.B., Jr. 1996. Determination of total sulphur in plant tissue using the HACH kit spectrophotometer technique. *Sulphur Agric.,* 19:58–62.

Jones, J.B., Jr. 1998a. Plant analysis, in *Plant Nutrition Manual*, CRC Press, Boca Raton, FL, 77–99.

Jones, J.B., Jr. 1998b. Field sampling procedures for conducting a plant analysis, in Y.P. Kalra, Ed., *Handbook of Reference Methods for Plant Analysis*, CRC Press, Boca Raton, FL, 25–35.

Jones, J.B., Jr. and V.W. Case. 1990. Sampling, handling, and analyzing plant tissue samples, in R.L. Westerman, Ed., *Soil Testing and Plant Analysis*, 3rd ed., SSSA Book Series Number 3, Soil Science Society of America, Madison, WI, 389–427.

Jones, J.B., Jr. and R.A. Issac. 1972. Determination of sulfur in plant material using a LECO sulfur analyzer. *J. Agric. Food Chem.,* 20:192–194.

Jones, J.B., Jr. and A. Wallace. 1992. Sample preparation and determination of iron in plant tissue samples. *J. Plant Nutr.,* 15:2085–2108.

Jones, J.B., Jr., R.L. Large, D.P. Pfeiderer, and H.S. Klosky. 1971. How to properly sample for a plant analysis. *Crops Soils,* 23:15–18.

Jones, J.B., Jr., H.V. Eck, and R. Voss. 1990. Plant analysis as an aid in fertilizing corn and grain sorghum, in R.L. Westerman, Ed., *Soil Testing and Plant Analysis*, 3rd ed., SSSA Book Series Number 3, Soil Science Society of America, Madison, WI, 521–547.

Jones, J.B., Jr., B. Wolf, and H.A. Mills. 1991. *Plant Analysis Handbook*, Micro-Macro Publishing, Athens, GA, 23–26.

Kalra, Y.P. and D.G. Maynard. 1998. Microwave digestion of plant tissue in an open vessel, in Y.P. Kalra, Ed., *Handbook of Reference Methods for Plant Analysis*, CRC Press, Boca Raton, FL, 63–67.

Kalra, Y.P., D.G. Maynard, and F.G. Aadford. 1989. Microwave digestion of tree foliage for multielement analysis. *Can. J. For. Res.,* 19:981–985.

Kane, P.F. 1987. Comparison of HgO and $CuSO_4/TiO_2$ as catalysts in manual Kjeldahl digestion for determination of crude protein in animal feed. Collaborative study. *J. Assoc. Off. Anal. Chem.*, 70:907–911.

Karamanos, R.E. 1998. Data processing in plant analysis, in Y.P. Kalra, Ed., *Handbook of Reference Methods for Plant Analysis*, CRC Press, Boca Raton, FL, 221–233.

Katyal, J.C. and B.D. Sharma. 1980. A new technique of plant analysis to resolve iron chlorosis. *Plant Soil,* 55:105–119.

Keeney, D.R. and D.W. Nelson. 1982. Nitrogen-inorganic forms, in A.L. Page, Ed., *Methods of Soil Analysis*, Part 2, 2nd ed., Agronomy Monograph Number 9, American Society of Agronomy, Madison, WI, 643–698.

Keller, P.N. 1992. An overview of analysis by inductively coupled plasma atomic emission spectrometry, in A. Montaser and D.W. Golightly, Eds., *Inductively Coupled Plasma–Atomic Emission Spectrometry*, VCH Publishers, New York, 601–629.

Kelling, K.A. and J.E. Matocha. 1990. Plant analysis as an aid in fertilizing forage crops, in R.L. Westerman, Ed., *Soil Testing and Plant Analysis*, 3rd ed., SSSA Book Series Number 3, Soil Science Society of America, Madison, WI, 603–641.

Kenworthy, A.L. 1961. Interpreting the balance of nutrient elements in leaves of fruit trees, in W. Reuther, Ed., *Plant Analysis and Fertilizer Problems*, Publication Number 8, American Institute of Biological Sciences, Washington, D.C., 28–43.

Kenworthy, A.L. 1969. Fruit, Nut, and Plantation Crops Deciduous and Evergreen: A Guide for Collecting Foliar Samples for Nutrient Element Analysis, Horticultural Rep. II, Michigan State University, East Lansing.

Kingston, K.M. and L.B. Jassie. 1986. Microwave energy for acid decomposition at elevated temperatures and pressures using biological and botanical samples. *Anal. Chem.,* 58:2534–2541.

Kingston, N.M. and L.B. Jassie. 1988. Monitoring and predicting parameters in microwave dissolution, in N.M. Kingston and L.B. Jassie, Eds., *Introduction to Microwave Sample Preparation*, American Chemical Society, Washington, D.C., 97–148.

Kirsten, W.J. and B.S. Nordenmark. 1987. Rapid, automated, high-precision method for micro, ultramicro, and trace determinations of sulfur. *Anal. Chim. Acta,* 196:59–68.

Knapp, G. 1985. Sample preparation techniques — an important part in trace element analysis for environmental research and monitoring. *Int. J. Environ. Anal. Chem.,* 22:71–83.

Koch, W.E. 1989. Ion chromatography and the certification of standard reference materials. *J. Chromatogr. Sci.,* 27:418–421.

Kovács, B., Z. Gyori, J. Prokisch, J. Loch, and P. Daniel. 1996. A study of plant sample preparation and inductively coupled plasma emission spectrometry parameters. *Commun. Soil Sci. Plant Anal.*, 27:1177–1198.

Kowalenko, C.G. 1985. A modified apparatus for quick and versatile sulphate–sulphur analysis using hydriodic acid reduction. *Commun. Soil Sci. Plant Anal.*, 16:289–300.

Kowalenko, C.G. 1998. Sulfate–sulfur determination in plant tissue, in Y.P. Kalra, Ed., *Handbook of Reference Methods for Plant Analysis*, CRC Press, Boca Raton, FL, 103–109.

Kowalenko, C.G. and C.J. van Laerhoven. 1998. Total sulfur determination in plant tissue, in Y.P. Kalra, Ed., *Handbook of Reference Methods for Plant Analysis*, CRC Press, Boca Raton, FL, 93–102.

Kowalenko, C.G. and L.E. van Lowe. 1972. Observations on the bismuth sulfide colorimetric procedure for sulfate analysis in soil. *Commun. Soil Sci. Plant Anal.*, 3:79–86.

Kubota, J. and V.A. Lazar. 1971. X-ray emssion spectrograph: techniques and uses for plant and soil analysis studies, in L.M. Walsh, Ed., *Instrumental Methods for Analysis of Soils and Plant Tissue*, Soil Science Society of America, Madison, WI, 67–83.

Labanauskas, C.K. 1968. Washing citrus leaves for leaf analysis. *Calif. Agric.,* 22:12–14.

Labanauskas, C.K. and M.F. Handy. 1973. Time, temperature and sample size effects in dry ashing of plant tissue, in *Summaries, Advances and Innovations in Progress for Soil, Plant, and Water Analysis*, University of California, Berkeley.

Lea, R. and C.C. Wells. 1980. Determination of extractable sulfate and total sulfur from plant and soil material by an autoanalyzer. *Commun. Soil Sci. Plant Anal.*, 11:507–516.

Lilleland, O. and J.G. Brown. 1943. Phosphate nutrition of fruit trees. *Proc. Am. Soc. Hortic. Sci.,* 41:1–10.

Liu, L. 1998a. Determination of ammonium–nitrogen in plant tissue, in Y.P. Kalra, Ed., *Handbook of Reference Methods for Plant Analysis*, CRC Press, Boca Raton, FL, 89–91.

Liu, L. 1998b. Determination of chloride in plant tissue, in Y.P. Kalra, Ed., *Handbook of Reference Methods for Plant Analysis*, CRC Press, Boca Raton, FL, 111–113.

Liu, L. and B.J. Shelp. 1992. Nitrogen partitioning in greenhouse-grown broccoli in response to varying $NH_4^+:NO_3^-$ ratios. *Commun. Soil Sci. Plant Anal.,* 24:45–60.

Lockman, R.B. 1970. Plant sample analysis as affected by sample decomposition prior to laboratory processing. *Commun. Soil Sci. Plant Anal.,* 1:13–19.

Losche, G. 1982. Microanalytical Azomethine-H method for boron determination in plant tissue. *Commun. Soil Sci. Plant Anal.,* 13:127–134.

Ludwick, A.E., Ed. 1990. *Western Fertilizer Handbook,* Horticulture ed., Interstate Publishers, Danville, IL.

Machold, O. and W.W. Stephen. 1969. Functions of iron porphyrin in chlorophyll biosynthesis. *Phytochemistry,* 8:2189–2192.

Macy, P. 1936. The quantitative mineral nutrient requirement of plants. *Plant Physiol.,* 11:749–764.

Maples, R.L., W.N. Miley, and T.C. Keisling. 1990. Nitrogen recommendations for cotton based on petiole analysis: strengths and limitations, in W.N. Miley and D.M. Oosterhuis, Eds., *Nitrogen Nutrition of Cotton, Practical Issues,* American Society of Agronomy, Madison, WI, 59–63.

Market, B. 1994. *Progress Report on the Element Concentration Cadaster Project ECCE of INTECOL/IUBS*, International Institute of Advanced Ecological and Economic Studies, IHI Zittau, Markt 23, 02763 Zittau, Germany.

Martin-Prevel, P., J. Gagnard, and P. Gautier., Eds. 1987. *Plant Analysis as a Guide to the Nutrient Requirements of Temperate and Tropical Crops*, Lavosier Publishers, New York.

Matrai, P.A. 1989. Determination of sulfur in ocean particulates by combustion fluorescence. *Mar. Chem.,* 26:227–238.

McGeehan, S.L. and D.V. Naylor. 1988. Automated instrumental analysis of carbon and nitrogen in plant and soil samples. *Commun. Soil Sci. Plant Anal.*, 19:493–505.

McLaughlin, R.D., M.S. Hunt, D.L. Murphy, and C.R. Chen. 1979. Instrumentation in environmental monitoring, in D.L. Hemphill, Ed., *Trace Elements in Environmental Health — XIII*, University of Missouri, Columbia, 224–245.

Mehrotta, S.C., C.P. Sharma, and S.C. Agarwala. 1985. A search for extractants to evaluate the iron status of plants. *Soil Sci. Plant Nutr.,* 31:155–162.

Milham, P.J., A.S. Awad, R.E. Paul, and J.H. Bull. 1970. Analysis of plants, soils, and waters for nitrate by using an ion-selective electrode. *Analyst,* 95:751–759.

Millard, P., G.S. Sharp, and N.M. Scott. 1985. The effect of sulphur deficiency on the uptake and incorporation of nitrogen in ryegrass. *J. Agric. Sci.* (Cambridge), 105:501–504.

Miller, R.O. 1998a. Determination of dry matter content of plant tissue: gravimetric moisture, in Y.P. Kalra, Ed., *Handbook of Reference Methods for Plant Analysis*, CRC Press, Boca Raton, FL, 51–52.

Miller, R.O. 1998b. High temperature oxidation: dry ashing, in Y.P. Kalra, Ed., *Handbook of Reference Methods for Plant Analysis*, CRC Press, Boca Raton, FL, 53–56.

Miller, R.O. 1998c. Microwave digestion of plant tissue in an open vessel, in Y.P. Kalra, Ed., *Handbook of Reference Methods for Plant Analysis*, CRC Press, Boca Raton, FL, 69–73.

Miller, R.O. 1998d. Nitric-perchloric acid wet digestion in an open vessel, in Y.P. Kalra, Ed., *Handbook of Reference Methods for Plant Analysis*, CRC Press, Boca Raton, FL, 57–61.

Miller, R.O. 1998e. Extractable nitrate in plant tissue: ion-selective electrode method, in Y.P. Kalra, Ed., *Handbook of Reference Methods for Plant Analysis*, CRC Press, Boca Raton, FL, 85–88.

Miller, R.O. 1998f. Extractable chloride, nitrate, orthophosphate, potassium, and sulfate-sulfur in plant tissue: 2% acetic acid extraction, in Y.P. Kalra, Ed., *Handbook of Reference Methods for Plant Analysis*, CRC Press, Boca Raton, FL, 115–118.

Mills, H.A. 1980. Nitrogen specific ion electrode for soil, plant, and water analysis. *J. Assoc. Off. Anal. Chem.*, 63:797–801.

Mills, H.A. and J.B. Jones, Jr. 1996. *Plant Nutrition Manual II*, MicroMacro Publishing, Athens, GA.

Miner, G.S. and M.R. Tucker. 1990. Plant analysis as an aid in fertilizing tobacco, in R.L. Westerman, Ed., *Soil Testing and Plant Analysis*, 3rd ed., SSSA Book Series Number 3, Soil Science Society of America, Madison, WI, 645–657.

Mitchell, R.L. 1964. The Spectrochemical Analysis of Soils, Plants and Related Materials, Technical Communication Number 44A, Commonwealth Agricultural Bureaux, Farnham Royal, Bucks, England.

Miyazawa, M., M.A. Paven, and M.F.M. Block. 1984. Determination of Ca, Mg, K, Mn, Cu, Zn, Fe, and P in coffee, soybean, corn, sunflower and pasture grass leaf tissues by HCl extraction method. *Commun. Soil Sci. Plant Anal.*, 15:141–147.

Morries, P. 1983. A century of Kjeldahl (1883–1983). *J. Assoc. Public Anal.*, 21:53–58.

Morrison, G.H. 1979. Elemental trace analysis of biological materials. *Crit. Rev. Chem.*, 8:287–320.

Munsinger, R.A. and R. McKinney. 1982. Modern Kjeldahl systems. *Am. Lab.*, 14:76–79.

Munson, R.D. 1998. Principle of plant analysis, in Y.P. Kalra, Ed., *Handbook of Reference Methods for Plant Analysis*, CRC Press, Boca Raton, FL, 1–24.

Munson, R.D. and W.L. Nelson. 1990. Principles and practices in plant analysis, in R.L. Westerman, Ed., *Soil Testing and Plant Analysis*, 3rd ed., SSSA Book Series Number 3, Soil Science Society of America, Madison, WI, 359–387.

Munter, R.C. and R.A. Grande. 1981. Plant analysis and soil extracts by ICP-atomic emission spectrometry, in R.M. Barnes, Ed., *Developments in Atomic Plasma Spectrochemical Analysis*, Heyden and Son, Ltd., London, England, 653–672.

Munter, R.C., T.L. Halverson, and R.D. Anderson. 1984. Quality assurance for plant tissue analysis by ICP-AES. *Commun. Soil Sci. Plant Anal.*, 15:1285–1322.

Murdrock, A. and O. Murdock. 1977. Analysis of plant material by x-ray fluorescence spectrometry. *X-ray Spectrosc.*, 6:215–217.

Nelson, D.W. and L.L. Sommers. 1980. Total nitrogen analysis of soil and plant tissues. *J. Assoc. Off. Anal. Chem.*, 63:770–778.

Nelson, P.B. and J.W. Boodley. 1965. An error involved in the prepartion of plant tissue for analysis. *Proc. Am. Soc. Hortic. Sci.*, 86:712–716.

Nickolas, D.J.D. 1956. An appraisal of the use of chemical tissue tests for determining the mineral status of crop plants, in T. Wallace, Ed., *Plant Analysis and Fertilizer Problems*, IRHO, Paris, France, 119–139.

Novozamsky, I., R. van Eck, J.J. van der Lee, V.J.G. Houba, and E.J.M. Temminghoff. 1986. Determination of total sulphur and extractable sulphate in plant materials by inductively-coupled plasma atomic emission spectrometry. *Commun. Soil Sci. Plant Anal.*, 17:1147–1157.

Oertli, J.J. 1994. Non-homogeneity of boron distribution in plants and consequences for foliar diagnosis. *Commun. Soil Sci. Plant Anal.*, 25:1133–1147.

Ohki, K. 1975. Lower and upper critical zinc levels in relation to cotton growth and development. *Plant Physiol.*, 35:96–110.

Ohki, K. 1981. Manganese critical levels for soybean growth and physiological processes. *J. Plant Nutr.*, 3:271–284.

Ohki, K. 1984. Zinc nutrition related to critical deficiency and toxicity levels for sorghum. *Agron. J.*, 76:253–256.

Ohki, K. 1987. Critical nutrient levels related to plant growth and some physiological processes. *J. Plant Nutr.*, 10:1583–1590.

Okamoto, K. and K. Fuwa. 1984. Low-contamination digestion bomb method using a Teflon double vessel for biological materials. *Anal. Chem.*, 56:1756–1760.

Parkenson, J.A. and S.E. Allen. 1975. A wet oxidation procedure suitable for the determination of nitrogen and mineral nutrients in biological material. *Commun. Soil Sci. Plant Anal.*, 6:1–11.

Perrott, K.W., B.E. Kerr, M.J. Kear, and M.M. Sutton. 1991. Determination of total sulphur in soil using inductively coupled plasma-atomic emission spectrometry. *Commun. Soil Sci. Plant Anal.*, 22:1477–1487.

Peverill, K.I. 1993. Soil testing and plant analysis in Australia, *Aust. J. Exp. Agric.*, 33:963–971.

Piper, C.S. 1942. *Soil and Plant Analysis*, Hassell Press, Adelaide, Australia.

Pinkerton, A. and P.J. Randall. 1995. Sulfur requirement during early growth of *Trifolium balansae*, *Trifolium substerraneum*, *Medicago murex*, and *Phalaris aquatica. Aust. J. Exp. Stn.*, 35:199–208.

Powers, R.F., D.L. van Gent, and R.F. Townsend. 1981. Ammonia electrode analysis of nitrogen in micro-Kjeldahl digests of forest vegetation. *Commun. Soil Sci. Plant Anal.*, 12:19–30.

Prevot, P. and M. Ollagnier. 1961. Law of the minimum and balanced nutrition, in W. Reuther, Ed., *Plant Analysis and Fertilizer Problems*, Publication Number 8, American Institute of Biological Sciences, Washington, D.C., 257–277.

Pritchard, M.W. and J. Lee. 1984. Simultaneous determinations of boron, phosphorus and sulphur in some biological and soil materials by inductively-coupled plasma emission spectrometry. *Anal. Chim. Acta,* 157:313–326.

Randall, P.J. and K. Spencer. 1980. Sulfur content of plant material: a comparison of methods of oxidation prior to determination. *Commun. Soil Sci. Plant Anal.*, 11:257–266.

Reuter, D.J. and J.B. Robinson, Eds. 1997. *Plant Analysis: An Interpretation Manual,* 2nd ed., CSIRO Publishing, Collingwood, Australia.

Reuter, D.J., J.B. Robinson, K.I. Peverill, and G.H. Price. 1986. Guidelines for collecting, handling, and analyzing plant materials, in D.J. Reuter and J.B. Robinson, Eds., *Plant Analysis: An Interpretation Manual*, Inkata Press, Melbourne, Australia, 11–35.

Reuther, W. 1961. Limitation of plant analysis as a research diagnostic tool, in W. Reuther, Ed., *Plant Analysis and Fertilizer Problems*, Publication Number 8, American Institute of Biological Sciences, Washington, D.C., 443–454.

Righetti, T.L., K.L. Wilder, and G.A. Cummings. 1990. Plant analysis as an aid in fertilizing orchards, in R.L. Westerman, Ed., *Soil Testing and Plant Analysis*, 3rd ed., SSSA Book Series Number 3, Soil Science Society of America, Madison, WI, 563–601.

Sabbe, W.E. and L.J. Zelinski. 1990. Plant analysis as an aid in fertilizing cotton, in R.L. Westerrnan, Ed., *Soil Testing and Plant Analysis*, 3rd ed., SSSA Book Series Number 3, Soil Science Society of America, Madison, WI, 469–493.

Sah, R.N. and R.O. Miller. 1992. Spontaneous reaction for acid dissolution of biological tissue in closed vessels. *Anal. Chem.,* 64:230–233.

Sahrawat, K.L. 1980. A rapid nondigestion method for determination of potassium in plant tissue. *Commun. Soil Sci. Plant Anal.,* 11:753–757.

Sahrawat, K.L. 1987. Determination of calcium, magnesium, zinc, and manganese in plant tissue using a dilute HCl extraction method. *Commun. Soil Sci. Plant Anal.,* 18:947–962.

Sayre, J.D. 1952. Accumulation of Radio Isotopes in Corn Leaves, Ohio Agriculture Experiment Station Bulletin 723.

Sayre, J.D. 1958. The accumulation of boron in margins of corn leaves, in C.B. Lamb, Ed., *Trace Elements*, Academic Press, New York, 245–250.

Scaife, A. and I.G. Burns. 1986. The sulphate-S/total S ratio in plants as an index of their sulphur status. *Plant Soil,* 91:61–71.

Scarf, N. 1988. One hundred years of the Kjeldahl method for nitrogen determination. *Arch. Acker. Pflanzenbau Bodenkd.,* 32:321–332.

Schmitter, B.M. and T. Rihns. 1989. Evaluation of a macrocombustion method for total nitrogen determination in feedstuffs. *J. Agric. Food Chem.,* 37:992–994.

Scott, N.M., M.E. Watson, K.S. Caldwell, and R.H.E. Inkson. 1983. Response of grassland to the application of sulphur at two sites in northeast Scotland. *J. Sci. Food Agric.,* 34:357–361.

Scott, N.M., P.W. Dyson, J. Ross, and G.S. Sharp. 1984. The effect of sulphur on the yield and chemical composition of winter barley. *J. Agric. Sci.* (Cambridge), 103:699–702.

Shear, C.B. and M. Faust. 1980. Nutritional ranges in deciduous tree fruits and nuts. *Hortic. Rev.,* 2:142–163.

Shuman, G.E. and F. Ruazi. 1981. Microwave drying of rangeland forage samples. *J. Range Manage.,* 34:426–428.

Simonne, E.N., J.B. Jones, Jr., H.A. Mills, D.A. Smittle, and C.G. Hussey. 1993. Influence of catalyst, sample weight, and digestion conditions on Kjeldahl N. *Commun. Soil Sci. Plant Anal.*, 24:1609–1616.

Simonne, E.N., H.A. Mills, J.B. Jones, Jr., D.A. Smittle, and C.G. Hussey. 1994. Comparison of analytical methods for nitrogen analysis in plant tissues. *Commun. Soil Sci. Plant Anal.*, 25:943–954.

Smith, F.W. and J.F. Loneragan. 1997. Interpretation of plant analysis: concepts and principles, in D.J. Reuter and J.B. Robinson, Eds., *Plant Analysis: An Interpretation Manual*, 2nd ed., CSIRO Publishing, Melbourne, Australia, 1–33.

Smith, J.H., D.L. Carter, M.J. Brown, and C.L. Parks. 1968. Differences in chemical composition of plant sample fractions resulting from grinding and screening. *Agron. J.,* 60:149–151.

Smith, M. and J.B. Storey. 1976. The influence of washing procedure on surface removal and leaching of certain elements from trees. *Hortic. Sci.,* 14:718–719.

Smith, P.F. 1962. Mineral analysis of plant tissues. *Annu. Rev. Plant Physiol.,* 13:81–108.

Smith, V.R. 1980. A phenol-hypochlorite manual determination of ammonium-nitrogen in Kjeldahl digests of plant tissue. *Commun. Soil Sci. Plant Anal.,* 11:709–722.

Soltanpour, P.N., C.W. Johnson, S.M. Workman, J.B. Jones, Jr., and R.O. Miller. 1998. Advances in ICP emission and ICP mass spectrometry. *Adv. Agron.,* 64:28–113.

Soon, Y.K., Y.P. Kalra, and S.A. Abboud. 1996. Comparison of some methods for the determination of total sulfur in plant tissues. *Commun. Soil Sci. Plant Anal.*, 27:809–818.

Sonneveld, C. and P.A. van Dijk. 1982. The effectiveness of some washing procedures on the removal of contaminants from plant tissue of glasshouse crops. *Commun. Soil Sci. Plant Anal.,* 13:487–496.

Spencer, K., Jr., R. Freney, and M.B. Jones. 1978. Diagnosis of sulphur deficiency in plants, in A.R. Ferguson et al., Eds., *Proceedings 8th International Colloquium Plant Analysis and Fertilizer Problems*, New Zealand Department of Science of International Research, Wellington, New Zealand, 507–513.

Steckel, J.E. and R.L. Flannery. 1971. Simultaneous determinations of phosphorus, potassium, calcium, and magnesium in wet digestion solutions of plant tissue by autoanalyzer, in L.M. Walsh, Ed., *Instrumental Methods for Analysis of Soils and Plant Tissue*, Soil Science Society of America, Madison, WI, 83–96.

Sterrett, S.B., C.B. Smith, M.P. Mascianica, and K.T. Demchak. 1987. Comparison of analytical results from plant analysis laboratories. *Commun. Soil Sci. Plant Anal.*, 18:287–299.

Stevens, R.J. 1985. Evaluation of the sulphur status of some grasses for silage in Northern Ireland. *J. Agric. Sci.* (Cambridge), 105:581–585.

Steyn, W.J.A. 1959. Leaf analysis. Errors involved in the preparative phase. *J. Agric. Food Chem.,* 7:344–348.

Steyn, W.J.A. 1961. The errors involved in the sampling of citrus and pineapple plants for leaf analysis purposes, in W. Reuther, Ed., *Plant Analysis and Fertilizer Problems*, Publication Number 8, American Institute of Biological Sciences, Washington, D.C., 409–430.

Stika, J.W.A. and G.H. Morrison. 1981. Analytical methods for mineral content of human tissues. *Fed. Proc.,* 40:2115–2120.

Stripp, R.A. and D. Bogen. 1989. The rapid decomposition of biological materials by using microwave acid digestion bomb. *J. Anal. Toxicol.,* 13:57–59.

Sung, J.F.C., A.E. Nevussu, and F.B. Dewalle. 1984. Simple digestion of sewage and sludge for multielement analysis. *J. Environ. Sci. Health,* A19:959–972.

Sweeney, R.A. 1989. Generic combustion method for determination of crude protein in feeds: collaborative study. *J. Assoc. Off. Anal. Chem.,* 72:770–774.

Tabatabai, M.A. 1982. Sulfur, in A.L. Page, Ed., *Methods of Soil Analysis,* Part 2, *Chemical and Microbiological Properties*, 2nd ed., Agronomy Monograph Number 9, American Society of Agronomy, Madison, WI, 501–538.

Tabatabai, M.A., H.T. Basta, and N.J. Pirela. 1988. Determination of total sulfur in soils and plant materials by ion chromatography. *Commun. Soil Sci. Plant Anal.*, 19:1701–1714.

Tel, D.A and C. Heseltine. 1990. Chloride analyses of soil leachate, using the *TRAACS 800* analyzer. *Commun. Soil Sci. Plant Anal.,* 21:1689–1693.

Thomas, W. 1945. Foliar diagnosis. *Soil Sci.,* 59:353–374.

Tolg, G. 1974. The basis of trace analysis, in E. Korte, Ed., *Methodium Chimicum,* Vol. 1, *Analytical Methods,* Part B, *Micromethods, Biological Methods, Quality Control, Automation*, Academic Press, New York, 698–710.

Topper, K. and J. Kotuby-Amacher. 1990. Evaluation of a closed vessel acid digestion method for plant analyses using inductively coupled plasma spectrometry. *Commun. Soil Sci. Plant Anal.*, 21:1437–1455.

Tucker, M.R. 1974. A modified heating block for plant tissue digestion. *Commun. Soil Sci. Plant Anal.,* 5:539–546.

Ulrich, A. 1948. Plant analysis — methods and interpretation of results, in H.B. Kitchen, Ed., *Diagnostic Techniques for Crops and Soils,* American Potash Institute, Washington, D.C., 157–198.

Ulrich, A. 1952. A physiological basis for assessing the nutritional requirements of plants. *Ann. Rev. Plant Physiol.,* 3:207–228.

Ulrich, A. and F.J. Hills. 1990. Plant analysis as an aid in fertilizing sugarbeet, in R.L. Westerman, Ed., *Soil Testing and Plant Analysis*, 3rd ed., SSSA Book Series Number 3, Soil Science Society of America, Madison, WI, 429–447.

Ulrich, A., D. Ririe, F.J. Hills, A.G. George, and M.D. Morse. 1959. I. Plant Analysis: A Guide for Sugar Beet Fertilization, California Agricultural Experiment Station Bulletin 766.

Ulrich, J.M. 1984. Pulsing air enhances Wiley mill utility. *Commun. Soil Sci. Plant Anal.,* 15:189–190.

van der Lee, J.J., I. Walinga, P.K. Mangeki, V.J.G. Houba, and I. Novozamsky. 1987. Determination of boron in fresh and dried plant material by plasma emission spectrometry after extraction with HF-HCl. *Commun. Soil Sci. Plant Anal.,* 18:789–802.

Viets, F.G., Jr., L.C. Boawn, and C.L. Crawford. 1954. Zinc contents and deficiency symptoms of 26 crops grown in a zinc deficient soil. *Soil Sci.,* 78:305–316.

Vigler, M.S., A.W. Varnes, and H.A. Strecker. 1980. Sample preparation techniques for AA and ICP spectroscopy. *Anal. Lab.*, 12:21–34.

Wall, L.L., C.W. Gehrke, and J. Suzuki. 1980. An automated turbidimetric method for total sulfur in plant tissue and sulfate sulfur in soils. *Commun. Soil Sci. Plant Anal.,* 11:1087–1103.

Wallace, A., J. Kinnear, J.W. Cha, and E.M. Romney. 1980. Effect of washing procedures on mineral analysis and their cluster analysis for orange leaves. *J. Plant Nutr.,* 2:1–9.

Wallace, A., J. Kinnear, J.W. Cha, and E.M. Romney. 1982. Influence of washing of soybean leaves on identification of iron deficiency by leaf analysis. *J. Plant Nutr.,* 5:805–810.

Walsh, L.M., Ed. 1971. *Instrumental Methods for Analysis of Soils and Plant Tissue*, Soil Science Society of America, Madison, WI.

Watson, M.E. 1998. Analytical instrumentations for determination of elements in plant tissue, in Y.P. Kalra, Ed., *Handbook of Reference Methods for Plant Analysis*, CRC Press, Boca Raton, FL, 137–152.

Watson, M.E. and R.A. Isaac. 1990. Analytical instruments for soil and plant analysis, in R.L. Westerman, Ed., *Soil Testing and Plant Analysis*, 3rd ed., SSSA Book Series Number 3, Soil Science Society of America, Madison, WI, 691–740.

Weetman, G.F. and C.G. Wells. 1990. Plant analysis as an aid in fertilizing forests, in R.L. Westerman, Ed., *Soil Testing and Plant Analysis*, 3rd ed., SSSA Book Series Number 3, Soil Science Society of America, Madison, WI, 659–690.

Westfall, D.G., D.A. Whitney, and D.M. Brandon. 1990. Plant analysis as an aid in fertilizing small grains, in R.L. Westerman, Ed., *Soil Testing and Plant Analysis*, 3rd ed., SSSA Book Series Number 3, Soil Science Society of America, Madison, WI, 495–519.

White, R.T., Jr. and G.E. Douthit. 1985. Use of microwave oven and nitric acid-hydrogen peroxide digestion to prepare botanical materials for elemental analysis by inductively coupled argon plasma emission spectroscopy. *J. Assoc. Off. Anal. Chem.,* 68:766–769.

Wikner, B. 1986. Pretreatment of plant and soil samples: a problem in boron analysis. I. Plants. *Commun Soil Sci. Plant Anal.,* 17:1–25.

Wolf, B. 1982. A comprehensive system of leaf analyses and its use for diagnosing crop nutrient status. *Commun. Soil Sci. Plant Anal.*, 13:1035–1059.

Wright, W.D. and D.R. Wilkinson. 1993. Automated micro-Kjeldahl nitrogen determination: a method. *Am. Environ. Lab.,* 2:30–33.

Zarcinas, B.A., B. Cartwright, and L.P. Spauncer. 1987. Nitric acid digestion and multielement analysis plant material by inductively coupled plasma spectrometry. *Commun. Soil Sci. Plant Anal.,* 18:131–147.

Zasoski, A.J. and A.G. Burau. 1977. A rapid nitric perchloric acid digestion method of multi-element tissue analysis. *Commun. Soil Sci. Plant Anal.,* 3:425–436.

Zhao, F., S.P. McGrath, and A.R. Crosland. 1994. Comparison of three wet digestion methods for the determination of plant sulphur by inductively coupled plasma atomic emission spectroscopy (ICP-AES). *Commun. Soil Sci. Plant Anal.*, 25:407–418.

Chapter 4

Tissue Testing

A. Introduction

The two primary objectives for conducting a tissue test are to (1) identify quickly the nutritional status of the plant for verification of an apparent nutrient element insufficiency, and (2) determine by evaluating the current nutrient-element status of the plant whether additional fertilizer is needed to ensure that the desired yield goal is obtained.

Factors that distinguish a tissue test from a plant analysis are (1) a tissue test is conducted in the field rather than on collected tissue that is sent to a laboratory for analysis, and (2) a tissue test is conducted on extracted sap, whereas a plant analysis is the determination of the total elemental content or determinations are made by extraction on oven-dried, ground plant tissue.

In general, a tissue test is conducted using chemically treated papers or test strips, test tubes or vials, and specially prepared reagents. The development of a color and its intensity are used to identify the presence of an element (normally its ion) and its concentration, respectively, or by a change in color with the addition (by drop count or pipette volume) of a reagent for concentration determination. Unfortunately, kits are not available at this time that can be used to conduct such tests; therefore, users must make their own kit materials. Instructions for preparing the reagents and test papers are given in this chapter. For those interested in seeing how a tissue test kit can be used in the field, a video by Jones (1994) is available. Tissue testing procedures and test kit use are described in the book chapters by Jones (1998) and Jones and Slovacek (1998).

There have been new developments in instrumentation that can also be used to conduct a tissue test, such as handheld, battery-operated specific ion meters for the determination of the NO_3^- anion, and the cations K^+ and Ca^{2+}.

A more recent development has been the use of a chlorophyll meter that measures the *greenness* of plant tissue, readings that can be correlated with the nutrient-element status of the plant, primarily for the assessment of the N status of a plant (Schepers et al., 1992a, b; 1998; Piekielek et al., 1993; Peterson et al., 1993; Blackmer et al., 1993; Blackmer and Schepers, 1995).

B. Testing Kits

The type of tests to be conducted and selected methodology will determine to a considerable degree what procedure will be used to assay the collected tissue; therefore, only general instructions can be specified.

A distinction is made between plant (leaf) analysis and tissue testing, the former being a laboratory analysis of a prepared (dried and ground) plant tissue sample, whereas a tissue test is an assessment of the elemental content of sap from fresh tissue, usually carried out in the field using special test papers, vials, reagents, and color charts.

Krantz et al. (1948) provided instructions for the field testing of corn, cotton, and soybean plants using sap pressed from fresh tissue for the semi-quantitative determination of NO_3, PO_4, and K employing test papers, vials, reagents, and color charts. Wickstrom (1984) has discussed the use of such tissue tests for field diagnosis. Syltie et al. (1972) have given procedural details for conducting tissue tests in the field for the crops corn and soybeans for the elements N, P, K, Mg, and Mn. They give instructions for the preparation of reagents and techniques for conducting the tests. Scaife and Stevens (1983) have found the use of "Merckoquant" test strips suitable for NO_3 determination in the field for assessing the N status of vegetable crops.

Iron is an element that can be determined by a tissue test conducted in the field, a procedure first developed by Bar-Akiva et al. (1978) and modified by Bar-Akiva (1984). Peroxidase activity is measured by floating leaf disks in a reactive solution; development of a blue color indicates adequate Fe in the plant tissue.

The ability to perform tissue tests in the field is considered by some of significant advantage in terms of immediate test results and low cost when compared with that required for a laboratory-conducted plant analysis. It should be remembered that most of the tests themselves are not entirely quantitative, but provide the tester with a qualitative "yes" or "no" evaluation of a crop; that is, the element being evaluated by the tissue test is either present or its presence is not at the desired concentration level. It requires practical experience with these test procedures and repeated observations to feel confident when making an interpretation of a test result. Tissue tests and their interpretation are considered by some as an "art" rather than a strict quantitative analytical science. However, it should be remembered that the test procedures themselves are based on sound analytical chemistry. It is their utilization and interpretation that require skill gained only by repeated practical experience.

Combining field observations with tests of soil and plant tissue by means of quick tests (field conducted) has been coined "The Diagnostic Approach," a procedure of observation, testing, and evaluation discussed in some detail in a special issue of *Better Crops* (Armstrong et al., 1984).

C. Preparation of Reagents for Conducting Tissue Tests Using Filter Paper

Reagent Preparation

Nitrate–Nitrogen (NO_3–N)

Reagent Nitrate Powder:
100 g dry barium sulfate ($BaSO_4$)
10 g manganese sulfate ($MnSO_4 \cdot H_2O$)
2 g finely powdered Zn
75 g citric acid
4 g sulfanic acid
2 g of ∂-naphthylamine
Finely grind as separate portions with a mortar and pestle, then thoroughly mix and store in a blackened container.
Reaction:
Any degree of red color produced on reaction with plant sap indicates the presence of NO_3.

Phosphorus (P)

Reagent A:
10 g ammonium molybdate [$(NH_4)_6Mo_7O_{24} \cdot 4H_2O$] dissolved in 85 mL water
Reagent B:
Mix 16 mL water with 170 mL concentrated hydrochloric acid (HCl).
Concentrated Solution:
Mix Solutions A and B and add 2 g boric acid (H_3BO_3) per 50 mL of the mixed solution.
Reagent C:
Dilute the concentrated solution 10 times with water.

Reductant Suspension:

Place a small amount of tin chloride ($SnCl_2{\cdot}2H_2O$) in a small dropping bottle and add water. Shake vigorously to mix and just prior to use.

Potassium (K)

Solution A:

Add 0.6 g dipicrylamine (2,2′,4,4′,6,6′-hexanitrodiphenylamine) and 0.6 g sodium carbonate (Na_2CO_3) to 25 mL water and boil for 10 min.

Solution B:

Dilute 8 mL of solution A to 25 mL with water.

Solution C:

Dilute 10 mL of solution B to 15 mL with water.

Preparation of Filter Paper

Place three separate 8-mm-diameter spots, one from each solution, A, B, and C, on a filter paper and allow to dry.

D. Sampling Techniques

To conduct most tissue tests successfully, a sufficient quantity of cell sap must be obtained to conduct the test. What are commonly selected are conductive tissues, such as leaf petioles, leaf midribs, or the plant stalk itself. It is from the recently mature leaves that the petiole or midrib tissues are collected. When the plant stalk is the test tissue, the stalk section at the base of the plant or the midsection is the portion of the stalk selected.

The time of sampling is determined by the purpose for the tissue test. For diagnostic evaluation — when dealing with a suspected nutrient-element insufficiency — the time would be when the first symptoms of stress are visually evident. For determining nutrient-element status — when the need for supplemental fertilization is to be determined — the time of sampling is based on a specific development period in the life cycle of the plant.

Here are some general instructions to be followed when collecting plant tissue for testing:

- Collect tissue between 8:00 A.M. and 5:00 P.M.
- Do not collect tissue immediately after a rain.
- Collect tissue from a range of plants, young plants to those near maturity.
- Do not collect tissue from plants during drought or when the plants are under some stress condition.

E. Testing Procedures

The type of tests to be conducted and the methods selected will determine to a considerable degree what procedure will be used to assay the collected tissue; therefore, only general instructions can be specified.

1. General Test Procedures for Paper and Vial-Type Kits

From a collected petiole or leaf midrib, using pliers, an aliquot of sap is squeezed onto the test paper in the area marked "Phosphorus," and then additional aliquots of sap are squeezed onto each of the three orange spots below "Potassium."

a. Nitrate–nitrogen test

A short section of midrib or petiole is placed across the end of the test paper marked "Nitrate–Nitrogen," a small aliquot of Nitrate Powder sprinkled along the piece of tissue, the paper folded over, and the area squeezed with the pliers, squeezing until sufficient to wet that area on the test paper with sap. If NO_3 is present, the powder will turn red, with the speed of color development and its intensity (pink–low; red–high) an indication of the concentration of NO_3 present.

Reading
No color or white — Very low
Pink — Low
Light red — Medium
Cherry red — High

b. Phosphorus test

On the sap-wetted spot on the test paper designated "Phosphorus," two drops of *Phosphorus Reagent C* are placed followed by two drops of *Phosphorus Reductant Suspension*. The appearance of a blue color, its speed of development, and intensity (pale blue–low; dark blue–high) indicate the presence of P and its probable concentration, respectively.

Reading
No color — Very low
Light blue — Low
Medium blue — Medium
Intense blue — High

c. Potassium test

Using *Phosphorus Reagent C*, place two drops on each of the three potassium orange dots on the test paper. If an orange precipitate remains, this indicates the presence of K. Of the three tests possible with this kit, the K test is the only defined quantitative test.

Reading

Orange left on sapspots, all 3 dots — High
Orange left on Medium and Low sapspots — Medium
Orange left on Low sapspot — Low
No orange color left — Very low

d. Nitrate–nitrogen stalk test

For a crop plant such as corn, cut a 3- to 4-in. section of stalk from the base of the plant, cut the stalk section in half, and then place some Nitrate Powder on the open cut. Put the two halves together, moving them in a way to mix the powder with the exposed stalk cut. After about 1 min, open the two halves. If NO_3 is present, the intensity of the red color (pink–low; red–high) is a measure of its probable concentration.

e. Phosphorus stalk test

For a crop plant such as corn, cut a 3- to 4-in. section of stalk at the base. Cut the stalk section in half, and place two to three drops of Phosphorus Reagent C on the open cut, followed by two to three drops of Phosphorus Reductant Suspension. Put the two halves together, moving them in a way to mix the added reagents with the exposed stalk cut. In about a minute, open the two halves. If P is present at a minimum concentration, a blue color will indicate its presence and the intensity of the blue color (pale blue–low; dark blue–high) will indicate its probable concentration.

f. Other tests

Similar tests can be conducted using vials, other types of test papers, and electronic meters. Some of these test procedures have an advantage since they are quantitative, a requirement that is essential when a supplemental fertilizer application may be made based on the concentration found at a specified period in the growth of the plant. Two examples would be the determination of the NO_3–N content at the base of the wheat plant stem, and the petiole NO_3–N content in a specifically selected cotton plant petiole. The NO_3–N content found would then determine how much additional N fertilizer means to be applied.

F. Methods of Interpretation

Interpretation of a tissue test result can be difficult for the inexperienced. Therefore, it is important to gain some practical experience in the field — testing plants in various stages of development and nutrient-element stress — before venturing forth under critical situations when a test result will form the basis for corrective action. An example of how tissue tests can be used to determine N and P fertilization needs is described by Beverly (1994).

With most tissue procedures and instruments, instructions are provided with the kit along with some interpretation information based on obtained test results. Before this interpretative information is used, some verification may be required based on trial tests by the user, as was stated above.

1. When and How to Use Tissue Tests

The procedures given below are applicable to any tissue testing procedure, no matter what test kit or device is used to conduct the test.

a. Use tissue test results along with all other available information — soil tests, past history, visual observations, current fertilizer use, etc. — to determine adequacy or inadequacy of nutrient element supplies.
b. Look for the one factor that is most limiting plant growth. Be careful — it may not be N, P, or K.
c. Use tissue tests to increase knowledge of plant nutrition.
d. Remember that the plant is a dynamic biological system, and that the nutrient elements (particularly NO_3–N and K) can be present in adequate amounts today only to be short a month from now because the soil could not supply them fast enough.

Sampling and Interpretation Chart

Plant test		Part to sample	Minimum level to avoid hidden hunger
		Corn	
Under 15 in.	NO_3	Midrib, basal leaf	High
	PO_4	Midrib, basal leaf	Medium
	K	Midrib, basal leaf	High

(continued)

Sampling and Interpretation Chart (continued)

Plant test		Part to sample	Minimum level to avoid hidden hunger
15 in. to ear showing	NO_3	Base of stalk	High
	PO_4	Midrib, 1st mature leaf[a]	Medium
	K	Midrib, 1st mature leaf	High
Ear to very early dent	NO_3	Base of stalk	
	PO_4	Midrib, leaf below ear	Medium
	K	Midrib, leaf below ear	Medium
		Soybean	
Early growth to midseason	NO_3	Not tested	
	PO_4	Pulvinus (swollen base of	High
	K	petiole), 1st mature leaf	High
Midseason to good pod development	PO_4	Pulvinus, 1st mature leaf	Medium
	K	Pulvinus, 1st mature leaf	Medium
		Cotton	
Early bloom	NO_3	Petiole, basal leaf	High
	PO_4	Petiole, basal leaf	High
	K	Petiole, basal leaf	High
Bloom to boll set	NO_3	Petiole, 1st mature leaf	High
	PO_4	Petiole, 1st mature leaf	Medium
	K	Petiole, 1st mature leaf	High
Boll set to early maturity	NO_3	Petiole, 1st mature leaf	Medium
	PO_4	Petiole, 1st mature leaf	Medium
	K	Petiole, 1st mature leaf	Medium
		Alfalfa	
Before 1st cutting	PO_4	Middle section of stem	High
	K	Middle section of stem	High
		Small Grains	
Shoot stage to milk stage	NO_3	Lower stem	High
	PO_4	Lower stem	Medium
	K	Lower stem	Medium

[a] 1st Mature Leaf — Avoid the young leaves at the top of the plant. Take the first fully matured but recently formed leaf as one goes down the plant.

Figure 4.1
Cardy Nitrate Specific Ion Meter. (Courtesy of Spectrum Technologies, Inc., Plainfield, IL.)

G. Use of a Specific-Ion Nitrate Meter

A relatively quick method for determining the NO_3–N level in petiole cell sap is with the use of a specific-ion meter, one type being the Cardy meter as shown in the Figure 4.1. The procedure for determining NO_3–N is as follows:

- Collect a representative sample of leaf or petiole tissue.
- Using a sap press (garlic press), squeeze an aliquot of sap onto a clean smooth plastic surface.
- Transfer an aliquot of the sap directly onto the meter sensor and read the NO_3–N concentration.

Using a reference source relating NO_3–N content with N plant status, compare the meter reading obtained with the reference to determine if the concentration found is within the sufficiency range for the plant part being tested, type of crop, and stage of crop development (see Chapter 3, Section I.2; Ludwick, 1990).

Although the NO_3–N meter is the most commonly available in use today, rapid developments are being made with other types of meters that can be used for the determination of other ions, such as K, Na, and Ca.

H. Sources for Kits and Instruments

The major suppliers of test kits for conducting tissue tests and other similar types of assays are as follows:

- HACH Chemical Company, P.O. Box 389, Loveland, CO 80539
- LaMotte Chemical Company, P.O. Box 329, Chestertown, MD 21620
- Spectrum Technologies, 23839 West Andrew Road, Plainfield, IL 60544

For specific instruments used in tissue testing, such as the Cardy Ion Meters and chlorphyll meters, contact:

- Spectrum Technologies, 23839 West Andrew Road, Plainfield, IL 60544

References

Armstrong, D., Ed. 1984. *The Diagnostic Approach to Maximum Economic Yield,* Potash & Phosphate Institute, Norcross, GA.

Bar-Akiva, A. 1984. Substitutes for benzidine as N-donors in the peroxidase assay for rapid diagnosis of iron in plants. *Commun. Soil Sci. Plant Anal.,* 15:929–934.

Bar-Akiva, A., D.N. Maynard, and J.E. English. 1978. A rapid tissue test for diagnosis of iron deficiencies in vegetable crops. *HortSciences,* 13:284–285.

Beverly, R.B. 1994. Stem sap testing as a real-time guide to tomato seedling nitrogen and potassium fertilization. *Commun. Soil Sci. Plant Anal.,* 25:1045–1056.

Blackmer, T.M. and J.S. Schepers. 1995. Use of a chlorophyll meter to monitor N status and schedule fertigation of corn. *J. Prod. Agric.,* 8:5–60.

Blackmer, T.M., J.S. Schepers, and M.F. Vigil. 1993. Chlorophyll meter readings in corn as affected by plant spacing. *Commun. Soil Sci. Plant Anal.,* 24:2507–2516.

Jones, J.B., Jr. 1994. *Tissue Testing* (VHS video). On Sat One, Anderson, SC.

Jones, J.B., Jr. 1998. Tissue Testing, in *Plant Nutrition Manual,* St. Lucie Press, Boca Raton, FL, 5.1–5.9.

Jones, J.B., Jr. and D. Slovacek. 1998. Tissue testing kits and procedures for nutrient element assessment in plant tissue, in Y.P. Kalra, Ed., *Handbook of Reference Methods for Plant Analysis*, CRC Press, Boca Raton, FL, 119–128.

Krantz, B.A., W.L. Nelson, and L.F. Burkhart. 1948. Plant-tissue tests as a tool in agronomic research, in H.B. Kitchen, Ed., *Diagnostic Techniques for Soils and Crops*, American Potash Institute, Washington, D.C., 137–156.

Ludwick, A.E., Ed. 1990. *Western Fertilizer Handbook,* Horticultural ed., Interstate Publishers, Danville, IL.

Peterson, T.A., T.M. Blackmer, D.D. Francis, and J.S. Schepers. 1993. Using a Chlorophyll Meter to Improve N Management, Nebguide G93–1171A, Cooperative Extension Service, University of Nebraska, Lincoln.

Piekielek, W.P. and R.H. Fox. 1992. Use of a chlorophyll meter to predict sidedress nitrogen requirements for maize. *Agron. J.,* 84:59–65.

Scaife, A. and K.L. Stevens. 1983. Monitoring sap nitrate in vegetable crops: comparison of test strips with electrode methods and effects of time of day and leaf position. *Commun. Soil Sci. Plant Anal.,* 14:761–771.

Schepers, J.S., D.D. Francis, M. Vigil, and F.E. Below. 1992a. Comparison of corn leaf nitrogen concentration and chlorophyll meter readings. *Commun. Soil Plant Anal.,* 23:2173–2187.

Schepers, J.S., T.M. Blackmer, and D.D. Francis. 1992b. Predicting N fertilizer needs for corn in humid regions: using chlorophyll meters, in D.R. Dock and K.R. Kelley, Eds., Predicting Fertilizer Needs for Corn in Humid Regions, Bulletin Y-226, National Fertilizer and Environmental Research Center, Tennessee Valley Authority, Muscle Shoals, AL, 105–114.

Schepers, J.S., T.M. Blackmer, and D.D. Francis. 1998. Chlorophyll meter method for estimating nitrogen content in plant tissue, in Y.P. Kalra, Ed., *Handbook of Reference Methods for Plant Analysis*, CRC Press, Boca Raton, FL, 129–135.

Syltie, P.W., S.W. Melsted, and W.M. Walker. 1972. Rapid tissue tests as indication of yield, plant composition, and fertility of corn and soybeans. *Commun. Soil Sci. Plant Anal.,* 3:37–49.

Wickstrom, G.A. 1967. Use of tissue testing in field diagnosis, in G.W. Hardy et al., Eds., *Soil Testing and Plant Analyis*, Part I, SSSA Special Publication Number 2, Soil Science Society of America, Madison, WI, 109–112.

Chapter 5

Principles of Instrumental Analysis

A. Introduction

The analyst today has a wide range of analytical instrumentation from which to choose; the factors that affect the choice are described by McLaughlin et al. (1979) and van Loon (1985). Hislop (1980) has developed analytical criteria for selection of an analysis technique based on accuracy, precision, limit of detection, elemental coverage, single or multielement, and determined chemical form. More recently, Sturgeon (2000) states that sample analysis is characterized by the interplay of at least three domains of activity: sample preparation, sample introduction, and instrumentation. Additional constituents exerting an effect are calibration, data reduction and analysis, and effective quality assurance/quality control parameters. The more practical considerations of instrument availability, cost, and sample form and quantity may become the governing factors rather than basing the choice on the criteria given by Hislop (1980). Analysts themselves, their skill and experience, can also be factors when selecting a particular instrument when more than one is available. The issues of accuracy and precision are significant considerations. Accuracy is the ability to obtain the "true" value and is dependent to a large degree on the availability and use of reliable standards. Precision, on the other hand, is a measure of the degree of variability of an obtained result determined by repeated analyses of the same sample through all the steps from sample preparation to the final obtained result. Horwitz (1982) also has evaluated various analytical procedures by assigning levels of performance based on his years of use in the determination of various elements and substances in various materials. Further consideration of these important factors is beyond the scope of this laboratory guide; the above references are provided to alert the reader to the factors that should be considered when selecting an analytical procedure.

Those in search of a suitable analytical procedure may find the articles by Morrison (1979) and Stika and Morrison (1981) helpful as they compare the relative sensitivity and precision of various analytical procedures. In those instances in which a nutrient element concentration in either a soil extract or plant tissue digest is high, detection limit considerations are of less significance than those factors that affect precision. In addition, with large varying concentrations of elements in an analyte — typical of what occurs in most soil extracts and plant tissue digests — interelement effects can be significant, which may either eliminate or specify a particular method of analysis, and/or require separation of a determined element from the matrix, or require the use of a matrix modifier.

A review of analytical procedures for the assay of soil extracts and plant tissue digests is given in the book edited by Walsh (1971), the MAFF/ADAS book (1986), the book edited by Carter (1993), and the book chapter written by Watson and Isaac (1990); for soil analysis in the book edited by Smith (1991); and for plant tissue digest assay in the book chapter written by Watson (1998). Two earlier treatises on soil and plant analysis procedures by Piper (1942) and Jackson (1958) provide useful analytical method procedures that have relevance today.

Advances in analytical chemistry and instrumental analytical chemistry in the past 2 decades have significantly improved, making elemental determinations in soil extracts and plant tissue digests easier, faster, and less challenging for the analyst. For most of the elemental analysis procedures, the more traditional wet chemistry procedures have been replaced by various analytical instrumental procedures that employ either emission or absorption spectrophotometry.

The elemental concentration in prepared soil extracts and plant digests can be determined by a number of instrumental analytical procedures, with some elements determinable by more than one technique as is shown in the following table:

Instrumental Method of Analysis

Element	UV-VIS	Emission			Atomic	Specific-ion
	Spectrophotometry	Flame	Spark	ICP	absorption	electrode
Boron (B)	good	na	good	excel	na	na
Calcium (Ca)	good	fair	good	excel	excel	poor
Copper (Cu)	good	na	good	excel	excel	na
Iron (Fe)	fair	na	good	excel	excel	na
Potassium (K)	poor[a]	excel	excel	excel	good	na

Instrumental Method of Analysis (continued)

Element	UV-VIS Spectrophotometry	Emission Flame	Emission Spark	Emission ICP	Atomic absorption	Specific-ion electrode
Magnesium (Mg)	fair	fair	excel	excel	excel	na
Manganese (Mn)	good	na	excel	excel	excel	na
Molybdenum (Mo)	good	na	poor	good	good[b]	na
Sodium (Na)	na	excel	excel	excel	good	na
Phosphorus (P)	excel	na	excel	excel	na	na
Zinc (Zn)	good	na	excel	excel	excel	na
Ammonium (NH_4)	good	na	na	na	na	good
Chloride (Cl)	good	na	na	na	na	good
Fluoride (F)	na	na	na	na	na	good
Nitrate (NO_3)	excel	na	na	na	na	good
Sulfate (SO_4)	good[a]	na	na	na	na	na

Key: na = not applicable; excel = excellent (high sensitivity with minimal interference); good = moderate sensitivity with some interference; fair = reasonable sensitivity but with matrix effects; poor = reasonable sensitivity with significant matrix effects.

[a] Turbidity.

[b] Flameless AA.

In the above table, all but the specific-ion electrode procedure involve some form of spectrophotometry, the utilization of a specific wavelength of light and its intensity to determine elemental concentration. The trend today is toward greater analytical sophistication, to multielement computer-controlled analytical instrumentation, resulting in a lessened understanding of the analytical principles involved, and "black box" concepts of instrument calibration, maintenance, and operation. Technicians are less knowledgeable about the analytical procedures they are using and more concerned about which button to push to carry an analysis forward.

Preparation of the analyte and suitable adaptation for a method of analysis require an understanding, by the analyst, of the principle of the method as well as its requirements and limitations. Adequate testing is usually required before putting the method to use, following procedures such as those that have been adapted by the Association of Official Analytical Chemists (AOAC) (McLain, 1982).

It is not possible to cover in adequate detail every method of analysis suited for the elemental concentration determination in soil extracts and plant tissue digests. Therefore, the objective here is to provide sufficient background information to guide the reader.

B. UV-VIS Spectrophotometry (Colorimetry)

UV-VIS spectrophotometry (in the past referred to as colorimetry) has had a long history of application and use for elemental determination in soil extracts and plant tissue digests (Piper, 1942; Jackson, 1958; Chapman and Pratt, 1982). Paul (1998) states that UV-VIS instruments continue to evolve. The technique has good selectivity and sensitivity for many of the elements (or ions) found in soils and plants, such as:

Boron (B)	Manganese (Mn)
Copper (Cu)	Molybdenum (Mo)
Iron (Fe)	Phosphorus (P)
Magnesium (Mg)	Zinc (Zn)

In addition, the two ionic forms of N, NH_4 and NO_3, are determinable by UV-VIS spectrophotometry.

A UV-VIS spectrophotometer is a relatively inexpensive instrument and easy to use. The main components in a UV-VIS spectrophotometer are a light source, a means of obtaining a monochromatic beam of light, a sample holder or cell, and a detector. The more sophisticated spectrophotometers employ either a prism or a grating in lieu of the interference filter to obtain a specific wavelength (monochromatic) of light, thereby improving their performance.

UV-VIS spectrophotometry is based on the principle of light absorption by a complex, with the amount absorbed (or the percentage of light transmitted) correlated with the presence and concentration of a particular element or ion in solution. This relationship is defined by Beer's law:

$$\log_{10}(\mathrm{PO/P}) = \mathrm{abs} \quad \text{or} \quad A = abc$$

where transmittance [$\log_{10}(\mathrm{PO/P})$], or absorbance (A), is a function of the constant (a) specific to the substance, the thickness (b), and concentration (c) of the relative number of colored ions or molecules in the light path. According to Beer's law, there is a linear relationship between absorbance (A) and concentration (c) when monochromatic light is used.

Determining the spectral properties of a formed complex, the wavelength of maximum absorption with the least interference is normally chosen. For example, the wavelength for the determination of P in soil extracts by the molybdenum-blue procedure is 882 nm (Knudsen and Beegle, 1988; Rodriquez et al., 1994; Kuo, 1996; Frank et al., 1998). Care is needed to ensure that the sample solution is of the proper pH and that interfering substances are absent or that compensation is made for their presence. For some elements and solutions, separation of the element(s) of interest may be needed to ensure removal of interfering substances and/or to concentrate the element of interest prior to analysis. Such procedures may be required when determining a micronutrient or trace element in the presence of high concentrations of other elements.

Turbidimetric spectrophotometry is an analytical technique in which the complex formed is a precipitate rather than a true solution, with the spectrophotometer used to determine the density of the suspended precipitate in the solution. Considerable care is required to ensure that a uniform particle-size precipitate is formed, which will remain in suspension during the measurement of absorption. The two elements that can be determined by this analyical technique are K and S, as the SO_4^{2-} anion. The same general conditions must be met regarding the properties of the test solution in terms of its pH and elemental composition in order for the precipitate to be successfully formed. The only turbidimetric procedure used to any great extent is the determination of SO_4–S (Schulte and Eik, 1988; Singh et al., 1995; Jones, 1996; Combs et al., 1998).

It has been the automation of colorimetric and turbidimetric procedures that has kept these techniques in use today; first, the invention of the Technicon AutoAnalyzer (Coakley, 1981; Smith and Scott, 1991) and, more recently, injection flow analyzers (Ranger, 1981; Ruzieka and Hansen, 1988; Smith and Scott, 1991), has kept spectrophotometry an important analytical procedure. In the late 1950s, Technicon Corporation introduced the AutoAnalyzer (registered trade name), a continuous-flow system of analysis suitable for a wide range of analytical applications for the assay of solutions. The AutoAnalyzer was quickly adapted for use in soil testing (Isaac and Jones, 1970; Flannery and Markus, 1972; 1980; Tel and Heseltine, 1990a, b; Smith and Scott, 1991) and plant analysis (Isaac and Jones, 1970; Steckel and Flannery, 1971; Flannery and Markus, 1980) laboratories. Its rapid acceptance, not only for this application, but for a wide range of other analytical applications (Knopp and McKee, 1983), is based on its excellent ability to perform repetitive analytical processes with minimal technical assistance at relatively good speed and excellent control, producing high-quality analytical results.

The principle of operation is that of a continuous-flow system, with standard solutions and the unknown analyte introduced intermittently into a flowing stream of reagents, mixed together by constant inversion through glass mixing coils and segmented by air bubbles equally spaced from one another. These air bubbles establish and maintain sample integrity, promote mixing of reagent and sample streams, and provide a visual check of the stream-flow characteristics for monitoring the behavior of the system.

The AutoAnalyzer is a train of interconnected modules, with its flowing stream directed through tubing from module to module. As a continuous flow system, reference and sample solutions enter the analytical stream and are segmented into discrete liquid increments, or slugs. As the slug is circulated through the system, reagents are added and physical manipulations (mixing, extractions, etc.) continuously take place. As a sequential system also, each event takes place one step at a time with each slug treated as an individual sample.

Although the AutoAnalyzer has been a revolutionary development in analytical chemistry and its use has standardized many colorimetric procedures (Smith and Scott, 1991), autoanalyzers are relatively slow and quite wasteful of reagents, and, in addition, the concentration range of detection is usually limited to 2 decades. By comparison, flow-injection analysis (Ranger, 1981; Ruzieka and Hansen, 1988; Smith and Scott, 1991) instruments are rapid, low-volume systems in which the analytical measurement is not always made at the equilibrium point of the reaction. By using micropump tubes and rapidly moving solutions, response time is in seconds and reagent use minimal, and the concentration range of detection can be greater than 2 decades.

C. Emission Spectrophotometry

This technique of analysis is based on the property of excited atoms deficient in shell electrons that release absorbed energy in the form of radiation when electrons are captured to give a ground-state atom (all electron shells are complete). This phenomenon can be demonstrated by the simple experiment of placing a platimum wire, after it is dipped into a solution of sodium chloride (NaCl), into a gas flame. A bright yellow color is seen, the result of emitted light energy emanating from the Na atoms due to shell electron movement. The wavelength of radiation emitted is characteristic of the element, and the intensity of the emitted radiation is correlated with concentration. Wavelength tables giving principal emission lines by element can be found in most handbooks of chemistry and/or physics (Dean, 1973).

Emission spectrophotometry has had a long history of application and use in the agricultural field (Mitchell, 1964). Prior to the 1950s, elemental excitation was by either AC or DC arc discharge, and the emission was recorded with the use of spectrographs, with the emitted radiation from elemental excitation recorded on a photographic plate. Since the 1950s, there has been a "revolution" in instrumention development and analytical technique. Today, high-energy plasma excitation and direct-reading polychromators make rapid, high-sensitivity, and high-precision analyses possible (Hieftje et al., 1976; Lajunen, 1992).

An emission spectrometer consists of an excitation source, a means of introducing the sample into the excitation source, a spectrometer, and a detector. With the excitation source either an AC or DC arc, a powdered form of the sample is placed into a crater of an electrode (usually made of graphite) and then an arc struck between the sample-containing electrode and another electrode.

For solution samples, an AC spark rotating disk electrode excitation source was devised and applied for the analysis of plant tissue digests (Jones and Warner, 1969). This excitation technique is not well suited for soil extracts because of insufficient sensitivity.

Whether an arc or spark discharge, the emitted radiation is focused through a slit into a spectrometer that divides it into discrete fines after passage through a prism or reflection from a grating. The wavelength of emitted radiation identifies the element, and its intensity, the elemental concentration.

1. Flame Emission Spectrophotometry

The first development in emission spectrophotometry that had a significant impact on soil testing and plant analysis was the flame photometer (Mavrodineanu, 1970; Isaac and Kerber, 1971; Ure, 1991; Wright and Stuczynski, 1996; Horneck and Hanson, 1998), which first appeared in the 1950s, making K and Na easy determinations. The flame photometer is a spectrophotometer in which the light source is replaced by a flame with the analyte (soil extract or plant tissue digest) carried by an air stream into the flame by means of an aerosol generated by a cross-flow nebulizer. The flame can be either mixtures of acetylene (C_2H_2) with air, oxygen (O_2), or nitrous oxide (N_2O), or agron (Ar) and hydrogen (H) gases in entrained air, with C_2H_2 and air the more frequently used.

The alkaline earth elements Ca and Mg can be determined by flame emission, but with difficulty. The presence of other ions (elements) in solution cause interference that must be either removed or compensated for if accurate determinations are to be made. The reason K and Na are so easily determined by flame excitation is in large part due to their relatively low excitation potentials; that is, there is substantial shell electron movement in the relatively low temperature (1000 to 1500 K) flame. By contrast, the extent of electron movement for the elements Ca and Mg in the same flame is considerably less because of their higher excitation potentials. Therefore, relatively less radiant energy is produced.

Another factor that limits the flame emission techniques is self-absorption. Atoms of the same element that are not in an excited state will absorb the emitted radiation when an atom of the same element is going from the excited to the ground state. Since in the excitation source the number of atoms in the ground state far exceeds those in the excited state for most elements, absorption by these ground state atoms significantly reduces the emission to be detected and measured. However, it is this property of self-absorption that led to the development of atomic absorption spectrophotometry, an analytical technique that will be discussed later.

For biological samples, the use of an internal standard, usually Li, is highly recommended. Multielement capability is possible, although most commercial instruments have but one or two detectors with or without an internal standard channel. Pickett and Koirtyohann (1969) have compared the detection limits for 62 elements determinable by flame emission and atomic absorption spectrophotometry.

2. Atomic Absorption Spectrophotometry

Atomic absorption spectrophotometry (frequently referred to as AAS) was introduced in the early 1960s, and at the time, it revolutionized elemental analytical chemistry. With AAS, the analyst was able easily to determine the elements Ca, Mg, Cu, Fe, Mn, and Zn (Hanlon, 1998) whose alternative assay methods were frequently difficult and tedious.

The principle of operation is the reverse of that for emission in that absorption by ground-state atoms of emitted radiation from the same element is utilized. Therefore, a source of radiant energy of the element of interest must be provided, as well as a means of bringing into the emitted radiation source the sample containing the same element whose concentration is to be determined.

The radiant energy source is provided by means of a hollow cathode lamp in which the cathode is made of the element to be determined. The sample in liquid form containing the element of interest is carried into a flame, commonly an acetylene (C_2H_2)–oxygen (O_2) flame, as an aerosol in the same manner as for analysis by flame emission spectrophotometry. The radiant energy from the hollow cathode lamp is passed through the flame and the reduction in emitted radiation (as a result of absorption by the same element as the cathode present in the analyte) determined. The degree of reduction is correlated with the concentration of the element in the analyte.

The essential components of an AAS spectrophotometer are the hollow cathode lamp, cross-flow nebulizer, and spray chamber, burner, monochromator, and detector.

Only one element at a time can be determined since most AAS spectrophotometers have a single analytical channel. However, this is not to be confused with AAS spectrophotometers that are single- or dual-beam instruments, with one beam the analytical channel and the other the reference or background channel. There is some setup time required between elemental sets as the proper hollow cathode lamp must be put into place, the system optically aligned and calibrated. Most AAS spectrophotometers have some type of processor control to assist the analyst in calibrating the spectrophotometer. However, compared with other techniques that have polychromators, analysis by AAS is relatively slow and tedious, although the relative performance in terms of accuracy and precision is comparable.

Flame AAS is not totally interference free, requiring the use of "releasing agents" and the addition of an excess of a "competing ion" to minimize interference effects. Lanthanum is the most common releasing agent, and K can be added as the competing ion. The addition of chelating agents, such as 8-hydroxyquinoline or the Na salt of ethylenediaminetetraacetic acid (EDTA), is also effective in reducing interfering effects.

Operating conditions and settings for an AAS are frequently provided by the manufacturer, and operating procedures for the assay of soil extracts and plant tissue digests are given in a number of reference sources (Willis, 1970; Isaac and Kerber, 1971; Buttgereit, 1974; Isaac, 1980; Baker and Shur, 1982; Ure, 1991; Wright and Stucsznski, 1996; Hanlon, 1998).

The common operating mode flame AAS is the introduction of the analyte into the flame as a water aerosol, but for those elements that form volatile hydrides, such as As, Bi, Pb, Se, and Sn, the hydride itself can be introduced directly into the flame (Thompson et al., 1978; Soon, 1998). The advantages are reduced interference and increased sensitivity at the nanogram level (Ure, 1991).

Flameless AAS by gaseous absorption of vaporized atoms from samples placed on a electrothermally heated tantalum boat (Hwang et al., 1972) or carbon rod (Sturgeon and Chakrabarti, 1978) increases detection limits substantially over flame AAS techniques (Jenke and Woodriff, 1982). There are various designs for the graphite furnace (carbon rod technique); the Massmann design (Massmann, 1968) with the L'vov platform (Routh et al., 1982) is considered the best. Techniques for the analysis of biological samples using the graphite furnance have been well described (Routh et al., 1982; Hinderberger et al., 1982; Gupta, 1998). Although the flameless ASS technique offers excellent sensitivity, sample preparation and handling as well as matrix interference can present difficult problems to the analyst.

Although flame and flameless AAS are very useful analytical techniques, AAS instruments are single-element analyzers not well suited for handling multielement assays. Flameless AAS is best for an elemental assay when high sensitivity is required for small-sized samples.

3. Inductively Coupled Plasma Emission Spectrometry

a. Introduction

A whole new era for emission spectrometry came into existence with the development and commercialization of plasmas as an excitation source (Fassel and Kniseley, 1974), providing a powerful multielement analysis technique with excellent sensitivity and emission stability. Although DC plasma was first introduced in the early 1970s, it is the inductively coupled plasma, frequently referred to as ICP, ICAP, or ICP-AES, that has been the more useful of the two sources. The very high temperature of the plasma (8000 to 10,000 K) results in high emissions, minimizing the effects due to self-absorption and other interference. Calibration curves are usually linear over several decades of concentration making for easy calibration using a two-point technique. Direct-reading polychromators and sequentially operated spectrometers with complete or partial computer control provide the analyst with a variety of operating options.

Most elements found in soil extracts and plant tissue digests, whether in macro-, micro-, or trace concentrations, can be easily determined by the ICP-AES technique. As with flame emission, the liquid sample is introduced as an aerosol into the plasma torch whose design and operating principles have been described by Scott et al. (1976). Samples with divergent elemental concentrations can be easily handled without the need to dilute or concentrate. The prominent ICP-AES emission lines for 70 elements have been identified (Winge et al., 1979) along with detection limits (Wolnik et al., 1981). Details for the use of ICP-AES for the analysis of a range of biological materials have been described by a number of authors (Jones, 1977; Dalquist and Knoll, 1978; Munter and Grande, 1981; Sharp, 1991; Soltanpour, 1991; Soltanpour et al., 1996; 1998; Isaac and Johnson, 1998). Petterelli (1995) has evaluated the axial viewing of the plasma for determinations of trace metals.

In addition to the ICP-AES, there is the DC plasma jet, an entirely different source that suffers from a general lack of high sensitivity and long-term stability. However, it is a suitable source for a number of applications (Skogerhoe et al., 1976; Coleman et al., 1998) and has been used for plant tissue digest elemental assay (DeBolt, 1980).

b. Spectrometer designs

Sequential spectrometer. As the name implies, one element at a time is assayed by physically moving either the grading or the detector in the spectrometer. This spectrometer configuration works best for the assay of substances that have varying elemental compositions, the analyst selecting those elements to be determined for each substance assayed. This spectrometer configuration is not well suited for the repeated assay of large numbers of samples for the same suite of elements, or for the determination of many elements (greater than ten, for example) in samples. For large numbers of elements, analysis time is slow and a considerable volume of analyte is required. Depending on the number of elements determined and the analysis time, calibration monitoring may be required on a fairly frequent interval of time.

Simultaneous spectrometer. The spectrometer is a polychromator as each element to be determined has its own exit slit and detector; therefore, all the elements so installed in the spectrometer are determined simultaneously. This configuration is best for the repetitious assay of samples for the suite of installed elements. Since only those elements with exit slits and detectors can be determined with this spectrometer configuration, some manufacturers will install a single detector spectrometer so that any element not present in the polychromator array may be determined with this added spectrometer. Unfortunately, this added feature may not be sufficient because of the limitations of the added spectrometer itself — poor sensitivity, inadequate dispersion, etc. Another option is to have the plasma physically placed between two spectrometers, one simultaneous, the other a polychromator, an expensive option but workable for situations that require such versatility for dealing with the requirements of both varied and high-volume analytical requirements.

Photodiode array spectrometer. This is the latest configuration that has promise of being the spectrometer of the future, replacing both sequential and simultaneous spectrometers by the use of a photodiode array as the detector. Recent significant advances have been made with these types of detectors, with considerable advancements yet to be made. The main advantage is the marked reduction in size of the spectrometer, meaning that the ICP-AES technique can be truly a desktop analyzer similar in physical size to most AAS spectrophotometers. This spectrometer design offers some unique features for doing "approximate" assays and for allowing the analyst to design a wide range of analytical programs and systems that are not possible with the current sequential or simultaneous spectrometers.

4. Operating Characteristics of an ICP-AES

a. Advantageous characteristics

ICP-AES is a multielement, high-speed, and very sensitive analytical procedure suitable for elemental content determination of many types of aqueous liquids. Sensitivity is in the ≤ μg/L (parts per billion, or ppb) range for many elements because of the very high temperature of the plasma discharge (8000 to 10,000 K). Its linear 3- to 5-decade range of elemental concentration reading range is because relatively few stable atoms (which absorb radiation) are present in the plasma, making the plasma itself what is termed *optically thin.*

b. Disadvantages

Not all elements exhibit the same emission characteristics in all parts of the plasma; for example, for the elements K and Na, maximum emission occurs at the top and along the edge of the plasma discharge, whereas for most of the heavy metals maximum intensity exists at the base of the plasma discharge. Axial and total viewing of the plasma discharge by variable positioning of the viewing height are possible means to compensate for these effects.

Samples containing high levels of organic substances are not easily assayed unless specific operating characteristics are used.

Samples must be free of suspended materials as clogging or impaired flow rates through the nebulizer will significantly affect the analysis results. Changing solution viscosity will alter the flow rate as well as the positioning of the sample container and depth of solution, if pneumatic lift is used to bring the analyte into the nebulizer. Peristaltic pumping of the analyte into the nebulizer minimizes all these effects and is highly recommended.

5. Standard Preparation

Calibration standards need to conform exactly to the matrix of the assayed samples as well as the elements present plus concentration range. The plasma discharge (temperature profile) itself will conform to the presence of the dominant element, which will affect the emission characteristcs for most elements.

For primary standards, it is best to use commercially prepared elemental standards (1000 or 10,000 mg/L, or ppm) rather than relying on standards prepared from laboratory reagents. When preparing "working standards," compatibility and interelemental factors should be considered. Generally, it is not possible for multielement assays to have only one standard containing all the elements to be determined.

It is well to have standards from known sources (such as NIST, etc.) available to verify accuracy. Multielement-containing standards can be obtained from commercial sources and should be used for verification of prepared working standards.

Details on the preparation of standards are given in Appendix B.

6. Calibration Techniques

The key to success for the assay of any unknown (particularly soil extracts and plant tissue digests) is the setting of the "zero" when using a two-point calibration routine. Therefore, the number of standards, their elemental composition, and the selection of the blank are critical decisions. Once the spectrometer is calibrated, its calibration should be verified by running the standards as "unknowns" following the calibration routine in addition to a quality assurance sample(s).

Accuracy verification should be conducted by using standards from known sources as described above. During an analytical run, a quality assurance sample(s) should begin and terminate the run, and it may also be interspersed several times among the unknowns during the run. Any variance from the known concentration of an element(s) warns of the need to adjust the analytical result or void the analytical run results.

To check for calibration curve linearity and possible interelement interference, high-concentration (1000 mg/L, or ppm, or higher concentrations), single-element commercial standards should be assayed following calibration, observing the determined concentration of the element itself plus that of the other elements in the suite of elements being determined. Lack of linearity for the calibration element and the "apparent" presence of other elements may need evaluation and correction before the assay of unknowns commences.

The use of an internal standard is not commonly recommended for the ICP-AES analysis technique; however, there are considerable advantages if the elemental concentrations among the unknowns are high and varied, which is particularly true for plant tissue digests. An internal standard would partially compensate for changing sample characteristics that would affect the flow rate (nebulization characteristics) of the unknown(s), and if a wide range in elemental concentration exists from one sample unknown to another. To obtain the maximum value from the use of an internal standard, the emission characteristics should closely match those of the element(s) being determined; for example, Li would be a suitable internal standard element for the determination of K and Na, while the elements In or Y would be best suited for many of the "heavy metals." The internal standard must be added to both standards and unknowns at the same concentration.

For samples with relatively high concentrations of an element, the "memory effect" should be checked to ensure that, before the next sample emission is integrated, no significant carryover effect is still apparent. To minimize such an effect, it may be necessary to have an extended wash time between samples, or unknowns can be positioned in the analytical run in a way that minimizes the influence of a carryover effect.

It also may be desirable to have the wash liquid between samples similar to that of the unknowns (for example, if a particular soil extractant is being assayed, use the extractant as the wash solution), or use the next sample, if sufficient quantity is available, as the wash liquid. Such simple "tricks" may significantly affect the analytical results obtained in terms of improved precision.

There is a stabilization (warm-up period) time required (usually 20 to 30 min) from the time the plasma is formed to the time the spectrometer is ready for calibration. The ICP-AES should be placed in a room isolated from other types of laboratory activity and where there is reasonable control of temperature and humidity. The discharge from the plasma should be directly vented outside.

7. Common Operating Problems

Irregular nebulizer flow, changing optical alignment, agron quality, and electronic irregularities and failures are commonly occurring problems.

Irregular nebulizer flow can occur as a result of changing viscosity of samples, pump tube wear, irregular pumping rate, partial plugging of the nebulizer intake capillaries, varying argon flow, or varying backpressure in the nebulizer chamber.

Optical alignment is essential for high-quality performance, ensuring proper positioning of the plasma image on the entrance slit of the spectrometer as well as the alignment of the entrance slit to the exit slits within the spectrometer. Alignment procedures to be followed will vary with the type of ICP-AES; therefore, the analyst must become thoroughly familiar with the procedures as specified by the manufacturer. With time and change in temperature, alignment will shift. Therefore, the analyst should check the alignment as frequently as required based on evaluation or past experience.

Without high-purity agron, formation of the plasma may be difficult, if not impossible. Even if there is just a slight change (purity and flow rate) of the agron from the supply tank, the temperature profile within the plasma will change, thereby affecting the assay result.

Electrical and electronic failures occur with continued use of the various components in an ICP-AES. For most ICP-AESs, the RF generator is the unit more likely to fail with greater frequency as compared with the other electric/electronic components. Having a competent and timely repair service and a supply of spare parts available is essential to avoid long periods of downtime. Periodic maintenance of the major electrical/electronic components is not cost-effective unless the failure of a component would result in a long period awaiting repair. It should be remembered that many electric/electronic components may not be easily evaluated for their potential for failure within an anticipated period of use time. A new component may be just as likely to fail as one that has been in operation for many months.

8. Important General Points

As with most complex analytical instruments, an ICP-AES should be operated only by a well-trained, experienced operator if reliable analytical data are to be obtained. There is no substitute for experience as most ICP-AESs seem to have a "character of their own" that responds to the care taken by the analyst in all their functions.

The analyst should maintain daily operating logs on the ICP-AES, noting repairs, service conducted (changing pump tubes, replacing the torch, etc.), standardization data, replacement standards, QC samples, etc. If optical intensity measurements are made, these values should be recorded. The analyst should continuously monitor the established QC procedures, making adjustments for drift or changing values for QC samples included in the analytical run.

Each ICP-AES has its own set of performance characteristics that may vary from the norm as specified by the manufacturer. For example, elemental sensitivity levels (usually identified as twice the standard deviation), saturation concentrations for elements, degree of linearity for calibration curves, level of precision by element, interelement interference (apparent and real), etc., are all specific characteristics that will vary among instruments. Therefore, tests should be conducted periodically to observe and record such performance characteristics for the ICP-AES. Also, any significant shift in any of these operating parameters should serve as a warning sign that some operating factor has changed.

Although the elemental detection limit is an important operating characteristic that is frequently used to identify performance for comparative purposes among ICP-AES instruments, those using the ICP-AES technique for the assay of soil extracts and plant tissue digests should focus primarily on precision and long-term stability of the operating system, for these are the characteristics that are highly desirable for this application since most of the elemental concentration levels in soil extracts and plant tissue digests are at levels considerably above detection limits. In fact, linearity of calibration curves between 100 and 1000 mg/L (ppm), saturation points of the detection sensors, and the apparent or real interelement effects among elements are far more important instrument characteristics than detection limits.

A comparison of aqueous detection limits (ng/L) for atomic optical spectrophotometry methods is shown in the following table:

Element	FAAL	FAFL	FAE	RFICP	ETA-AA
Aluminum (Al)	100	100	3.0	0.2	0.1
Arsenic (As)	30	100	10,000	40	0.8
Boron (B)	2500	—	50	5.0	20
Calcium (Ca)	2.0	20	0.1	0.02	0.04
Cadmium (Cd)	1.0	0.001	800	2.0	0.008
Chromium (Cr)	2.0	5.0	2.0	0.3	0.2
Copper (Cu)	4.0	0.5	10	0.1	0.06
Iron (Fe)	4.0	8.0	5.0	0.3	1.0
Lead (Pb)	10	40	50	2.0	0.4
Magnesium (Mg)	3.0	0.1	70	0.05	0.004
Manganese (Mn)	0.8	1.0	1.0	0.06	0.02
Molybdenum (Mo)	30	500	200	0.2	0.3
Nickel (Ni)	5.0	3.0	20	0.4	0.9
Phosphorus (P)	10	10	100	2.0	0.2
Potassium (K)	3.0	—	0.05	—	4.0
Sodium (Na)	0.08	—	0.5	0.2	—
Zinc (Zn)	1.0	0.02	10,000	2.0	0.003

Key: FAAL = flame atomic absorption with line source; FAFL = flame atomic fluorescence with line source; FAE = flame atomic emission; RFICP = radiofrequency inductively coupled plasma; ETA-AA = electrothermal atomization–atomic absorption.

9. Spectrophotometry Terms

Absorbance, *A*. Negative logarithm to the base 10, of the transmittance (T): $A = -\log_{10}(T)$.

Absorptively, a. Absorbance divided by the product of the sample path length (b) and the concentration of the absorbing substance (c). $a = A/bc$.

Absorptivity, molar, e. Product of the absorptivity (a) and molecular weight of the absorbing substance.

Analysis. Ascertainment of the identity and/or the concentration of the constituents or components of a sample.

Angstrom, Å. Unit of length = 10^{-10} m or 0.1 nm.

Beer's law. Absorptivity of a substance is a constant with respect to changes in concentration.

Concentration, c. Quantity of the substance contained in a unit quantity of sample.

Determination. The ascertainment of the quantity or concentration of a specific substance in a sample.

Frequency, v. Number of cycles per unit time.

Micrometer, μm. Unit of length = 10^{-6} m.

Nanometer, nm. Unit of length = 10^{-9} m.

D. pH and Specific-Ion Electrodes

pH and specific-ion electrodes work on the same principle as that of the glass electrode for measuring H ion concentration (Fisher, 1984; Talibudeen, 1991). These electrodes employ liquid or solid ion-exchange membranes, or solid membranes composed of single crystals, or precipitates compressed into a plug or dispersed in a matrix such as silicone rubber. The instrument components are the specific-ion and reference electrodes (they may be seperate electrodes or one combination electrode) and the recording meter. The advantages of specific-ion electrodes are their speed of determination and a general simplicity of operation. However, they are not without limitations, such as lack of sensitivity when the analyte of interest is in low concentration and interference. A good description of their uses in soil and plant analysis applications is given by Milham et al. (1970), Carlson and Keeney (1971), Mills (1980), Talibudeen (1991), and Miller (1998).

Specific-ion electrodes are available for a number of elemental determinations as shown below.

Ion		Type of electrode
Direct	Indirect	Membrane
Cations		
Ammonium		Liquid (in polymer)
Calcium		Liquid (in polymer)
Copper		Solid state (crystal)
Potassium		Solid state (glass)
		Liquid (in polymer)
Sodium		Solid state (glass)
		Solid state (crystal)
		Liquid (in polymer)

(continued)

Ion		Type of electrode
Direct	Indirect	Membrane
Anions		
Bromide		Solid state (crystal)
Chloride		Solid state (crystal)
Cyanide		Solid state (crystal)
Fluoride	Aluminum	Solid state (crystal)
Fluoroborate	Boron	Liquid (free flowing)
Iodide		Solid state (crystal)
Nitrate		Liquid (in polymer)
Sulfide		Solid state (crystal)

The most widely used specific-ion electrode is for the determination of the NO_3 anion in soil extracts (Dahnke and Vasey, 1973; Mills, 1980; Gelderman and Beegle, 1998) and plant tissue (Baker and Thompson, 1992). The NO_3 electrode consists of an electrode body and a replaceable pretested sensing module.

The sensing module contains a gelled internal filling solution, a membrane saturated with a liquid ion exchanger, and a reservoir of liquid ion exchanger. The membrane separates sample from the electrode filling solution and is selective for the NO_3 anion. A potential develops across the membrane dependent on the NO_3 concentration in the unknown solution and is measured against a constant reference potential with a specific solution as described by the Nernst Equation:

$$\mathbf{E} = \mathbf{Eo} - S \log (A)$$

where **E** = measured electrode potential, **Eo** = reference potential (a constant), A = nitrate concentrated in unknown solution, S = electrode slope. A, the activity or "effective concentration," is related to the NO_3 concentration, C, by the ion activity coefficient, y (γ), where $A = yC$.

Ionic activity coefficients are variable and largely dependent on total ionic strength. If background ionic strength is high and constant, relative to the sensed ion concentration, the activity coefficient is constant and activity is directly proportional to concentration.

An ionic strength adjuster is normally added to all NO_3 standards and samples so that the background ionic strength is high and constant relative to the variable concentration of NO_3. The same basic principle is followed for other cation and anion determinations.

Specific-ion electrodes are useful for some applications, but they have been primarily limited to the determination of NO_3, Cl, and F in soil extracts, and NH_4 in soil and plant digests (Eastin, 1976; Gallaher and Jones, 1976; Watson and Isaac, 1990; Talibudeen, 1991). Taboada-Castro et al. (2000) compared an ion-selective electrode (ISE) procedure vs. other methods of analysis (inductively coupled plasma emission mass spectrometry, or ICP-MS, and capillary electrophoresis, or CE) for water samples for the determination of NO_3^-, Cl^-, Ca^{2+}, K^+, and Na^+, finding that only the Na^+ determinations were not well correlated.

E. Ion Chromatography

The anions F^-, Cl^-, NO_2^-, PO_4^{3-}, NO_3^-, and SO_4^{2-}, which can be present in both soil extracts and plant tissue digests, can be determined by ion chromatography (Mosko, 1984; Ettre, 1980; Watson and Isaac, 1990; Tabatabai and Basta, 1991; Tabatabai and Frankenberger, 1996). The anions are listed in the order of appearance on an anion chromatogram. The time required to obtain a typical chromatogram that includes these six anions will take about 20 min. Sensitivity and concentration range of determination will vary with ion-exchange column and instrument characteristics.

In addition, the monovalent cations Li^+, Na^+, K^+, Rb^+, and Cs^+ (listed in their order of appearance on the cation chromatogram) can be determined, and the divalent cations Mg^{2+}, Ca^{2+}, Sr^{2+}, and Ba^{2+} (listed in their order of appearance on the cation chromatogram) can also be determined by ion chromatography, but require different ion-exchange columns and operating conditions. Anions and the mono- and divalent cations cannot be determined simultaneously using the same ion-exchange column, but require different systems and operating conditions.

Although this method has considerable appeal because of its ability to determine more than one ion species per determination, sample preparation procedures and matrix effects are considerable, limiting its use, as has been pointed out by Tabatabai and Basta (1991). Frequently, separation of the ion species of interest must be made prior to analysis. In addition, the time required to perform a complete ion determination ranges from 12 to 20 min, which makes this method of limited use if large numbers of samples are to be assayed within a specified time period.

Karmarkar (1998) described a method of combining sequential flow injection analysis with specific ion electrodes to determine various anions.

Ion chromatography is still in its infancy and future developments in ion-exchange column characteristics and instrumental procedures may make this method of multianion/cation determination of value for the assay of soil extracts and plant tissue digests.

References

Baker, D.L. and N.H. Suhr. 1982. Atomic absorption and flame emission spectometry, in A.L. Page, Ed., *Methods of Soil Analysis,* Part 2, *Chemical and Microbiological Properties*, Agronomy Monograph Number 9, America Society of Agronomy, Madison, WI, 13–27.

Baker, W.H. and T.L. Thompson. 1992. Determination of nitrate-nitrogen in plant samples by selective ion electrode, in C.O. Plank, Ed., Plant Analysis Reference Procedures for the Southern Region of the United States, Southern Cooperative Series Bulletin 368, University of Georgia, Athens, 25–31.

Buttgereit, G. 1974. Atomic absorption, flame emission and atomic fluorescence spectroscopy, in F. Korte, Ed., *Methodicum Chimicum,* Vol. 1, *Analytical Methods,* Academic Press, New York, 710–725.

Carlson, R.M. and R.D. Keeney. 1971. Specific-ion electrodes: techniques and uses in soil, plant and water analysis, in L.M. Walsh, Ed., *Instrumental Methods for Analysis of Soil and Plant Tissue,* Soil Science Society of America, Madison, WI, 39–66.

Carter, M.R., Ed. 1993. *Soil Sampling and Methods of Analysis*, Lewis Publishers, Boca Raton, FL.

Chapman, H.D. and P. Pratt. 1982. *Methods of Analysis for Soils, Plants and Water*, Priced Publication 4034, University of California, Division of Agricultural Sciences, Berkeley.

Coakley, W.A. 1981. *Handbook of Automated Analysis-Continuous Flow Techniques*, Marcel Dekker, New York.

Coleman, G.N., D.E. Miller, and R.W. Starek. 1998. A new look at direct current plasma spectrometry. *Am. Lab.,* 30(4):33R-33Y.

Combs, S., J. Denning, and K.D. Frank. 1998. Sulfate-sulfur, in J.R. Brown, Ed., Recommended Chemical Soil Test Procedures for the North Central Region, North Central Regional Research Publication Number 221 (revised), Missouri Agricultural Experiment Station SB 1001, University of Missouri, Columbia, 35–40.

Dahnke, W.C. and E.H. Vasey. 1973. Testing soils for nitrogen, in L.M. Walsh and J.D. Beaton, Eds., *Soil Testing and Plant Analysis*, Soil Science Society of America, Madison, WI, 97–114.

Dalquist, R.L. and J.W. Knoll. 1978. Inductively (coupled plasma-atomic emission spectroscopy: analysis of biological materials and soil for major, trace, and ultra-trace elements. *Appl. Spectrosc.*, 32:1–30.

Dean, J.A. 1973. *Lange's Handbook of Chemistry*, Section 8–19, 11th ed., McGraw-Hill, New York.

DeBolt, D.C. 1980. Multielement emission spectroscopic analysis of plant tissue using DC agron plasma source. *J. Assoc. Off. Anal. Chem.,* 63:803–805.

Eastin, E.F. 1976. Use of ammonia electrode for total nitrogen determination in plants. *Commun. Soil Sci. Plant Anal.,* 7:477–481.

Ettre, L.S. 1980. Evolution of liquid chromatography: a historical overview, in C. Horvath, Ed., *High Performance Liquid Chromatography: Advances and Perspectives*, Vol. 1, Academic Press, New York, 1–74.

Fassel, V.A. and R.N. Kniseley. 1974. Inductively coupled plasmas. *Anal. Chem.,* 46:1155A-1164A.

Fisher, J.E. 1984. Measurement of pH. *Am. Lab.,* 16:54–60.

Flannery, R.L. and D.K. Markus. 1972. Determination of phosphorus, potassium, calcium and magnesium simultaneously in North Carolina, ammonium acetate, and Bray PI soil extracts by Auto Analyzer, in L.M. Walsh, Ed., *Instrumental Methods for Analysis of Soils and Plant Tissue*, Soil Science Society of America, Madison, WI, 97–112.

Flannery, R.L. and D.K. Markus. 1980. Automated analysis of soil extracts for phosphorus, potassium, calcium, and magnesium. *J. Assoc. Off. Anal. Chem.,* 63:779–787.

Frank, K., D. Beegle, and J. Denning. 1998. Phosphorus, in J.R. Brown, Ed., Recommended Chemical Soil Test Procedures for the North Central Region, North Central Regional Research Publication Number 221 (revised), Missouri Agricultural Experiment Station SB 1001, University of Missouri, Columbia, 21–29.

Frankenberger, W.T., Jr., H.C. Mehra, and D.T. Gjerde. 1990. Environmental applications of ion chromatography. *J. Chromatogr.,* 504:211–245.

Gallaher, R.N. and J.B. Jones, Jr. 1976. Total extractable, and oxalate calcium and other elements in normal and mouse ear pecan tree tissues. *J. Am. Soc. Hortic. Sci.,* 101:692–696.

Gelderman, R.H. and D. Beegle. 1998. Nitrate-nitrogen, in J.R. Brown, Ed., Recommended Chemical Soil Test Procedures for the North Central Regal, North Central Regional Publication No. 221 (revised), Missouri Agricultural Experiment Station SB 1001, Columbia, 17–20.

Gupta, U.C. 1998. Determination of boron, molybdenum, and selenium in plant tissue, in Y.P. Kalra, Ed., *Handbook of Reference Methods for Plant Analysis*, CRC Press, Boca Raton, FL, 171–182.

Hanlon, E.A. 1992a. Determination of potassium, calcium, and magnesium in plants by atomic absorption techniques, in C.O. Plant, Ed., Plant Analysis Reference Procedures for the Southern Region of the United States, Southern Cooperative Series Bulletin 368, University of Georgia, Athens, 33–36.

Hanlon, E.A. 1992b. Determination of total manganese, iron, copper, and zinc in plants by atomic absorption technqiues, in C.O. Plant, Ed., Plant Analysis Reference Procedures for the Southern Region of the United States, Southern Cooperative Series Bulletin 368, University of Georgia, Athens, 49–51.

Hanlon, E.A. 1998. Elemental determination by atomic absorption spectrophotometry, in Y.P. Kalra, Ed., *Handbook of Reference Methods for Plant Analysis*, CRC Press, Boca Raton, FL, 157–164.

Hieftje, G.M., T.R. Copeland, and D.R. de Olivares. 1976. Flame emission, atomic absorption, and atomic fluorescence spectrometry. *Anal. Chem.*, 48:142R-174R.

Hinderberger, E.J., M.L. Kaiser, and S.R. Koirtyohann. 1982. Furnace atomic absorption analysis of biological samples using the L'vov platform and matrix modification. *Atom. Spectrosc.*, 2:1–7.

Hislop, J.S. 1980. Choice of the analytical method, in P. Bratter and P. Schramel, Eds., *Trace Element Analytical Chemistry in Medicine and Biology,* DeGrutyer, Berlin, Germany, 747–767.

Horneck, D.A. and D. Hanson. 1998. Determination of potassium and sodium by flame emission spectrophotometry, in Y.P. Kalra, Ed., *Handbook of Reference Methods for Plant Analysis*, CRC Press, Boca Raton, FL, 153–155.

Horwitz, W. 1982. Evaluation analytical methods used for regulation of foods and drugs. *Anal. Chem.*, 54:67A-76A.

Hwang, J.Y., C.J. Mokeler, and P.A. Ulluci. 1972. Maximization of sensitivities in tantalum ribbon flameless atomic absorption spectrometry, in *Progress in Analytical Atomic Spectrometry,* Vol. 1, Pergamon, New York, 5–50.

Isaac, R.A. 1980. Atomic absorption methods for analysis of soil extracts amd plant tissue digests. *J. Assoc. Off. Anal. Chem.*, 63:788–796.

Isaac, R.A. and W.C. Johnson, Jr. 1998. Elemental determination by inductively coupled plasma atomic emission spectrometry, in Y.P. Kalra, Ed., *Handbook of Reference Methods for Plant Analysis*, CRC Press, Boca Raton, FL, 165–170.

Isaac, R.A. and J.B. Jones, Jr. 1970. Auto-analyzers systems for the analysis of soil and plant tissue extracts, in *Advances in Automated Analysis, Technicon Congress Proceedings,* Technicon Corp., Tarrytown, NY, 57–64.

Isaac, R.A. and J.D. Kerber. 1971. Atomic and flame photometry: techniques and uses in soil, plant, and water analysis, in L.M. Walsh, Ed., *Instrumental Methods for Analysis of Soils and Plant Tissue*, Soil Science Society of America, Madison, WI, 17–38.

Jackson, M.L. 1958. *Soil Chemical Analysis*, Prentice-Hall, Englewood Cliffs, NJ.

Jenke, D.R. and R. Woodriff. 1982. Application of the Woodriff constant temperature graphite furnace atomizer to atomic spectroscopy. *Am. Lab.,* 14:14–26.

Jones, J.B., Jr. 1977. Elemental analysis of soil extracts and plant tissue ash by plasma emission spectroscopy. *Commun. Soil Sci. Plant Anal.,* 8:345–365.

Jones, J.B., Jr. 1996. Determination of total sulphur in plant tissue using the HACH kit spectrophotometer technique. *Sulphur Agric.,* 19:58–62.

Jones, J.B., Jr. and M.H. Warner. 1969. Analysis of plant-ash solutions by spark-emission spectroscopy, in E.L. Grove and H.D. Perkins, Eds., *Developments in Applied Spectroscopy,* Vol. 7A, Plenum Press, New York, 152–160.

Kammin, W.R., S. Cull, R. Knox, J. Ross, M. Mclntosh, and D. Thomson. 1995. Labware cleaning protocols for the determination of low-level metals by ICP-MS. *Am. Environ. Lab.,* 11/95–12/95:1–3.

Karmarkar, S.V. 1998. Enhanced ion chromatography with sequential flow injection analysis: determination of common anions and nitrite. *Am. Environ. Lab.*, 10:16–7.

Knopp, J.F. and D. McKee. 1983. Methods for Chemical Analysis of Water and Wastes, EPA-60014-79-020, Environmental Protection Agency, Cincinnati, OH.

Knudsen, D. and D. Beegle. 1988. Recommended phosphorus tests, in W.C. Wahnke, Ed., Recommended Chemical Soil Test Procedures for the North Central Region, North Central Regional Publication Number 221 (revised), North Dakota Agricultural Experiment Station Bulletin Number 499 (revised), North Dakota State University, Fargo, 12–15.

Kuo, S. 1996. Phosphorus, in R.L. Sparks, Ed., *Methods of Soil Analysis,* Part 3, *Chemical Methods*, SSSA Book Series Number 5, Soil Science Society of America, Madison, WI, 869–919.

Lajunen, L.H.J. 1992. *Spectrochemical Analysis by Atomic Absorption and Emission*, The Royal Society of Chemistry, Cambridge, U.K.

MAFF/ADAS. 1986. The Analysis of Agricultural Materials, 3rd ed., Reference Book 427, Her Majesty's Stationery Office, London.

Massmann, H. 1968. Vergleich von Atomabsorption und Atomfluoreszenze in der Graphitkuvette. *Spectrochim. Acta,* 23B:215–226.

Mavrodineanu, R., Ed. 1970. *Analytical Flame Spectroscopy*, Springer-Verlag, New York.

McLain, D.B., Ed. 1982. *Handbook for AOAC Members*, 5th ed., Association of Official Analytical Chemists, Arlington, VA.

McLaughlin, R.D., M.S. Hunt, D.L. Murphy, and C.R. Chen. 1979. Instrumentation in environmental monitoring, in D.L. Hemphill, Ed., *Trace Elements in Environmental Health — XIII*, University of Missouri, Columbia, 224–245.

Milham, P.J., A.S. Awad, R.E. Paul, and J.H. Bull. 1970. Analysis of plants, soils, and waters for nitrate by using an ion-selective electrode. *Analyst,* 95:751–759.

Miller, R.O. 1998. Extractable nitrate in plant tissue: ion-selective electrode method, in Y.P. Kalra, Ed., *Handbook of Reference Methods for Plant Analysis*, CRC Press, Boca Raton, FL, 85–88.

Mills, H.A. 1980. Nitrogen specific-ion electrodes for soil, plant, and water analysis. *J. Assoc. Off. Anal. Chem.*, 63:797–801.

Mitchell, R.L. 1964. The Spectrographic Analysis of Soil, Plants, and Related Materials, Technical Communication No. 44A of the Commonwealth Bureau of Soils, Harpenden, Commonwealth Agricultural Bureaux, Farnham Royal, Bucks, England.

Morrison, G.H. 1979. Elemental trace analysis of biological materials. *Crit. Rev. Anal. Chem.,* 8:287–320.

Morrison, G.H. and T.H. Risby. 1979. Elemental trace element analysis of biological materials. *CRC Crit. Rev. Anal. Chem.,* 8:287–320.

Mosko, J.A. 1984. Automated determination of inorganic anions in water by ion chromatography. *Anal. Chem.,* 56:629–633.

Munter, R.C. and R.A. Grande. 1981. Plant tissue and soil extract analysis by ICP-atomic emission spectrometry, in R.M. Batmes, Ed., *Developments in Atomic Plasma Spectrochemical Analysis,* Heyden and Son, London, 653–672.

Paul, S. 1998. UV-VIS instruments continue to evolve. *Today's Chem. Work,* 7(9):21–25.

Petterelli, M.N. 1995. Axial ICP for trace metal analysis. *Environ. Test Anal.*, 13:50–54.

Pickett, E.E. and S.R. Koirtyohann. 1969. Emission flame photometer: a new look at an old method. *Anal. Chem.,* 41:28A-30A.

Piper, C.S. 1942. *Soil and Plant Analysis*, The University of Adelaide, Adelaide, Australia.

Ranger, C. 1981. Flow injection analysis — principles, techniques, applications, design. *Anal. Chem.,* 53:20A-26A.

Rodriquez, J.B., J.R. Self, and P.N. Soltanpour. 1994. Optical conditions for phosphorus analysis by the ascorbic acid-molybdenum blue method. *Soil Sci. Soc. Am. J.,* 58:866–870.

Routh, M.W., P.S. Doidge, J. Chidzey, and B. Fray. 1982. Advances in graphite furnace atomization. *Am. Lab.,* 14:80–95.

Ruzieka, J. and E.H. Hansen. 1988. *Flow Injection Analysis*, 2nd ed., John Wiley & Sons, New York.

Schulte, E.E. and K. Eik. 1988. Recommended sulfate–sulfur tests, in W.C. Dahnke, Ed., Recommended Chemical Soil Test Procedures for the North Central Region, North Central Regional Publication Number 221 (revised), North Dakota Agricultural Experiment Station Bulletin Number 499 (revised), North Dakota State University, Fargo, 17–19.

Scott, R.N., V.A. Fassel, R.N. Kniseley, and D.E. Dixon. 1976. Inductively coupled plasma optical emission analytical spectroscopy. *Anal. Chem.,* 46:75–80.

Sharp, B.L. 1991. Inductively coupled plasma spectrometry, in K.A. Smith, Ed., *Soil Analysis: Modern Instrumental Techniques*, 2nd ed., Marcel Dekker, New York, 63–109.

Singh, R., D.K. Bhumbia, and R.F. Keefer. 1995. Recommended soil sulfate-S tests, in J.T. Sims and A.M. Wolf, Eds., Recommended Soil Testing Procedures for the Northeastern United States, Northeast Regional Publications Number 493 (revised), University of Delaware, Newark, 46–50.

Skogerhoe, R.K., J.T. Urasa, and G.N. Coleman. 1976. Characteristics of a dc plasma as an ion excitation source for multielement analysis. *Appl. Spectrosc.*, 30:500–507.

Smith, K.A., Ed. 1991. *Soil Analysis. Modern Instrumental Techniques*, 2nd ed., Marcel Dekker, New York.

Smith, K.A. and A. Scott. 1991. Continuous-flow and discrete analysis, in K.A. Smith, Ed. *Soil Analysis. Modern Instrumental Techniques*, 2nd ed., Marcel Dekker, New York, 183–227.

Soltanpour, P.N. 1991. Determination of nutrient element availability and elemental toxicity by the AB–DTPA soil tests and ICPS. *Adv. Soil Sci.,* 16:165–190.

Soltanpour, P.N., G.W. Johnson, S.M. Workman, J.B. Jones, Jr., and R.O. Miller. 1996. Inductively coupled plasma emission spectrometry and inductively coupled plasma-mass spectrometry, in D.L. Sparks, Ed., *Methods of Soil Analysis,* Part 3. *Chemical Methods*, SSSA Book Series Number 5, Soil Science Society of America, Madison, WI, 91–139.

Soltanpour, P.N., C.W. Johnson, S.M. Workman, J.B. Jones, Jr., and R.O. Miller. 1998. Advances in ICP emission and ICP mass spectrometry. *Adv. Agron.,* 64:28–113.

Soon, Y.K. 1998. Determination of arsenic and mercury in plant tissue, in Y.P. Kalra, Ed., *Handbook of Reference Methods for Plant Analysis*, CRC Press, Boca Raton, FL, 183–192.

Steckel, J.E. and R.L. Flannery. 1971. Simultaneous determinations of phosphorus, potassium, calcium, and magnesum in wet digestion solutions of plant tissue by autoanalyzer, in L.M. Walsh, Ed., *Instrumental Methods for Analysis of Soils and Plant Tissue*, Soil Science Society of America, Madison, WI, 83–96.

Stika, J.W.A. and G.H. Morrison. 1981. Analytical methods for mineral content of human tissues. *Fed. Proc.,* 40:2115–2120.

Sturgeon, R.E. 2000. Current practice and recent developments in analytical methodology for trace element analysis of soils, plants, and water. *Commun. Soil Sci. Plant Anal.,* 31:1479–1512.

Sturgeon, R.E. and C.L. Chakrabarti. 1978. Recent advances in electrothermal atomization in graphite furnace atomic absorption spectrometry, in *Progress in Analytical Atomic Spectroscopy,* Vol. 1, Pergamon, New York, 5–50.

Tabatabai, M.A. and N.T. Basta. 1991. Ion chromatography, in K.A. Smith, Ed., *Soil Analysis: Modern Instrumental Techniques*, 2nd ed., Marcel Dekker, New York, 229–259.

Tabatabai, M.A. and W.T. Frankenberger, Jr. 1996. Liquid chromatography, in R.L. Sparks, Ed., *Methods of Soil Analysis,* Part 3, *Chemical Methods*, SSSA Book Series Number 5, Soil Science Society of America, Madison, WI, 225–245.

Taboada-Castro, T., A. Diéguez, B. López, and A. Paz-González. 2000. Comparison of conventional water testing methods with ion-selective electrodes techniques for NO_3^-, Cl^-, Ca^{2+}, K^+, and Na^+. *Commun. Soil Sci. Plant Anal.,* 31:1993–2005.

Talibudeen, O. 1991. Ion-selective electrodes, in H.A. Smith, Ed., *Soil Analysis: Modern Instrumental Techniques*, 2nd ed., Marcel Dekker, New York, 111–182.

Tel, D.A. and C. Heseltine. 1990a. The analyses of KCl soil extracts for nitrate, nitrite, and ammonium using a TRAACS 800 analyzer. *Commun. Soil Sci. Plant Anal.,* 21:1681–1688.

Tel, D.A. and C. Heseltine. 1990b. Chloride analyses of soil leachate using the TRAACS 800 analyzer. *Commun. Soil Sci. Plant Anal.,* 21:1689–1693.

Thompson, M., B. Pahlavanpour, and S.J. Walton. 1978. Simultaneous determination of trace concentrations of arsenic, antimony, bismuth, selenium, and tellurium in aqueous solution by introduction of the gaseous hydrides into an inductively coupled plasma source for emission spectrometry. I. *Analyst* (London), 103:568–579.

Ure, A.M. 1991. Atomic absorption and flame emission spectrometry, in K.A. Smith, Ed., *Soil Analysis: Modern Instrumental Techniques*, 2nd ed., Marcel Dekker, New York, 1–62.

van Loon, J.C. 1985. *Selected Methods of Trace Metal Analysis. Biological and Environmental Samples*, John Wiley & Sons, New York.

Walsh. L.M., Ed. 1971. *Instrumental Methods for Analysis in Soils and Plant Tissue*, Rev. ed., Soil Science Society of America, Madison, WI.

Watson, M.E. 1998. Analytical instruments for the determination of elements in plant tissue, in Y.P. Kalra, Ed., *Handbook of Reference Methods for Plant Analysis*, CRC Press, Boca Raton, FL, 137–152.

Watson, M.E. and R.A. Isaac. 1990. Analytical instruments for soil and plant analysis, in R.L. Westerman, Ed., *Soil Testing and Plant Analysis,* 3rd ed., SSSA Book Series Number 3, Soil Science Society of America, Madison, WI, 691–740.

Willis, J.B. 1970. Atomic absorption spectrometry, in R. Mavodineau, Ed., *Analytical Flame Spectroscopy*, Springer-Verlag, New York, 525–598.

Winge, R.F., V.J. Peterson, and V.A Fassel. 1979. Inductively coupled plasma-atomic emission spectroscopy: prominent lines. *Appl. Spectrosc.,* 33:206–219.

Wolnik, K.A., R.W. Kuennen, and R.L. Fricke. 1981. Determination of toxic and nutritional elements in raw agricultural crops using ICAP spectroscopy, in R.M. Barnes, Ed., *Developments in Atomic Plasma Spectrochemical Analysis*, Heyden, London, 685–696.

Wright, R.J. and T.I. Stuczynski. 1996. Atomic absorption and flame emission spectrometry, in D.L. Sparks, Ed., *Methods of Soil Analysis,* Part 3, *Chemical Methods*, SSSA Book Series Number 5, Soil Science Society of America, Madison, WI, 65–90.

Chapter 6

Quality Assurance/Quality Control in the Laboratory

A. Introduction

Quality assurance (QA) in an analytical laboratory is an essential management tool to ensure reliable performance. Criteria for implementation of a QA program have been established by the Association of Official Analytical Chemists (AOAC). The book by Garfield (1984) listed these elements of a QA program as follows:

- Administration
- Personnel management
- Management of equipment and supplies
- Records maintenance
- Sample analysis
- Proficiency testing
- Audit procedures
- Design and safety of facilities

All these factors will not be covered in detail in this chapter. Users of this guide should refer to *Soil Analysis Handbook of Reference Methods* (Anonymous, 1999) for specific QA procedures in a soil testing laboratory, and *Handbook of Reference Methods for Plant Analysis* (Kalra, 1998) for a plant analysis laboratory. For a more general review of this topic, refer to the books by Garfield (1984), Dux (1986), Taylor (1987), and Garner and Barge (1988).

Aldenhoff and Ernest (1983) defined QA as:

> The sum total of activities that document and maintain the quality of monitoring data as a means of assuring that only proven methods are used, that instruments are properly calibrated and maintained, that uniform operating procedures are established and followed, and that the performances are documented and audited.

Another aspect of laboratory performance is quality control (QC), which Aldenhoff and Ernest (1983) define as:

> The routine application of activities and procedures designed and used to ensure that quality and reliability of laboratory results, whose requirements consist of trained technicians, precision and accuracy testing, and a visual display of results as final assurances that analyses are in control.

Both these definitions imply that QC is measured by performance, whereas QA focuses on the implementation of those management criteria needed to ensure reliable performance.

The basis for reliable performance in an analytical laboratory has been called "Good Laboratory Practices," the subject of a *Federal Registry* entry in 1979 (Anonymous, 1979) and described in some detail by Fischbeck (1980). Good laboratory practices involve organization and personnel, standard operating procedures, study protocol, and study conduct, record keeping, and final reports.

For those engaged in conducting soil and plant analyses, obtaining accurate results rests on two important criteria: the use of acceptable methods and standards.

The handbooks, manuals, and books referenced in this instruction guide that describe procedures for the elemental assay of soils and plant tissues are sources of reliable and acceptable methods of analysis. Matching the method of analysis with the sample type is very important, particularly for soil analysis. Horwitz (1982), as a result of his many years of association with AOAC, has discussed the practical limits of acceptable variability in methods of analysis, focusing on the important aspects of reliability, reproducibility, repeatability, systematic error of bias, specificity, and limit of reliable measurement. An additional criterion is the "ruggedness" factor, the limits to any change in the steps of the analysis procedure that must not be exceeded. Examples are how much variation from temperature and time factors, reagent characteristics, etc. given in the procedure can be tolerated and still maintain the validity of the method. Unfortunately, most soil and plant analysis procedures have not been so described. However, Houba et al. (2000) did set the specific temperature (20°C; 68°F) for the 0.01 *M* $CaCl_2$

extraction procedure, and it has been found that if the pH of the Olsen P extraction reagent (see Chapter 2, Section I.2.c) varies from 8.5, the results obtained will be significantly affected.

B. Accuracy and Precision

The issues of accuracy and precision are significant considerations for the analyst and can be determining factors in method selection. Accuracy is the ability to obtain the "true" value and is dependent to a large degree on the availability and use of reliable standards. Precision, on the other hand, is a measure of the degree of variability of an obtained result determined by repeated analyses of the same sample through all the steps from sample preparation to the final obtained result. Hislop (1980) has an excellent article on the requirements for obtaining accurate and precise analytical results. Horwitz (1982) also has evaluated various analytical procedures by assigning levels of performance based on years of use in the determination of elements and substances in a wide range of materials. Rayment et al. (2000) looked at the percent coefficients of variation for a number of commonly performed soil analysis procedures, found that the pH determination had the lowest (<5%), that determinations for Cl, organic C, nitrate-N, DTPA Zn and Mn, and exchangeable Ca, K, and Mg ranged from 10 to 20%, that exchangeable Na was between 20 and 30%, and that Olsen P and Bray P were greater than 30%. They also looked at intralaboratory RSDs (%), finding that the greatest precision was obtained for the determination of pH (RSD of 1.6%), that determinations for ammoniun-N, Bray P1, and organic carbon had RSDs of 6.2, 6.3, and 8.3%, respectively, and that nitrate-N (Cd reduction) and Olsen P had RSDs of 14.5 and 11.8%, respectively. All these evaluations would suggest that there may be an inherent variance in a particular method and/or that the methodology associated with these assay procedures should be more carefully examined to determine the source of variance.

C. Standards

Various kinds of standards are required in the laboratory to monitor analytical procedures, calibrate instruments, standardize reagents, and verify the final assay result. Standard solutions and reagents are readily available from commercial sources or they can be prepared by the analyst (see Appendix B).

In normal laboratory routine, standards may be used not only to monitor the analytical procedure, but also, when placed periodically into the sequence of unknowns, to serve a dual purpose as a "marker" as well as a standard.

At the end of the analytical run, the analyst can determine if an unknown was skipped or duplicated, and if there was a significant shift in the calibration and instrument performance during the analytical run.

Depending on the established QA program, the placement of standards may be known to the analyst, or standards may serve as "blind" unknowns. Each version has its advantages. Knowing the placement and value for the standard may cause the analyst to focus on the standard at the expense of the total analytical run, forcing the procedure to give a result that "fits" the standard. However, having the placement of the standard known to the analyst can save valuable time if there is a problem in the analysis routine that can be spotted quickly and corrected before an entire analysis run is completed. If the analyst does not know either the placement or the value for the standard, the analyst's skill in performing the analysis can be evaluated. Therefore, the "ideal" would be to have both systems of standard use in place.

When it comes to verifying the accuracy of the final obtained analytical result, good standards are not always available. In soil testing, for example, there are few "standard" soils. Most soil testing laboratories prepare their own by taking a volume of soil, mixing to obtain reasonable homogeneity, and assaying over a period of time. After a period of repeated assays, the prepared sample becomes the "standard" that is used to monitor laboratory procedures. Such a standard may contribute significantly to maintaining good precision, but may not ensure good accuracy. To obtain some measure of accuracy, the preparing laboratory may submit its "standard" soil to other laboratories for analysis. Depending upon the extent of agreement, a value is assigned for the assay performed. Unfortunately, the system for preparing standard soil samples sounds easy and efficient, but it is not. The lack of agreement among testing laboratories and the insufficient homogeneity of the soil sample itself are frequent difficulties. At the present time, the Soil and Plant Analysis Council (621 Rose Street, Lincoln, NE 68502-2040) is making available soils that have been assayed as a part of the North American Proficiency Testing Program (see Appendix D) for use as standards.

For plant analysis, there are National Institute of Science and Technology (NIST) plant tissues available for verification purposes. These Standard Reference Materials, referred to as SRMs (Uriano, 1979), also have some limitations to their usefulness. Unfortunately, not every element of current interest to the analyst is certified, thereby limiting their usefulness. In addition, there has been disagreement among some analysts regarding some of the certified values for certain elements. Some of this discrepancy relates to both the method of sample preparation, preparation for analysis by either wet acid digestion or high-temperature dry ashing (see Chapter 3, Section G), and the element assay procedure employed for determining the certified value, which may be quite different from the procedure used by the analyst (i.e., UV-VIS spectrophotometry vs. AAS or ICP-ES). Lack of homogeneity is also a factor that has limited certification of some elements in some SRMs. More recently, Ihnat (1998) has prepared a comprehensive list of plant tissues varying widely in plant type and element(s) certified, as well as source, that can be obtained and used as standards.

D. Instrumentation

The analyst today has a wide range of instrumentation from which to choose, the choice being determined by a number of factors as described by McLaughlin et al. (1979) and, more recently, for trace element assays by Sturgeon (2000). A similar set of evaluation criteria of analytical techniques has been developed by Hislop (1980); that includes (1) accuracy, (2) elemental coverage, (3) precision, (4) single or multielement, (5) limit of detection, and (6) determined chemical form.

It may be the more practical considerations of instrument availability, cost, and sample form and quantity that become the governing factors rather than choices based on the criteria given by McLaughlin et al. (1979), Hislop (1980), and Sturgeon (2000). Analysts themselves and their skill and experience may be factors when selecting a particular instrument when more than one is available. For the assay of most soil extracts and plant tissue digests, sensitivity is not always the significant factor; rather, precision and long-term stability of the analytical instrument are critical. However, in some instances, particularly when dealing with trace element determinations, analytical instrumentation with excellent sensitivity is required; this instrumentation is listed in an article by Sturgeon (2000).

E. Laboratory Procedures

Good management practices (Anonymous, 1979) in an analytical laboratory include such factors as the employment of statistical control procedures, the use of control charts, the conduct of blind studies and continuous system audits, and the maintenance of documents, for example, the dates of preparation of reagents, service and maintenance records for analytical instruments, etc. The training and supervision of analysts, well organized and performed on a timely basis, are significant QC factors. A review of these procedures in the management of a soil testing laboratory may be found in the *Soil Analysis Handbook of Reference Methods* (Anonymous, 1999), and for a plant analysis laboratory in the *Handbook of Reference Methods for Plant Analysis* (Kalra, 1998).

F. Participation in Proficiency Testing Programs

Probably one of the most important aspects for ensuring reliable analytical results from soil and plant analysis laboratories is participation in a Proficiency Testing Program. Rayment et al. (2000) lists "Collaborative Networks," which exist in most areas of the world, as well as 16 proficiency testing programs worldwide. The two major proficiency testing programs for both soils and plants are the North American Proficiency Testing Program (NAPT) (see Appendix D), and the WEPL (ISE and IPE) program run by the Wageningen Agricultural University, Wageningen, the Netherlands (Houba et al., 1996). Reports from the WEPL program have been published by Houba et al. (1997a, b; 1998), and from the NAPT program by Wolf and Miller (1998) and Rayment et al. (2000).

One of the major advantages emerging from these programs is the evaluation of test procedures and the opportunity of laboratories to obtain assistance from groups that either are involved directly in the programs themselves or have experience in the conduct of soil and plant analyses, using the reports from the proficiency testing programs to determine where sources of errors exist. Yearly reviews are made of test methodology by the Soil and Plant Analysis Council and by the Soil Science Society of America, which manages the NAPT program (see Appendix D).

References

Aldenhoff, G.A. and A. Ernest. 1983. Quality assurance program at a municipal wastewater treatment plant laboratory. *J. Water Pollut. Control Fed.*, 55:1132–1137.

Anonymous 1979. Nonclinical Laboratory Studies, Good Laboratory Practices, Department of Health, Education, and Welfare, FDA, *Fed. Regis.*, Friday, December 22, 1979, Part II, 43(247).

Anonymous 1999. *Soil Analysis Handbook of Reference Methods*, CRC Press, Boca Raton, FL, 193–205.

Dux, J.P. 1986. *Handbook on Quality Assurance for the Analytical Laboratory,* Van Nostrand Reinhold, New York.

Fischbeck, R.D. 1980. Good laboratory practice regulations. *Am. Lab.,* 12:125–130.

Garfield, F.M., Ed. 1984. *Quality Assurance Principles for Analytical Laboratories,* Association of Official Analytical Chemists, Arlington, VA.

Garner, W.Y. and M.S. Barge, Eds. 1988. *Good Laboratory Practices,* ACS Symposium Series 369, American Chemical Society, Washington, D.C.

Hanlon, E.A. 1998. Quality control procedures for plant analysis, in Y.P. Karla, Ed., *Handbook of Reference Methods for Plant Analysis*, CRC Press, Boca Raton, FL, 199–207.

Hislop, J.S. 1980. Choice of the analytical method, in P. Bratter and P. Schramel, Eds., *Trace Element Analytical Chemisry in Medicine and Biology*, DeGruyter, Berlin, Germany, 747–767.

Horwitz, W. 1982. Evaluation of analytical methods used for regulation of foods and drugs. *Anal. Chem.*, 54:67A-76A.

Houba, V.J.G., J. Uttenbogaard, and P. Pellen. 1996. Wageningen evaluating programmes for analytical laboratorires (WEPAL), organization and purpose. *Commun. Soil Sci. Plant Anal.,* 27:421–429.

Houba, V.J.G., J. Uittenbogaard, P. Pellen, and D. van Dijk. 1997. International Soil-Analytical Exchange Report, 1996, Wageningen Evaluating Programmes for Analytical Laboratories, Department Soil Science Plant Nutrition, Wageningen Agricultural University, Wageningen, the Netherlands.

Houba, V.J.G., J. Uittenbogaard, D. Van Dijk, P. Pellen, and A. Brader. 1998. International Plant-Analytical Exchange Report, 1997, Wageningen Evaluating Programmes for Analytical Laboratories, Department Soil Science Plant Nutrition, Wageningen Agricultural University, Wageningen, the Netherlands.

Houba, V.J.G., E.J.M. Temminghoff, G.A. Gaikhorst, and W. van Vark. 2000. Soil analysis using 0.01 *M* $CaCl_2$ as extraction reagent. *Commun. Soil Sci. Plant Anal.,* 31:1299–1396.

Ihnat, M. 1998. Plant and related reference materials for data quality control of elemental content, in Y.P. Kalra, Ed., *Handbook of Reference Methods for Plant Analysis*, CRC Press, Boca Raton, FL, 235–284.

Kalra, Y.P., Ed. 1998. *Handbook of Reference Methods for Plant Analysis*, CRC Press, Boca Raton, FL, 199–207.

McLaughlin, R.D., M.S. Hunt, D.L. Murphy, and C.R. Chen. 1979. Instrumentation in environmental monitoring, in D.L. Hamphill, Ed., *Trace Elements in Environmental Health–XIII*, University of Missouri, Columbia, 224–234.

Miller, R.O., J. Kotuby-Amacher, and N.B. Dellevalle. 1996. A proficiency testing program for the agricultural laboratory industry: results for the 1994 program. *Commun. Soil Sci. Plant Anal.,* 27:451–461.

Rayment, G.E., R.O. Miller, and E. Sulaeman. 2000. Proficiency testing and other interactive measures to enhance analytical quality in soil and plant laboratories. *Commun. Soil Sci. Plant Anal.,* 31:1513–1530.

Sturgeon, R.E. 2000. Current practice and recent developments in analytical methodology for trace element analysis of soils, plants, and water, *Commun. Soil Sci. Plant Anal.* 31:1479–1512.

Taylor, J.K. 1987. *Quality Assurance of Chemical Measurements,* Lewis Publishers, Chelsea, MI.

Uriano, G.A. 1979. The NBS Standard Reference Materials Program: SRMs today and tomorrow. *ASTM Standardization News,* 7:8–13.

Wolf, A.M. and R.O. Miller. 1998. Development of a North American Proficiency Testing Program for soil and plant analysis. *Commun. Soil Sci. Plant Anal.,* 28:1685–1690.

Wolf, A.M., J.B. Jones, Jr., and T. Hood. 1996. Proficiency testing for improving analytical performance in soil testing proficiency testing programs. *Commun. Soil Sci. Plant Anal.,* 27:1611–1622.

Appendix A

List of Reagents, Standards, pH Buffers, Acids, and Indicators, and Preparation of Standard Acids, Bases, and Buffers Required in the Instruction Guide

A. Reagents

aluminum chloride ($AlCl_3 \cdot 6H_2O$)
aluminum sulfate [$Al_2(SO_4)_3$]
ammonium acetate ($NH_4C_2H_3O_2$)
ammonium bicarbonate (NH_4HCO_3)
ammonium chloride (NH_4Cl)
ammonium dihydrogen phosphate ($NH_4H_2PO_4$)
ammonium fluoride (NH_4F)
ammonium hydroxide (NH_4OH)
ammonium iron sulfate [$(NH_4)_2Fe(SO_4)_2 \cdot 6H_2O$]
ammonium molybdate [$(NH_4)_6Mo_7O_{24} \cdot 4H_2O$]
ammonium nitrate (NH_4NO_3)
ammonium sulfate [$(NH_4)_2SO_4$]
antimony metal (Sb)
antimony potassium tartrate [$K(SbO)C_4H_4O_6 \cdot \frac{1}{2}H_2O$]
ascorbic acid ($C_6H_8O_6$)
azomethine-H

barium chloride ($BaCl_2 \cdot 2H_2O$)
boric acid (H_3BO_3)
"Brij-35" — wetting agent
butanol

calcium acetate [$Ca(C_2H_3O_2)_2$]
calcium carbonate ($CaCO_3$)
calcium chloride ($CaCl_2 \cdot 2H_2O$)
calcium dihydrogen phosphate [$Ca(H_2PO_4)_2 \cdot 2H_2O$]
calcium hydroxide [$Ca(OH)_2$]
calcium nitrate [$Ca(NO_3)_2 \cdot H_2O$]
calcium oxide (CaO)
calcium sulfate ($CaSO_4 \cdot 5H_2O$)
CALGON (*see* sodium heximetaphosphate)
cesium chloride (CsCl)
citric acid ($C_6H_8O_7$)
copper sulfate ($CuSO_4 \cdot 5H_2O$)

Devada's alloy
DTPA — diethylenetriaminepentaacetic acid
DTPA — diethylenetriaminepentaacetic acid (penta sodium salt)
4,5-dihydroxy-2,7-naphthalene-disulfonic acid (disodium salt)
[$(HO)_2C_{10}H_4(SO_3Na)_2$] (CTA)
dipicrylamine (2,2′,4,4′,6,6′-hexanitrodiphenylamine)
disodium phosphate ($Na_2HPO_4 \cdot 12H_2O$)
disodium salt of ethylenedintrillo tetraacetic acid
disodium salt of nitrilotriacetic acid

ethanol (C_2H_5OH)
ethyl alcohol, denatured (C_2H_5OH)
EDTA — ethylenediaminetetraacetic acid
(ethylenedinitrillo) tetraacetic acid-tetrasodium salt

ferric nitrate [$Fe(NO_3)_3 \cdot H_2O$]
ferrous ammonium sulfate [$Fe(NH_4)_2SO_4 \cdot 6H_2O$]

Gum Acacia

Humic Acid Standard (Aldrich Chemical Company)
hydrogen peroxide (H_2O_2)
hydroxylamine-HCl

lanthanum nitrate hexahydrate [$La(NO_3)_3 \cdot 6H_2O$]

lanthanum oxide (La_2O_3)
lithium chloride (LiCl)

magnesium nitrate [$Mg(NO_3)_2 \cdot 6H_2O$]
magnesium perchlorate [$Mg(ClO_4)_2$]
manganese oxide (MnO_2)
mercuric thiocyanate [$Hg(CNS)_2$]

a-naphthylamine ($C_{10}H_9N$)
o-naphthyletyhylene-diamine dihydrochloride ($C_{12}H_{16}Cl_2N_2$)
nitrilotriacetic acid-disodium salt
p-nitrophenol ($HO \cdot C_6H_4 \cdot NO_2$)

paranitrophenol
phenol
phenyl acetate ($C_6H_5 \cdot O \cdot CO \cdot CH_3$)
potassium aluminum sulfate [$KAl(SO_4)_2 \cdot 12H_2O$]
potassium antimony tartrate ($KSbOC_4H_4O_6 \cdot 5H_2O$)
potassium chloride (KCl)
potassium dichromate ($K_2Cr_2O_7$)
potassium dihydrogen phosphate (KH_2PO_4)
potassium hydrogen phthalate ($KC_8H_5O_4$)
potassium hydroxide (KOH)
potassium nitrate (KNO_3)
potassium permanganate ($KMnO_4$)
potassium persulfate ($K_2S_2O_8$)
potassium sulfate (K_2SO_4)
propanol-2

selenium metal (Se)
silver sulfate (Ag_2SO_4)
sodium acetate ($NaC_2H_3O_2$)
sodium bicarbonate ($NaHCO_3$)
sodium borate ($Na_2B_4O_7 \cdot 10H_2O$)
sodium carbonate (Na_2CO_3)
sodium chloride (NaCl)
sodium dichromate ($Na_2Cr_2O_7$)
sodium glycerophosphate [$Na_2C_3H_5(OH)_2PO_4 \cdot 5\frac{1}{2}H_2O$]
sodium heximetaphosphate [$Na_6(PO_3)_6$]
sodium hydroxide (NaOH)
sodium hypochlorite (NaOCl)
sodium nitroprusside dihydrate {Na_2-[$Fe(CN)_5NO$]$\cdot H_2O$}

sodium potassium tartrate ($C_2H_4KNaO_6$)
sodium salicylate ($C_7H_5O_3Na$)
sodium tetraborate ($Na_2B_4O_7 \cdot 10H_2O$)
stannous chloride ($SnCl_2 \cdot 2H_2O$)
sulfamic acid (NH_2HSO_3)
sulfanilamide ($C_6H_8N_2O_2S$)

tin chloride ($SnCl_2 \cdot 2H_2O$)
titanium oxide (TiO_2)
triethanolamine (TEA)

zinc sulfate ($ZnSO_4 \cdot 7H_2O$)

B. Reagents for Preparation of Standards

ammonium dihydrogen phosphate ($NH_4H_2PO_4$)
arsenic (As) metal
arsenic oxide (As_2O_3)
barium chloride ($BaCl_2$)
barium carbonate ($BaCO_3$)
barium nitrate [$Ba(NO_3)_2$]
boric acid (H_3BO_3)
cadmium (Cd) metal
cadmium nitrate [$Cd(NO_3)_2 \cdot H_2O$]
cadmium oxide (CdO)
calcium carbonate ($CaCO_3$)
calcium nitrate [$Ca(NO_3)_2$]
chromium (Cr) metal
chromium chloride ($CrCl_3 \cdot 6H_2O$)
cobalt (Co) metal
cobalt chloride ($CoCl_2$)
copper (Cu) metal
copper oxide (CuO)
humic acid
iron (Fe) wire
iron oxide (Fe_2O_3)
lead (Pb) metal
lead nitrate [$Pb(NO_3)_2$]
lead oxide (PbO)
magnesium (Mg) ribbon
magnesium chloride ($MgCl_2$)

magnesium sulfate ($MgSO_4 \cdot 7H_2O$)
manganese (Mn) metal
manganese oxide (MnO_2)
molybdenum (Mo) metal
molybdic oxide (MoO_3)
nickel (Ni) metal
nickel chloride ($NiCl_2 \cdot 6H_2O$)
nickel nitrate [$Ni(NO_3)_2 \cdot 6H_2O$]
nickel oxide (NiO)
potassium carbonate (K_2CO_3)
potassium chloride (KCl)
potassium hydrogen phthalate ($KC_8H_5O_4$)
potassium sulfate (K_2SO_4)
selenium oxide (SeO_2)
sodium ammonium dihydrogen phosphate ($NaNH_4H_2PO_4$)
sodium carbonate (Na_2CO_3)
sodium chloride (NaCl)
sodium dihydrogen phosphate (NaH_2PO_4)
zinc (Zn) metal
zinc oxide (ZnO)
zinc sulfate [$Zn(NO_3)_2 \cdot 6H_2O$]

C. Reagents for Preparation of pH Buffers

boric acid (H_3BO_3)
citric acid ($C_6H_8O_7$)
disodium phosphate ($Na_2HPO_4 \cdot 12H_2O$)
monopotassium phosphate (KH_2PO_4)
potassium chloride (KCl)
potassium phthalate ($KHC_8H_4O_4$)
sodium hydroxide (NaOH)

D. Concentrated Acids

glacial acetic acid (CH_3COOH)
hydrochloric acid (HCl)
hydrofluoric acid (HF)
nitric acid (HNO_3)
perchloric acid ($HClO_4$)
phosphoric acid (H_3PO_4)
sulfuric acid (H_2SO_4)

E. Indicators

Bromocresol green
Bromocresol purple
Bromothymol blue
Erichrome Black T
malachite green oxalate
methyl red
methylene blue
phenolphathalein
n-phenylanthranilic acid

F. Standard Acids, Bases, and Buffers

To assist the user of this instruction guide, the following procedures to prepare standard acids, bases, and buffers are given.

1. Normality of Concentrated Acids

Acid	Specific gravity	Percent by weight	Normality
Conc. H_2SO_4	1.84	95.5–96.5	36.02
Conc. HCl	1.19	37.6	12.27
Conc. HNO_3	1.423	70.7	15.96
Conc. NH_4OH	0.899	28.8	7.39
Glacial CH_3COOH	1.049	100	17.47
Conc. H_3PO_4	1.689	86.3	44.62

2. Normal Acid Solution Preparation

To make various normal solutions, dilute aliquots of concentrated acid to 1000 mL with deionized water.

	mL of concentrated acid to make 1000 mL				
Acid	0.1 *N*	1.0 *N*	2.0 *N*	5.0 *N*	10.0 *N*
Conc. H_2SO_4	2.78	27.76	55.52	138.81	277.62
Conc. HC1	8.15	81.50	163.00	407.50	815.00
Conc. HNO_3	6.26	62.66	125.31	313.28	626.56

Acid	mL of concentrated acid to make 1000 mL				
	0.1 *N*	1.0 *N*	2.0 *N*	5.0 *N*	10.0 *N*
Conc. NH_4OH	13.53	135.32	270.63	676.59	—
Glacial CH_3COOH	5.72	57.24	114.48	286.2	572.4
Conc. H_3PO_4	2.24	22.41	44.82	112.06	224.11

3. Preparation of Buffer Solutions

a. Stock Solutions

Note: *Use reagent-grade chemicals and pure water.*

0.2 M Acid Potassium Phthalate ($KHC_8H_4O_4$)

Dry $KHC_8H_4O_4$ to constant weight at 110 to 115°C (230 to 239°F).

Weigh 40.836 g into a 1000-mL volumetric flask and bring to the mark with water.

0.2 M Monopotassium Phosphate (KH_2PO_4)

Dry KH_2PO_4 to constant weight at 110 to 115°C (230 to 239°F).

Weigh 27.232 g into a 1000-mL volumetric flask and bring to the mark with water.

0.2 M Boric Acid (H_3BO_3)–Potassium Chloride (KCl)

Dry H_3BO_3 to constant weight in desiccator over anhydrous calcium chloride ($CaCl_2$).

Dry KCl 2 days in an oven at 115 to 120°C (239 to 248°F).

Weigh 12.405 g H_3BO_3 and 14.912 g KCl into a 1000-mL volumetric flask and bring to the mark with water.

0.2 M Sodium Hydroxide (NaOH)

To 1 part NaOH, add 1 part water.

Mix to dissolve and let stand until clear (about 10 days).

Dilute 16 mL of this solution to 1000 mL with water.

Titrate against weighed amount of acid potassium phthalate ($KHC_8H_4O_4$).

Note: *0.04084 g* $KHC_8H_4O_4$ *= 1 mL 0.2 M NaOH. It is preferable to use factor with solution rather than try to adjust to exactly 0.2 M.*

b. Preparation of Buffer Solutions

Prepare standard buffer solutions from designated amounts of stock solutions and dilute to 200 mL with water.

	Phythalate–NaOH mixtures	
pH of mixture	0.2 *M* $KHC_8H_4O_4$, mL	0.2 *M* NaOH, mL
5.0	50	23.65
5.2	50	29.75
5.4	50	35.25
5.6	50	39.70
5.8	50	43.10
6.0	50	45.40
6.2	50	47.00

	KH_2PO_4–NaOH mixtures	
pH of mixture	0.2 *M* KH_2PO_4, mL	0.2 *M* NaOH, mL
5.8	50	3.66
6.0	50	5.64
6.2	50	8.55
6.4	50	12.60
6.6	50	17.74
6.8	50	23.60
7.0	50	29.54
7.2	50	34.90
7.4	50	39.34
7.6	50	42.74
7.8	50	45.17
8.0	50	46.85

Appendix B

Standards and Standard Preparation

A. Purpose

Calibration standards of precisely known composition are required to assay correctly soil extractants and plant tissue digests no matter what instrumental procedure is used. The three requirements for these standards are:

1. The elemental concentration must be certified or verified independently of the user.
2. The elemental concentration must be in the range bracket that is expected to be found in the unknowns.
3. The matrix composition must be the same or similar to that of the unknowns.

There are two types of standards, *primary standards* and *working standards*.

B. Primary Standards

Primary standards normally consist of an element at a specific concentration, such as 100, 1000, or 10,000 mg/L. They may be made by the analyst or purchased from a commercial company (see list at the end of this appendix). A primary standard may be made in the same matrix (such as the extraction reagent and digestion acids) as that of the unknowns for which the standard is to be used.

1. Preparation of Primary Standards

Note: *The elements, compounds, and acids used must be of reagent grade, free of adsorbed water, and the water used must be pure (Anonymous, 1997). The glassware used to prepare these standards should be carefully washed to minimize contamination (Kammin et al., 1995).*

Standards

Ammonium-Nitrogen Standard (1000 mg N/L)

Weigh 4.72 g ammonium sulfate [$(NH_4)_2SO_4$] into a 1000-mL volumetric flask and bring to volume with water.

Boron Standard (100 mg B/L)

Weigh 0.5716 g boric acid (H_3BO_3) into a 1000-mL volumetric flask and bring to volume with water.

Cadmium Standard (1000 mg Cd/L)

Weigh 1.000 g pure Cd metal into a 1000-mL volumetric flask.
Add 5 to 10 mL concentrated hydrochloric acid (HCl).
Evaporate almost to dryness and then dilute to 1000 mL with dilute (1:10) HCl.

Calcium Standard (1000 mg Ca/L)

Weigh 2.498 g calcium carbonate ($CaCO_3$) into a 1000-mL volumetric flask.
Add 50 mL water and add dropwise a minimum volume (approximately 20 mL) concentrated hydrochloric acid (HCl) to effect the complete solution of the $CaCO_3$.
Dilute to the mark with water.

Chloride Standard (1000 mg Cl/L)

Weigh 2.103 g potassium chloride (KCl) into a 1000-mL volumetric flask and bring to volume with water.

Note: *See Potassium Standard as this standard also contains 1103 mg K/L.*

Copper Standard (1000 mg Cu/L)

Weigh 1.000 g pure Cu metal into a 1000-mL volumetric flask.
Add in minimum amount concentrated nitric acid (HNO_3).
Add 5 mL concentrated hydrochloric acid (HCl).

Evaporate almost to dryness and dilute to 1000 mL with dilute (1:10) HCl.

Iron Standard (1000 mg Fe/L)

Weigh 1.000 g pure Fe wire into a 1000-mL volumetric flask.
Add 5 to 10 mL concentrated hydrochloric acid (HCl).
Evaporate to dryness.
Then bring to volume with dilute (1:10) HCl.

Magnesium Standard (1000 mg Mg/L)

Weigh 1.000 g Mg ribbon into a 1000-mL volumetric flask.
Dissolve in a minimum volume of dilute hydrochloric acid (HCl).
Dilute to 1000 mL with water.

Manganese Standard (1000 mg Mn/L)

Weigh 1.582 g manganese oxide (MnO_2) into a 1000-mL volumetric flask.
Add 5 mL concentrated hydrochloric acid (HCl).
Evaporate to dryness.
Then dilute to volume with dilute (1:10) HCl.

Phosphorus Standard (100 mg P/L)

Weigh 0.4394 g monobasic potassium phosphate (KH_2PO_4), which has been oven-dried at 100°C (212°F) into a 1000-mL volumetric flask. Bring to volume with water.

Note: *See Potassium Standard as this standard also contains 126.2 mg K/L.*

Potassium Standard (1000 mg K/L)

Weigh 1.9080 g potassium chloride (KCl) into a 1000-mL volumetric flask.
Bring to volume with water.

Note: *See Chloride Standard as this standard also contains 907.2 mg Cl/L.*

Nickel Standard (1000 mg Ni/L)

Weigh 1.000 g pure Ni metal into a 1000-mL volumetric flask.
Add a minimum amount of concentrated nitric acid (HNO_3).
Add 5 mL concentrated hydrochloric acid (HCl).
Evaporate almost to dryness and dilute to 1000 mL with dilute (1:10) HCl.

Nitrate–Nitrogen Standard (1000 mg NO_3–N/L)

Weigh 7.218 g potassium nitrate (KNO_3) into a 1000-mL volumetric flask.

Bring to volume with water.

Note: *See Potassium Standard as this standard also contains 2791.5 mg K/L.*

Sodium Standard (1000 mg Na/L)

Weigh 2.542 g sodium chloride (NaCl) into a 1000-mL volumetric flask.

Bring to volume with water.

Note: *See Chloride Standard as this standard contains 1541.7 mg Cl/L.*

Sulfate–Sulfur Standard (1000 mg S/L)

Weigh 5.434 g potassium sulfate (K_2SO_4) into a 1000-mL volumetric flask in about 400 mL ethylenediaminetetraacetic acid (EDTA) solution (weigh 5.84 g H_4EDTA into a 1000-mL volumetric flask, add 30 mL concentrated aqueous ammonia (NH_4OH), and bring to volume with water).

Note: *See Potassium Standard as this standard also contains 1219.3 mg K/L.*

Zinc Standard (1000 mg Zn/L)

Weigh 1.000 g pure Zn metal into a 1000-mL volumetric flask.

Add 5 to 10 mL concentrated hydrochloric acid (HCl).

Evaporate almost to dryness and dilute to 1000 mL with dilute (1:10) HCl.

Or weigh 4.3478 g zinc sulfate ($ZnSO_4 \cdot 7H_2O$) into a 1000-mL volumetric flask and bring to volume with water.

Ward (1978) has suggested the following reagents and solvents for preparing 1000 mg/L primary standards for use when an analyte is being assayed by ICP-AES.

Element	Reagents	Weight, g	Solvents
Aluminum (Al)	Al metal	1.0000	6 *M* HCl
	$AlCl_3 \cdot 6H_2O$	8.9481	1 *M* HCl
Arsenic (As)	As metal	1.0000	4 *M* HCl
	As_2O_3	1.3203	4 *M* HCl
Barium (Ba)	$BaCl_2$[a]	1.1516	Water
	$BaCO_3$[a]	1.4369	0.05 *M* HNO_3
	$Ba(NO_3)_2$	1.9029	Water

Element	Reagents	Weight, g	Solvents
Boron (B)	H_3BO_3	5.7191	Water
Cadmium (Cd)	Cd metal	1.0000	4 *M* HNO_3
	CdO	1.1423	4 *M* HNO_3
Calcium (Ca)	$CaCO_3$	2.4972	0.5 *M* HNO_3
	$Ca(NO_3)_2 \cdot 4H_2O$[a]	5.8920	Water
Chromium (Cr)	Cr metal	1.00011	4 *M* HCl
	$CrCl_3 \cdot 6H_2O$	5.1244	Water
Cobalt (Co)	Co metal	1.0000	4 *M* HCl
	$CoCl_2 \cdot 6H_2O$	4.0373	Water
Copper (Cu)	Cu metal	1.0000	4 *M* HNO_3
	CuO	1.2518	4 *M* HNO_3
Iron (Fe)	Fe metal	1.0000	4 *M* HCl
	Fe_2O_3	1.4297	4 *M* HCl
Lead (Pb)	Pb metal	1.0000	4 *M* HNO_3
	PbO	1.0772	4 *M* HNO_3
	$Pb(NO_3)_2$	2.6758	Water
Magnesium (Mg)	Mg metal	1.6581	0.5 *M* HCl
	$MgCl_2 \cdot 6H_2O$[a]	8.3621	Water
Manganese (Mn)	Mn metal	1.0000	4 *M* HNO_3
	MnO_2	1.5825	4 *M* HNO_3
Molybdenum (Mo)	Mo metal	1.0000	*Aqua regia*[b]
	MoO_3	1.5003	*Aqua regia*[b]
Nickel (Ni)	Ni metal	1.0000	4 *M* HCl
	NiO	1.2725	4 *M* HCl
	$NiCL_2 \cdot 6H_2O$	4.0489	Water
Phosphorus (P)	NaH_2PO_4	3.8735	Water
	$NaNH_4H_2PO_4$	3.7137	Water
Potassium (K)	KCl	1.9067	Water
	K_2CO_3	1.7673	1 *M* HCl
Selenium (Se)	SeO_2	1.4053	Water
Sodium (Na)	NaCl	2.5421	Water
	Na_2CO_3	2.3051	1 *M* HCl
Zinc (Zn)	Zn metal	1.0000	4 *M* HNO_3
	ZnO	1.2448	4 *M* HNO_3
	$Zn(NO_3)_2 \cdot 6H_2O$	4.5506	Water

[a] Not Specpure materials.
[b] *Aqua regia* (1 part concentrated HCl and 3 parts concentrated HNO_3).

C. Working Standards

Working standards are normally prepared from primary standards by diluting aliquots of the primary standard with either extraction reagents or digestion reagents to cover the anticipated range in concentration to be found in the soil extraction filtrate or plant tissue digest, and also to be within the operating range of the analytical procedure employed. Working standards may also be purchased from a commercial company (see list in Section G of this appendix).

Single-element-containing working standards are relatively easy to prepare. An example is the preparation of a series of calibration standards for the spectrophotometric (colorimetric) determination of P and No_3–N in a soil extract by the phosphomolybdate blue procedure and cadmium reduction, respectively. The working concentration range must be determined and standards prepared to cover that range. Normally, the useful concentration range for most spectrophotometric procedures (UV-VIS spectrophotometry, and flame emission and AAS) is 1 decade (0 to 10 mg/L), or possibly 2 (0 to 100 mg/L). In addition, it is necessary to determine, either by preliminary testing or from past experience, what elemental concentration range is expected in the unknowns. If the concentration range of unknowns is not within the working range of the spectrophotometric procedure chosen, then the unknowns must be diluted to bring them into that working range. For the UV-VIS spectrophotometric procedure, depending on the size of the cuvet, that is, the length of the light path through the cuvet, and the characteristics of the spectrophotometer, the working range may be decreased or increased.

A wider concentration range of elemental concentration (2 to 5 decades) for working standards is possible when ICP-AES is the elemental determination procedure.

The greater the number of working standards, the more precise the calibration curve will be. However, there are practical considerations that must be considered. Five working standards, including the zero standard (that is, a standard without the element of interest), are normally a sufficient number.

The same general procedure would be followed for multielement standards, standards required when the same analytical instrument can assay an unknown for more than one element. For example, standards for calibrating flame emission (for the elements K and Na) and AAS spectrophotometers (for the elements Ca and Mg, as well as Cu, Fe, Mn, and Zn) could contain more than one element, thereby simplifying their preparation and use. Normally, the working concentration range would be 1 to 2 decades, which either must be determined by actual testing or it may already be known. Then, the concentration range in the unknowns must be determined. Dilution of the unknowns may be necessary so that their elemental content will fall within the working range of the analytical procedure. When diluting samples, it is important that the obtained diluted sample be identical in matrix composition to the standards and that the diluting reagent be identical to that contained in the working standards. Diluting unknowns with water may alter the matrix effect, invalidating the assay.

For multielement working standards, it is not wise to have every element in the series at the same increasing concentration so that in one working standard all the elements are at the lowest and in another all are at the highest concentration. It may also be necessary to alternate the concentration of various elements in the working standards so that there is a minimum of possible interaction among the elements that might result in precipitation or some undesirable complex formation. The so-called zero standard, when the assay procedure is by either flame emission and AAS or ICP-AES, may be one in which there are elements at concentrations found in the unknowns. For example, Ca may be an element common in all the unknowns at a certain range in concentration; therefore, the zero standard for an element at relatively low concentration in the unknown, such as Zn, would contain Ca within that general range in concentration. This procedure in working standard preparation and composition would tend to minimize any matrix effect that Ca may have on emission or absorptive characteristics.

Finally, an essential requirement in the preparation of working standards is that they closely match the unknowns whose elemental content will be determined by comparison with the working standards. This means that for soil extractants, the working standards should be made in the extraction reagent. If fairly large aliquots of the primary standards are needed, they should also be made in the extraction reagent. The same requirement should be followed for the preparation of plant digest working standards. This requirement is intended to minimize what is known as the *matrix effect.*

D. Matrix Effects

The matrix effect is an influence that the chemical and physical characteristics of a standard or sample unknown can have on its assay. For example, the total ion content and type, pH, presence of organics, color, viscosity, etc. can contribute to the way a standard or unknown will react when assayed.

E. Matrix Modifiers

In some types of elemental determinations, most frequently assays by either AAS or GL-AAS, a matrix modifier is needed to ensure that the elemental species being detected or utilized is present. For example, in the determination of Ca by AAS, the presence of other companion ions, such as Al^{3+}, PO_4^{3-}, and SO_4^{2-}, as well as organic substances in the analyte, can combine with the element, reducing the concentration of the species (ground-state Ca atoms) being measured in the flame or gaseous material, resulting in an erroneous (normally low) assay result. The addition of a matrix modifier, in this case La, and sometimes Sr, will prevent this combining. For more details on this subject, refer to the book chapters by Watson and Isaac (1990) and Wright and Stuczynski (1996).

F. Blanks

A *blank* is obtained by carrying forward the analysis but without interaction with a sample. For example, carrying forward a soil extraction without interaction with a soil sample would produce a blank. A blank serves a very useful purpose by determining, if in the preparation of unknowns, an element(s) to be determined in the assay is being added systematically to the unknowns. A standard of matrix only (void of analyte elements) is not a blank and should not be so used. However, an assay of reagents used should be carefully checked for their freedom from the analyte(s) being determined. Blanks should be prepared when beginning a series of sample preparations and periodically thereafter, particularly when a change in the procedure occurs. High blank values indicate a significant source of contamination and may be sufficient to invalidate a determination. Some have suggested that working standards might be processed in the same manner as the unknown samples to contaminate both standards and unknowns equally. Such a procedure is dangerous since elimination of the source of contamination is the proper solution.

G. Commercial Sources for Primary and Working Standards

AccuStandard, Inc., 125 Market Street, New Haven, CT 06513 (800-442-5290)

Hawk Creek Laboratory, R.D. 1, Box 686, Simpson Road, Glen Rock, PA 17327 (800-637-2436; e-mail: sales@hawkcreeklab.com)

NSI Solutions, Inc., 2 Triangle Drive, Research Triangle Park, NC 27709 (800-234-7837)

Spex Chemical Division, 203 Horcross Ave., Metuchen, NJ 08840 (908-549-7145)

H. Soil and Plant Tissue Standards

The availability and use of soils and plant tissues of known composition are essential to ensure the accuracy of conducted assays. Sources and quality of such standards have been a problem in the past (primarily soil standards have not been readily available), but today a limited number of soil standards are obtainable from the Soil and Plant Analysis Council (621 Rose Street, Lincoln, NE 68502-2040), soils that have been assayed as a part of the North American Proficiency Testing Program (see Appendix D). However, a standard soil of a particular soil type (texture, organic matter content, pH, etc.) or analyte result, such as Bray P1 P, may not exist among the standard soils available from the council. Therefore, there is need for a wider range of soil standards representing a wider range of soil properties and assay results.

For plant tissues, the availability and range of materials of known composition available are considerable (Ihnat, 1998). For some plant tissues, the method of assay may be a factor in the use and reliability of these standards. For example, plant tissue standards, known as standard reference materials (SRMs) (Urinao, 1979; Alverez, 1980) from the National Institute of Science and Technology (NIST, formerly known as the National Bureau of Standards, NBS, Standard Reference Materials Program, Room 204, Building 202, Gaithersburg, MD 20899-0001) have reference values based on assay methods not commonly used in plant analysis laboratories; therefore, these reference values may not be the same as those that might be obtained using other assay procedures. Ihnat (1998) has listed plant materials available from various sources.

References

Alverez, R. 1980. NBS plant tissue standard reference materials. *J. Assoc. Off. Chem.*, 63:806–808.

Anonymous 1997. Water: what's in it and how to get it out. *Today's Chem.*, 6(1):16–19.

Ihnat, M. 1998. Plant and related reference materials for data quality control of element content, in Y.P. Kalra, Ed., *Handbook of Reference Methods for Plant Analysis*, CRC Press, Boca Raton, FL, 235–284.

Kammin, W.R., S. Cull, R. Know, J. Ross, M. McIntosh, and D. Thomson. 1995. Labware cleaning protocols for the determination of low-level metals by ICP-MS. *Am. Environ. Lab.*, 11/95–12/95:1–3.

Soltanpour, P.N., G.W. Johnson, S.M. Workman, J.B. Jones, and R.O. Miller. 1996. Inductively coupled plasma emission spectrometry and inductively coupled plasma-mass spectrometry, in D.L. Sparks, Ed., *Methods of Soil Analysis,* Part 3, *Chemical Methods*, SSSA Book Series Number 5, Soil Science Society of America, Madison, WI, 91–139.

Sturgeon, R.E. 2000. Current practice and recent developments in analytical methodology for trace element analysis of soils, plants, and water, *Commun. Soil Sci. Plant Anal.* 31:1479–1512.

Uriano, G.A. 1979. The NBD Standard Reference Materials Program: SRM's today and tomorrow. *ASTM Standardization News*, 7:9–13.

Ward, A.F. 1978. Stock standards preparation. *Jarrell-Ash Plasma Newslett.*, 1(2):14–15.

Watson, M.E. and R.A. Isaac. 1990. Analytical instruments for soil and plant analysis, in R.L. Westerman, Ed., *Soil Testing and Plant Analysis*, 3rd ed., SSSA Book Series Number 3, Soil Science Society of America, Madison, WI, 691–740.

Wright, R.J. and T.I. Stuczynski. 1996. Atomic absorption and flame emission spectrometry, in D.L. Sparks, Ed., *Methods of Soil Analysis*, Part 3, *Chemical Methods*, SSSA Book Series Number 5, Soil Science Society of America, Madison, WI, 65–90.

Appendix C

Extraction Reagents and Procedures

Notes: *(1) Use reagent-grade chemicals and pure water. (2) The elements or ions listed for determination by each Extraction Reagent are for those commonly determined by the procedure, and therefore may be incomplete since additional elements, mostly micronutrients and heavy metals, also have been determined by these Extraction Reagent procedures in more recent studies.*

Listed below are the extraction reagents and extraction procedures most commonly used to define the extractable elemental content of a soil; methods are given in fuller detail elsewhere in this guide.

Neutral Normal Ammonium Acetate

For extraction of Ca, K, Mg, and Na.

Extraction Reagent

1 N $NH_4C_2H_3O_2$, pH 7.0

Dilute 57 mL *glacial* acetic acid (CH_3COOH) with water to a volume of approximately 500 mL.

Then add 69 mL concentrated ammonium hydroxide (NH_4OH).

Caution: *Use a fumehood.*

Add sufficient water to obtain a volume of 990 mL.

After thoroughly mixing the solution, adjust the pH to 7.0 using either ammonium hydroxide (NH_4OH) or *glacial* acetic acid (CH_3COOH).

Dilute to a final volume of 1000 mL with water.

Alternate Method

Weigh 77.1 g ammonium acetate ($NH_4C_2H_3O_2$) in about 900 mL water in a 1000-mL volumetric flask.
After thoroughly mixing the solution, adjust the pH to 7.0 using either 3 *N* CH_3COOH or 3 *N* NH_4OH.
Bring to volume with water.

Extraction Procedure

Weigh 5 g or scoop 4.25 cm^3 air-dried <10-mesh-screened (2-mm) soil into a 50-mL extraction vessel.
Add 25 mL Extraction Reagent and shake for 5 min on a reciprocating shaker at a minimum of 180 oscillations/min.
Filter and collect the filtrate.

Mehlich No. 1

Formerly Double Acid or North Carolina Extractant, for extraction of Ca, Mg, P, K, Na, and Zn.

Extraction Reagent

0.05 N HCl in 0.025 N H_2SO_4

Pipette 4.3 mL concentrated hydrochloric acid (HCl) and 0.7 mL concentrated sulfuric acid (H_2SO_4) into a 1000-mL volumetric flask.
Bring to volume with water.

Extraction Procedure

Weigh 5 g or scoop 5 cm^3 air-dried <10-mesh-screened (2-mm) soil into a 50-mL extraction vessel.
Add 25 mL Extraction Reagent and shake for 5 min on a reciprocating shaker at a minimum of 180 oscillations/min.
Immediately filter and collect the filtrate.

Mehlich No. 3

For extraction of B, Ca, K, Mg, Mn, Na, P, and Zn.

Extraction Reagent

0.2 *N* acetic acid (CH_3COOH); 0.25 *N* ammonium nitrate (NH_4NO_3); 0.015 *N* ammonium fluoride (NH_4F); 0.13 *N* nitric acid (HNO_3); 0.001 *M* EDTA

Ammonium Fluoride–EDTA Stock Reagent

Add approximately 600 mL water to a 1000-mL volumetric flask.
Add 138.9 g ammonium fluoride (NH_4F) and dissolve.
Then add 73.05 g EDTA.
Dissolve the mixture and bring to volume with water.
Store in a plastic container.

Final Extraction Reagent Mixture

Add approximately 3000 mL water to a 4000-mL volumetric flask.
Add 80 g ammonium nitrate (NH_4NO_3) and dissolve.
Add 16 mL NH_4F–EDTA stock reagent (above) and mix well.
Add 46 mL *glacial* acetic acid (CH_3COOH) and 3.28 mL concentrated nitric acid (HNO_3).
Then bring to volume with water and mix thoroughly.
Achieve a final pH of 2.5 ± 0.1.
Store in a plastic container.

Extraction Procedure

Scoop 5 cm^3 air-dried <10-mesh-screened (2-mm) soil into an acid-washed 100-mL extraction vessel.
Add 50 mL Extraction Reagent and shake for 5 min on a reciprocating shaker.
Immediately filter and collect the filtrate and save for elemental content determination.
Store in a plastic container.

Note: *For the rationale of using a volume soil measure, refer to Mehlich (1973).*

Morgan

For extraction of B, Ca, Cu, Fe, K, Mg, Mn, P, and Zn.

Extraction Reagent

0.5 N $NaC_2H_3O_2 \cdot 3H_2O$

Weigh 100 g sodium acetate ($NaC_2H_3O_2 \cdot 3H_2O$) into a 1000-mL volumetric flask.
Add about 900 mL water.
Add 30 mL *glacial* acetic acid (CH_3COOH), adjust the pH to 4.8, and bring to volume with water.

Wolf Modification Extraction Reagent

Weigh 100 g sodium acetate ($NaC_2H_3O_2 \cdot 3H_2O$) into a 1000-mL volumetric flask and add about 300 mL water.
Add 30 mL *glacial* acetic acid (CH_3COOH) and 0.05 g DTPA (diethylenetriaminepentaacetic acid).
Dilute to 950 mL with water, adjust the pH to 4.8, and bring to volume with water.

Extraction Procedure

Scoop 5 cm^3 air-dried <10-mesh-sieved (2-mm) soil into a 50-mL extraction vessel.
Add 25 mL Extraction Reagent and shake for 5 min on a reciprocating shaker at a minimum of 180 oscillations/min.
Immediately filter and collect the filtrate.

Ammonium Bicarbonate–DTPA

For extraction of Cu, Fe, K, Mn, NO_3–N, P, and Zn.

Extraction Reagent

1 M NH_4HCO_3*–DPTA*

Obtain 0.005 *M* DTPA (diethylenetriaminepentaacetic acid) solution by adding 9.85 g DTPA (acid form) to 4500 mL water in a 5000-mL volumetric flask.
Shake for 5 h constantly to dissolve the DTPA.
Bring to 5000 mL with water. This solution is stable with regard to pH.

To 900 mL of the 0.005 *M* DTPA solution, add 79.06 g ammonium bicarbonate (NH_4HCO_3) gradually and stir gently with a rod to facilitate dissolution and to prevent effervescence when bicarbonate is added.

Dilute the solution to 1000 mL with the 0.005 *M* DTPA solution and mix gently with a rod.

Adjust the pH to 7.6 with slow agitation with a rod by adding 2 *M* hydrochloric acid (HCl).

Store the AB–DTPA solution under mineral oil.

Check the pH after storage and adjust with 2 *M* HCl dropwise, if necessary.

Note: *The cumulative volume of HCl added should not exceed 1 mL/L limit, after which a fresh solution should be prepared.*

Extraction Procedure

Weigh 10 g air-dried <10-mesh-sieved (2-mm) soil into a 125-mL conical flask.

Add 20 mL Extraction Reagent and shake on an Eberbach reciprocal shaker or an equivalent shaker for exactly 15 min at 180 cycles/min with flasks kept open.

Immediately filter the extracts through Whatman 42 filter paper.

Water

For extraction of Ca, K, Mg, and Na.

Extraction Reagent

Pure water

Extraction Procedure

Weigh 5 g or scoop 4.25 cm^3 air-dried <10-mesh-sieved (2-mm) soil into a 50-mL extraction vessel.

Add 25 mL water, seal the vessel with a stopper, and shake for 30 min on a reciprocating shaker at a minimum of 180 oscillations/min.

Allow to stand for 15 min to let the bulk of the soil settle.

Filter the supernatant liquid.

Note: *Discard the initial filtrate if it is turbid.*

Bray P1

For extraction of P.

Extraction Reagent

0.03 N NH_4F in 0.025 N HCl

Mix 30 mL 1 *N* ammonium fluoride (NH_4F) (weigh 37 g NH_4F into a 1000-mL volumetric flask and bring to volume with water; store in a polyethylene container and avoid prolonged contact with glass) with 50 mL 0.5 *N* HCl [dilute 20.4 mL concentrated hydrochloric acid (HCl) to 500 mL with water] in a 1000-mL volumetric flask and dilute to volume with water. Store in polyethylene.

Note: *This solution has a pH of 2.6 and is stable for more than 1 year.*

Extraction Procedure

Weigh 2.0 g or scoop 1.70 cm^3 air-dried <10-mesh-sieved (2-mm) soil into a 50-mL extraction vessel.

Add 20 mL Extraction Reagent and shake for 5 min on a reciprocating shaker at a minimum of 180 oscillations/min.

Immediately filter through Whatman No. 2 filter paper, limiting the filtration time to 10 min, and save the filtrate.

Bray P2

For extraction of P.

Extraction Reagent

0.03 N NH_4F in 0.1 N HCl

Mix 30 mL 1 *N* ammonium fluoride (NH_4F) (weigh 37 g NH_4F into a 1000-mL volumetric flask and bring to volume with water; store in a polyethylene container and avoid prolonged contact with glass) with 200 mL 0.5 *N* hydrochloric acid (HCl) (pipette 20.4 mL concentrated HCl into a 500-mL volumetric flask and dilute to volume with water) in a 1000-mL volumetric flask and bring to volume with water.

Note: *This solution has a pH of 2.6 and is stable for more than 1 year.*

Store in polyethylene.

Extraction Procedure

Weigh 2.0 g or scoop 1.70 cm^3 air-dried <10-mesh-sieved (2-mm) soil into a 50-mL extraction vessel.
Add 20 mL Extraction Reagent and shake for 5 min on a reciprocating shaker at a minimum of 180 oscillations/min.
Immediately filter through Whatman No. 2 filter paper, limiting the filtration time to 10 min, and save the filtrate.

Olsen

For extraction of P.

Extraction Reagent

0.5 N $NaHCO_3$

Weigh 42.0 g sodium bicarbonate ($NaHCO_3$) into a 1000-mL volumetric flask and bring to volume with water.
Adjust the pH to 8.5 with 50% sodium hydroxide (NaOH) or 0.5 *N* hydrochloric acid (HCl).
Add mineral oil to avoid exposure of the solution to air.
Store in a polyethylene container and check the pH of the solution before use and adjust if necessary.

Note: *Maintenance of the pH at 8.5 is essential.*

Extraction Procedure

Weigh 2.5 g or scoop 2 cm^3 air-dried <10-mesh-sieved (2-mm) soil into a 250-mL extraction vessel.
Add 50 mL Extraction Reagent and shake for 30 min on a reciprocating shaker at a minimum of 180 oscillations/min.
Immediately filter and collect the filtrate.

Caution: *Soil extraction is sensitive to temperature, changing 0.43 mg P/kg for each degree C for soils containing 5 to 40 mg P/kg.*

DTPA

For extraction of Cu, Fe, Mn, and Zn.

Extraction Reagent

DTPA (diethylenetriaminepentaacetic acid)

Weigh 1.96 DTPA $\{[(HOCOCH_2)_2NCH_2]_2NCH_2COOH\}$ into a 1000-mL volumetric flask.
Add 14.92 g triethanolamine (TEA).
Bring to approximately 950 mL with water.
Add 1.47 g calcium chloride ($CaCl_2 \cdot 2H_2O$).
Bring to 1000-mL with water while adjusting the pH to exactly 7.3 with 6 *N* hydrochloric acid (HCl) to achieve final concentration of 0.005 *M* DTPA (acid form), 0.1 *M* TEA, and 0.01 *M* $CaCl_2$.

Extraction Procedure

Weigh 10 g or scoop 8.5 cm^3 air-dried <10-mesh-sieved (2-mm) soil into a 125-mL extraction vessel.
Add 20 mL Extraction Reagent and shake on a reciprocating shaker for 2 h at a minimum of 180 oscillations/min.

Note: *Samples that are shaken longer than 2 h will give high results because a final equilibrium of the metal and soil is not reached in 2 h.*

Immediately filter and collect the filtrate.

2 *M* Potassium Chloride

For extraction of NH_4^- and NO_3–N).

Extraction Reagent

2 M KCl

Weigh 150 g potassium chloride (KCl) into a 1000-mL volumetric flask and bring to volume with water.

Extraction Procedure

Weigh 10 g air-dried <10-mesh-sieved (2-mm) soil into a 125-mL conical flask.
Add 50 mL Extraction Reagent and shake on an Eberbach reciprocal shaker or an equivalent shaker for exactly 15 min at 180 cycles/min.
Filter the slurry through Whatman 42 filter paper.

0.01 *M* Calcium Sulfate

For extraction of NO_3–N.

Extraction Reagent

0.01 M $CaSO_4 \cdot 2H_2O$

Weigh 1.72 g calcium sulfate ($CaSO_4 \cdot 2H_2O$) into a 1000-mL volumetric flask and bring to volume with water.

Extraction Procedure

Weigh 5 g air-dried <10-mesh-sieved (2-mm) soil into a 125-mL conical flask.
Add 50 mL Extraction Reagent and shake on an Eberbach reciprocal shaker or an equivalent shaker for exactly 15 min at 180 cycles/min.
Filter the slurry through Whatman 2 filter paper.

0.04 *M* Ammonium Sulfate

For extraction of NO_3–N.

Extraction Reagent

0.04 M $(NH_4)_2SO_4$

Weigh 5.28 g ammonium sulfate [$(NH_4)_2SO_4$] into a 1000-mL volumetric flask and bring to volume with water.

Extraction Procedure

Weigh 5 g air-dried <10-mesh-sieved (2-mm) soil into a 125-mL conical flask.

Add 50 mL Extraction Reagent and shake on an Eberbach reciprocal shaker or an equivalent shaker for exactly 15 min at 180 cycles/min.

Filter the slurry through Whatman 2 filter paper.

Calcium Phosphate (500 mg P/L)

For extraction of SO_4–S.

Extraction Reagent

$Ca(H_2PO_4)_2{\cdot}2H_2O$, 500 mg P/L

Weigh 2.03 g calcium phosphate [$Ca(H_2PO_4)_2{\cdot}2H_2O$] into a 1000-mL volumetric flask and bring to volume with water.

Extraction Procedure

Weigh 10 g air-dried <10-mesh-sieved (2-mm) soil into an extraction vessel.

Pipette into the vessel 25 mL Extraction Reagent and shake for 30 min at 180 cycles/min.

Add ¼ teaspoon (about 0.15 g) powdered charcoal and shake for an additional 3 min.

Filter and transfer a 10-mL aliquot into another flask.

0.5 *M* Ammonium Acetate in 0.25 *M* Acetic Acid

For extraction of SO_4–S.

Extraction Reagent

0.5 M $NH_4C_2H_3O_2$ in 0.25 M CH_3COOH

Weigh 39 g ammonium acetate ($NH_4C_2H_3O_2$) into a 1000-mL volumetric flask and bring to volume with 0.25 *M* acetic acid (CH_3COOH) (dilute 14.31 *glacial* CH_3COOH in 1000 mL water).

Extraction Procedure

Weigh 10 g air-dried <10-mesh-sieved (2-mm) soil into an extraction vessel.
Pipette 25 mL Extraction Reagent into the flask and shake for 30 min at 180 cycles/min.
Add ¼ teaspoon (about 0.15 g) of powdered charcoal and shake for an additional 3 min.
Filter and transfer a 10-mL aliquot into another flask.

0.01 *M* Calcium Chloride

For extraction of B, Cu, Fe, K, Mg, Mn, NO_3^- and NH_4–N, P, SO_4–S, Zn.

Extraction Reagent

0.01 M $CaCl_2 \cdot 2H_2O$

Weigh 1.47 g calcium chloride dihydrate ($CaCl_2 \cdot 2H_2O$) into a 1000-mL volumetric flask and bring to volume with water.

Comment: *$CaCl_2 \cdot 2H_2O$ may absorb water on standing and should be standardized by titration with EDTA at pH = 10 with Eriochrome Black T as an indicator.*

Extraction Procedure

Weigh 10.0 g air-dried <10-mesh-sieved (2-mm) soil into a 250-mL polystyrene bottle.
Add 100 mL 0.01 *M* $CaCl_2$ and shake mechanically for at least 2 h at room temperature (20°C; 68°F).
Decant about 60 mL of the slurry into a 100-mL centrifuge tube and centrifuge for 10 min at about 1800 *g*.
Carefully remove the supernatant for analysis.

Appendix D

The North American Proficiency Testing Program for Soil, Plant, and Water Analysis Laboratories (NAPT)

The goal of the North American Proficiency Testing Program (NAPT) is to assist soil, plant, and water testing laboratories in their performance through interlaboratory sample exchanges and statistical evaluation of the analytical data. The program was created to benefit the agricultural testing laboratory industry.

The program guidelines have been developed for the agricultural industry by representatives from groups familiar with and involved in standardizing methods and in developing nutrient recommendations for soil and plant analysis methods within the United States and Canada. It is operated as an activity of the Soil Science Society of America and overseen by an oversight committee comprising representatives of the above-mentioned groups. These include Regional Soil and Plant Analysis Workgroups; scientific organizations; state/provincial departments of agriculture; the Soil and Plant Analysis Council, and private and public soil and plant analysis laboratories.

Program Objectives

- Provide an external quality assurance program for agricultural laboratories
- Develop a framework for long-term improvement of quality assurance of the agricultural laboratory industry
- Identify variability of specific analytical methods

Specific Soil Analyses

Saturated Paste Percentage (%), pH, EC_e, HCO_3, Ca, Mg, Na, SAR, Cl, SO_4, and B
Soil pH: (1:1), (1:2) in water and 0.01 *M* $CaCl_2$
Buffer pH*
NO_3–N*
NH_4–N, KCl extractable
Extractable P*
Extractable K* and Al
Extractable Ca, Mg, and Na*
Extractable SO_4–S
Micronutrients: Zn, Mn, Fe, Cu, B, and Cl*
Soil organic matter*
Soil total organic carbon and nitrogen*
Inorganic carbon
Particle Size Analysis: sand, silt, and clay
CEC

Specific Plant Analyses

NO_3–N, PO_4–P, SO_4–S, and Cl
Total nitrogen (two methods)
Total P, K, S, Ca, Mg, Na, Al, B, Zn, Mn, Fe, Cu, and Mo (three methods)

Specific Water Analyses

pH, EC, Ca, Mg, Na, Cl, SAR, NO_3–N, HCO_3, CO_3, SO_4–S, B, total P, and cation/anion ratio

Program Basics

The NAPT program is based on the quarterly submission of five soil (600 g), three plant materials (8 g), and three water samples (200 mL) using reference methods of analysis described in Regional Soil Work Group publications of NEC-67, NCR-13, SERA-6, WCC-103, and Methods Manual for Forest Soil and Plant Analysis, Forestry Canada.

* Multiple methods.

Quarterly program results are statistically compiled and outlier values of individual laboratories identified. Warning limits are reported for each analysis based on the median and median absolute deviation. Results are kept confidential and provided within 30 days. An annual report is prepared with an overview of the program quantifying proficiency and reviewing inter- and intralaboratory precision. Laboratories enrolling in state/province certification are the programs of Iowa, Illinois, Minnesota, Missouri, Nebraska, and Ontario. The NAPT program provides quarterly summaries to the respective state program coordinator(s).

Fee Structure

The 2000 NAPT annual fees are:

Soils-only program	\$495
Soil and plant program	\$500
Soil and plant and water	\$620

Exchanges will occur in the months of March, May, August, and November.

To enroll in the NAPT Program, payment is made payable to SSSA:

Attn. 2000 NAPT Program
Soil Science Society of America
677 South Segoe Road
Madison, WI 53711-1086

For further information, contact the program coordinator:

Dr. Robert O. Miller
NAPT Program Coordinator
Soil and Crop Sciences Department
Colorado State University
Fort Collins, CO 80523

Phone: 970-686-5702
Cellular: 970-227-2549
Fax: 970-491-0564
E-mail: Rmiller@lamar.colostate.edu

To purchase reference soils and plant materials utilized in the program, contact:

Janice Kotuby-Amacher
USU Analytical Laboratory
Utah State University
Logan, UT 84332
Phone: 435-797-0008
Fax: 435-797-3376
E-mail: jkotuby@mendel.usu.edu

Acknowledgments

The NAPT is overseen by an oversight committee of the SSSA and acknowledges the contributions of the following groups: regional work groups NEC-67, NCR-13, SERA-6, and WCC-103; Soil and Plant Analysis Council; Soil Science Society of America; Canadian Society of Soil Science; The Minnesota Department of Agriculture; The Missouri Laboratory Testing Program; Nebraska Department of Agriculture; Iowa Department of Ag & Land Stewardship; Illinois Soil Testing Association; Missouri Soil Test Certification Program; Purdue FSA Program; Ministry of Agriculture Ontario; Michigan State FSA Program; Ohio FSA Program; Wisconsin FSA Program; USDA-NRCS; USEPA; and the commercial laboratory industry.

References

Miller. R.O., J. Kotuby-Amacher, and N.B. Dellevalle. 1996. A proficiency testing program for the agricultural laboratory industry: results for the 1994 program. *Commun. Soil Sci. Plant Anal.,* 27:451–461.

Rayment, G.E., R.O. Miller, and E. Sulaeman. 2000. Proficiency testing and other interactive measures to enhance analytical quality in soil and plant laboratories. *Commun. Soil Sci. Plant Anal.,* 31:1513–1530.

Wolf, A.M. and R.O. Miller. 1998. Development of a North American Proficiency Testing Program for soil and plant analysis. *Commun. Soil Sci. Plant Anal.,* 28:1685–1690.

Wolf, A.M., J. B. Jones, Jr., and T. Hood. 1996. Proficiency testing for improving analytical performance in soil testing proficiency testing programs. *Commun. Soil Sci. Plant Anal.,* 27:1611–1622.

Appendix E

ASI Extraction Reagent Method for Soil Analysis

Principle of Method

This method is effective for predicting fertilizer needs for soils varying in a wide range in pH (acid to alkaline) and in a range of soil conditions. First published in 1972 (ISFEIP, 1972), these methods are in use at the Agro Services International, Orange City, Florida, a laboratory providing soil analysis and interpretation services.

A. Determination of P, K, Cu, Fe, Mn, and Zn

Extraction Reagent

0.25 *N* sodium bicarbonate; 0.01 *M* EDTA; 0.01 *N* ammonium fluoride.

Weigh 21.0 g sodium bicarbonate ($NaHCO_3$), 3.72 g disodium EDTA, and 0.37 g ammonium fluoride (NH_4F) into 500 mL water in a 1000-mL volumetric flask.

Add 100 mL previously dissolved Superfloc 127 (weigh 5 g Superfloc into a 1000-mL beaker, add 5 mL methanol while stirring, bring to 1000 mL with water, and let stand for 12 h; store in a closed bottle).

Bring to volume with water.

Extraction Procedure

Scoop 2.5 cm^3 air-dried 2-mm soil in a 60-mL extraction vessel.
Add 25 mL Extracting Reagent and stir for 10 min.
Immediately filter the extracts through Whatman No. 1 or comparable filter paper.

Determination. P can be determined by UV-VIS spectrophotometry (colorimetry) or by plasma emission spectrometry; K by flame photometry; Cu, Fe, Mn, and Zn by AAS; and K, Cu, Fe, Mn, and Zn by plasma emission spectrometry.

B. Extractable Acidity and Determination of NH_4–N, Ca, Mg, and Na

Extraction Reagent 1N KCl

Weigh 74.5 g potassium (KCl) into a 1000-mL flask.
Add 0.025 g dissolved Superfloc 127 (weigh 5 g Superfloc into a 1000-mL beaker, add 5 mL methanol while stirring, bring to 1000 mL with water, and let stand for 12 h; store in a closed bottle).
Dilute to volume with water.

Extraction Procedure

Scoop 2.5 cm^3 of soil into a 60-mL extraction vessel.
Add 25 mL 1 *N* KCl and stir the mixture for 10 min.
Immediately filter through Whatman No. 1 or comparable filter paper.

Determination of extractable acidity. Take 10 mL of the filtrate, add 15 mL water, add about 3 to 4 drops of phenolphthalien, and titrate to pink end point with 0.01 *N* sodium hydroxide (NaOH).

Determination of NH_4

NaOH Reagent: Weigh 27 g sodium hydroxide (NaOH), 3 g disodium EDTA, and 5 g sodium actetate ($NaC_2H_3O_2$) in 1000-mL volumetric flask and dilute to volume with water; store in polyethylene bottle.

Phenol Reagent: On the day of use, add 12 mL 90% phenol to 250 mL NaOH Reagent.

NaOCl Reagent: Dilute 100 mL 5.25% sodium hypochlorite (NaOCl) (Clorox) to 400 mL with water.

Determination Procedure: Take 3 mL soil filtrate, add 4 mL Phenol Reagent and 10 mL NaOCl Reagent, let stand at least 20 min but no longer than 2 h, and determine the transmittance at 630 nm in a spectrophotometer.

Determination of Ca, Mg, and Na. Ca and Mg can be determined by AAS, Na by flame emission photometry, and Ca, Mg, and Na by plasma emission spectrometry.

C. Determination of S and B

Extraction Reagent

Weigh 2.02 g monobasic calcium phosphate [$CaH_4(PO_4)_2 \cdot H_2O$] into a 1000-mL volumetric flask, 1 mL concentrated hydrochloric acid (HCl), and 0.05 g Superfloc (weigh 5 g Superfloc into a 1000-mL beaker, add 5 mL methanol while stirring, bring to 1000 mL with water, and let stand for 12 h; store in a closed bottle).

Dilute to volume with water.

Extraction Procedure

Scoop 5.0 cm^3 soil into a 60-mL extraction vessel.
Add 25 mL Extraction Reagent and stir the mixture for 10 min.
Immediately filter through Whatman No. 1 or comparable filter paper.

Determination of sulfate (SO_4)

Acid Seed Reagent: Add 180 mL concentrated nitric acid (HNO_3) to about 500 mL water into a 2000-mL volumetric flask, add 400 mL *glacial* acetic acid (CH_3COOH), add 10 g polyvinyl prolidone (PVP-K30), which has been dissolved in about 300 mL water, 6 mL 1000 SO_4–S solution, and bring to volume with water.

Acetic Acid Reagent: Add 120 mL *glacial* acetic acid (CH_3COOH) to 1000 mL with water.

Turbidity Barium Chloride Reagent: On the day of use, weigh 37.5 g barium chloride ($BaCl_2 \cdot 2H_2O$) in 250 mL Acetic Acid Reagent.

Determination Procedure: Take 7 mL extract, add 9 mL Acid Seed Reagent, and 4 mL Turbidity Barium Chloride Reagent. Thoroughly mix and allow to stand at least 10 min but no longer than 30 min, and read the transmittance at 535 nm with a UV-VIS spectrophotometer.

Caution: *The determination solution should not be colder than 23°C (73°F).*

Determination of B. Boron can be determined by either UV-VIS spectrophotometry or plasma emission spectrometry.

D. Soil Organic Matter Determination

Extraction Reagent

0.5 N NaOH–0.01 M EDTA–2% methanol

In a 2000-mL plastic volumetric flask, add 1000 mL water, 16 g sodium hydroxide (NaOH), 7.44 g sodium EDTA, and dissolve.

Then add 200 mL methanol and bring to volume with water.

Procedure

Scoop 1 cm^3 of soil into a 60-mL extraction vessel.

Add 25 mL Extraction Reagent and stir the mixture for 10 min.

After stirring, add 25 mL Superfloc (weigh 5 g Superfloc into a 1000-mL beaker, add 5 mL methanol, while stirring bring to 1000 mL with water, and let stand for 12 h; store in a closed bottle). Stir just enough to mix.

Let stand undisturbed for 20 min.

Take a 2-mL aliquot of the supernatant and add 10 mL water.

Transfer to a 1-cm cuvette and read the transmittance at 420 nm with a UV-VIS spectrophotometer.

Standard curve

% Transmittance	% Organic matter
100	0
33	3.5
11	7.0

E. Interpretation Levels

Element	Critical level	Low	Optimum	Above
	Sandy Soils, meq 100/cm³			
Calcium (Ca)	1.3	1.6	10	20
Magnesium (Mg)	0.4	0.8	5	10
Potassium (K_2O)	0.2	0.4	2	4
	Sandy Soils, mg/kg			
Phosphorus (P_2O_5)	12	24	150	300
Sulfur (S)	12	24	150	300
Boron (B)	0.2	0.4	6	12
Copper (Cu)	1.0	2.0	30	50
Iron (Fe)	12	24	200	400
Manganese (Mn)	3.0	6.0	100	300
Zinc (Zn)	1.5	3.0	25	100
	Loams, Clays, Organic Soils, meq 100/cm³			
Calcium (Ca)	2.0	4.0	24	48
Magnesium (Mg)	1.0	2.0	12	24
Potassium (K_2O)	0.2	0.4	2	4
	Loams, Clays, Organic Soils, mg/kg			
Phosphorus (P_2O_5)	12	24	200	400
Sulfur (S)	12	24	200	400
Boron (B)	0.3	0.6	8	16
Copper (Cu)	1.5	3	25	75
Iron (Fe)	12	24	300	600
Manganese (Mn)	3	6	150	300
Zinc (Zn)	1.5	3	30	125

Source: International Soil Fertility Evaluation and Improvement Program (ISFEIP). 1972. Annual Report. Soil Science Department North Carolina State University, Raleigh.

Appendix F

Reference Texts

Soil Analysis

Anonymous 1970. Physical and Chemical Methods of Soil and Water Analysis, FAO Soil Bulletin Number 10, Food and Agriculture Organizations of the United Nations, Rome, Italy.

Brown, J.R., Ed. 1998. Recommended Chemical Soil Test Procedures for the North Central Region, North Central Regional Publication No. 221 (revised), Missouri Agricultural Experiment Station SB 1001, University of Missouri, Columbia.

Carter, R.L., Ed. 1993. *Soil Sampling and Methods of Analysis,* CRC Press, Boca Raton, FL.

Dahnke, W.C., Ed. 1988. Recommended Chemical Soil Test Procedures for the North Central Region. North Central Regional Publication No. 221 (revised), North Dakota Agricultural Experiment Station Bulletin No. 499 (revised), North Dakota State University, Fargo.

Davis, J. and F. Freitas. 1970. Physical and Chemical Methods of Soil and Water Analysis, Soils Bulletin 10, Food & Agriculture Organization, Rome, Italy.

Donahue, S.J., Ed. 1983. Reference Soil Test Methods for the Southern Region of the United States, Southern Cooperative Series Bulletin 289.

Donohue, S.J., Ed. 1992. Reference Soil and Media Diagnostic Procedures for the Southern Region of the United States, Southern Cooperative Series Bulletin No. 374, Virginia Agricultural Experiment Station, Blacksburg.

Hesse, R. 1971. *A Textbook of Soil Chemical Analysis*, Chemical Publishing Company, New York.

Jackson, M.L. 1958. *Soil Chemical Analysis*, Prentice-Hall, Englewood Cliffs, NJ.

Klute, A., Ed. 1996. *Methods of Soil Analysis, Physical, and Mineralogical Properties*, Part 1, 2nd ed., ASA Monograph Number 9, American Society of Agronomy, Madison, WI.

Loveday, J., Ed. 1974. Methods for Analysis of Irrigated Soils, Technican Communication Number 54, Commonwealth Bureau of Soils, Farnham Royal, Bucks, England.

Page, A.L., R.H. Miller, and D.R. Keeney, Eds. 1982. *Methods of Soil Analysis,* Part 2, *Chemical and Microbiological Properties*, 2nd ed., Soil Science Society of America, Madison, WI.

Radojevic, M. and V.N. Bashkin. 1999. *Practical Environmental Analysis*, The Royal Society of Chemistry, Cambridge, U.K.

Sims, J.T. and A.M. Wolf, Eds. 1991. Recommended Soil Testing Procedures for the Northeastern United States, Northeast Regional Publication Number 493, Agricultural Experiment Station, University of Delaware, Newark.

Smith, K.A., Ed. 1983. *Soil Analysis: Instrumental Techniques and Related Procedures*, Marcel Dekker, New York.

Smith, K.A., Ed. 1991. *Soil Analysis: Modern Instrumental Techniques,* 2nd ed., Marcel Dekker, New York.

Sparks, D.L., Ed. 1996. *Methods of Soil Analysis,* Part 3, *Chemical Method*, SSSA Book Series Number 5, Soil Science Society of America, Madison, WI.

Plant Analysis

Johnson, C.M. and A. Ulrich. 1959. Analytical Methods for Use in Plant Analysis, Bulletin 766, University of California Agricultural Experiment Station, Berkeley.

Kalra, Y.P., Ed. 1998. *Handbook of Reference Methods for Plant Analysis,* CRC Press, Boca Raton, FL.

Plank, C.O. 1992. Plant Analysis Reference Procedures for the Southern Region of the United States, Southern Cooperative Series Bulletin 368, University of Georgia, Athens.

Walinga, I., J.J. van der Lee, V.J.G. Houba, and I. Novozamsky, Eds. 1995. *Plant Analysis Manual,* Kluwer Academic Publishers, Dordrecht, the Netherlands.

Soil and Plant Analysis

Anonymous 1980. Soil and Plant Testing and Analysis, FAO Soil Bulletin Number 38/1, Food and Agriculture Organization of the United Nations, Rome, Italy.

Chapman, H.D. and P.F. Pratt, Eds. 1982. *Methods of Analysis of Soil, Plants, and Waters,* Priced Publication 4034, Division of Agricultural Sciences, University of California, Berkeley.

Gavlak, R.G., D.A. Horneck, and R.O. Miller, Eds. 1994. Plant, Soil, and Water Reference Methods for the Western Region, Western Regional Extension Publication WREP 125, University of Alaska Extension Service, Fairbanks.

Horwitz, W., Ed. 2000. *Official Methods of Analysis by the Association of Official Analytical Chemists. Chemicals, Contaminants, Drugs,* Vol. I, Association of Official Analytical Chemists, Arlington, VA.

Kalra, Y.P. and D.G. Maynard. 1991. Methods Manual for Forest Soil and Plant Analysis, Northwest Region, Information Report NOR-X-319, Forestry Canada, Edmonton, Canada.

Kitchens, H.B., Ed. 1948. *Diagnostic Techniques for Soils and Crops,* The American Potash Institute, Washington, D.C.

Mitchell, R.L. 1964. The Spectrochemical Analysis of Soils, Plants, and Related Materials, Technical Communications No. 44A, Commonwealth Agricultural Bureax, Farnham Royal, Bucks, England.

Mortvedt, J.J., F.R. Cox, L.M. Shuman, and R.M. Welch, Eds. 1991. *Micronutrients in Agriculture*, 2nd ed., SSSA Book Series Number 4, Soil Science Society of America, Madison, WI.

Piper, C.S. 1942. *Soil and Plant Analysis,* Hassell Press, Adelaide, Australia.

Walsh, L.M. and J.D. Beaton, Eds. 1972. *Soil Testing and Plant Analysis,* rev. ed., Soil Science Society of America, Madison, WI.

Westerman, R.L., Ed. 1990. *Soil Testing and Plant Analysis,* 3rd ed., SSSA Book Series Number 3, Soil Science Society of America, Madison, WI.

Interpretation Texts

Adriano, D.C. 1986. *Trace Elements in the Terrestrial Environment*, Springer-Verlag, New York.

Bennett, W.F., Ed. 1993. *Nutrient Deficiencies and Toxicities in Crop Plants,* APS Press, American Phytopathological Society, St. Paul, MN.

Bergmann, W. 1992. *Nutritional Disorders of Plants: Development, Visual and Analytical Diagnosis*, Gustav Fischer Verlag, Jena, Germany.

Beverly, R.B. 1991. *A Practical Guide to the Diagnosis and Recommendation Integrated System (DRIS)*, MicroMacro Publishing, Athens, GA.

Black, C.A. 1993. *Soil Fertility Control and Evaluation*, Lewis Publishers, Boca Raton, FL.

Brown, J.R., Ed. 1987. *Soil Testing: Sampling, Correlation, Calibration, and Interpretation,* SSSA Special Publication 21, Soil Science Society of America, Madison, WI.

Chapman, H.D., Ed. 1966. *Diagnostic Criteria for Plants and Soils*, University of California, Division of Agriculture, Berkeley.

Childers, N.F. 1966. *Nutrition of Fruit Crops, Temperate, Sub-Tropical, and Tropical,* Horticultural Publications, Rutgers, The State University of New Jersey, New Brunswick.

Cottenie, A. 1980. Soil and Plant Testing as a Basis of Ferilizer Recommendations, FAO Soils Bulletin Number 38/2, Food and Agriculture Organization of the United Nations, Rome, Italy.

Davidescu, D. and V. Davidescu. 1972. *Evaluation of Fertility by Plant and Soil Analysis*, Abacus Press, Kent, England.

Fairbridge, R.W. and C.W. Finkl, Jr. Eds. 1979. *The Encyclopedia of Soil Science, Physics, Chemistry, Biology, Fertility, and Technology*, Part 1, Dowden, Hutchinson & Ross, Stroudsburg, PA.

Foth, H.D. and B.G. Ellis. 1997. *Soil Fertility*, 2nd ed., Lewis Publishers, Boca Raton, FL.

Glass, A.D.M. 1989. *Plant Nutrition. An Introduction to Current Concepts*, Jones and Bartlett, Boston, MA.

Goodall, D.W. and P.G. Gregory. 1947. Chemical Composition of Plants as an Index of Their Nutritional Status, Imperial Bureau Horticultural Plantation Crops, Technical Communications No. 17, Ministry of Agriculture, London, England.

Grundon, N.J. 1987. *Hungry Crops. A Guide to Nutrient Deficiencies in Field Crops,* Queensland Department of Primary Industries, Brisbane, Australia.

Halliday, D.J. and M.E. Trenkel, Eds. 1992. *IFA World Fertilizer Use Manual,* International Fertilizer Industry Association, Paris, France.

Hardy, G.W., Ed. 1981. *Soil Testing and Plant Analysis. Soil Testing,* Part I, Special Publication No. 2, Soil Science Society of America, Madison, WI.

Hardy, G.W., Ed. 1967. *Soil Testing and Plant Analysis. Plant Analysis,* Part II, Special Publication Number 2, Soil Science Society of America, Madison, WI.

Jones, J.B., Jr. 1994a. *Plant Analysis* (VHS video). On Sat One, Anderson, SC.

Jones, J.B., Jr. 1994b. *Tissue Testing* (VHS video). On Sat One, Anderson, SC.

Jones, J.B., Jr. 1998. *Plant Nutrition Manual*, CRC Press, Boca Raton, FL.

Ludwick, A.E., Ed. 1998. *Western Fertilizer Handbook*, 2nd Horticultural ed., Interstate Publishers, Danville, IL.

Marschner, H. 1986. *Mineral Nutrition of Higher Plants,* Academic Press, New York.

Martin-Prevel, P., J. Gagnard, and P. Gauier. 1987. *Plant Analysis as a Guide to the Nutrient Requirements of Temperate and Tropical Crops*, Lavoisier Publishing, New York.

Maynard, D.N. and G.J. Hochmuth. 1997. *Knott's Handbook for Vegetable Growers*, 4th ed., John Wiley & Sons, New York.

Mengel, K. and E.A. Kirkby. 1982. *Principles of Plant Nutrition*, 3rd ed., International Potash Institute, Berne, Switzerland.

Mills, H.A. and J.B. Jones, Jr. 1996. *Plant Analysis Handbook II*, MicroMacro Publishing, Athens, GA.

Pais, I. and J.B. Jones, Jr. 1996. *Handbook on Trace Elements in the Environment,* St. Lucie Press, Boca Raton, FL.

Peverill, K.I., L.A. Sparrow, and D.J. Reuter, Eds. 1999. *Soil Analysis: An Interpretation Manual,* CSIRO Publishing, Collingwood, Australia.

Peck, T.R., J.T. Cope, Jr., and D.A. Whitney, Eds. 1977. *Soil Testing: Correlating and Interpreting the Analytical Results,* ASA Special Publication Number 29, American Society of Agronomy, Madison, WI.

Rechcigl, J.E., Ed. 1995. *Soil Amendments and Environmental Quality,* Lewis Publishers, Boca Raton, FL.

Reid, K., Ed. 1998. Soil Fertility Handbook, Ontario Ministry of Agriculture, Food and Rural Affairs, Queen's Printer for Ontario, Toronto, Canada.

Reuter, D.J. and J.B. Robinson. 1997. *Plant Analysis: An Interpretation Manual,* 2nd ed., CSIRO Publishing, Collingwood, Australia.

Scaife, A. and M. Turner. 1984. *Diagnosis of Mineral Disorders in Plants,* Vol. 2, *Vegetables,* Chemical Publishing, New York.

Sumner, M.E., Ed. 2000. *Handbook of Soil Science*, Lewis Publishers, Boca Raton, FL.

General Related Texts

Adriano, D.C. 1996. *Trace Elements in the Terrestrial Environment*, Springer-Verlag, New York.

Adriano, D.C., Ed. 1996. *Biogeochemistry of Trace Elements,* Lewis Publishers, Boca Raton, FL.

Anonymous 1996. *Soil Fertility Manual*, Potash & Phosphate Institute, Norcross, GA.

Barber, S.A. 1995. *Soil Nutrient Bioavailability: A Mechanistic Approach,* 2nd ed., John Wiley & Sons, Chichester, U.K.

Bould, C., E.J. Hewitt, and P. Needham. 1984. *Diagnosis of Mineral Disorders in Plants: Principles*, Vol. 1, Chemical Publishing Co., New York.

Childers, N.F., Ed. 1968. *Fruit Nutrition: Temperate to Tropical*, Horticultural Publications, Rutgers, The State University of New Jersey, New Brunswick.

Kabata-Pendias, A. 2000. *Trace Eelements in Soil and Plants,* 3rd ed., CRC Press, Boca Raton, FL.

Nickolas, D.J.D. and A.R. Egan, Eds. 1975. *Trace Elements in Soil–Plant–Animal Systems,* Academic Press, New York.

Sillanpää, M. 1972. Trace Elements in Soils and Agriculture, Soils Bulletin Number 17, FAO of the United Nations, Rome, Italy.

Sillanpää, M. 1990. Micronutrient Assessment at the Country Level: An International Study, Soils Bulletin Number 63, FAO of the United Nations, Rome, Italy.

Sillanpää, M. and H. Jansson. 1992. Status of Cadmium, Lead, Cobalt, and Selenium in Soils and Plants of Thirty Countries, Soil Bulletin Number 65, FAO of the United Nations, Rome, Italy.

Sprague, H.B., Ed. *Hunger Signs in Crops,* 3rd ed., David McKay Company, New York.

Tan, K.H. 1998. *Principles of Soil Chemistry,* 3rd ed., Marcel Dekker, New York.

Analytical Texts

Anonymous 1986. The Analysis of Agricultural Materials, 3rd ed., Reference Book 427, Her Majesty's Stationery Office, London, England.

Anonymous 1989. *Standard Methods for the Examination of Water and Wastewater*, 17th ed., American Public Health Association, Washington, D.C.

Boumans, W.J.M., Ed. 1987. *Inductively Coupled Plasma Emission Spectroscopy, Part II: Applications and Fundamentals,* John Wiley & Sons, New York.

Ewing, G.W., Ed. 1990. *Analytical Instrumentation Handbook,* Marcel Dekker, New York.

Horwitz, W., Ed. 2000. *Official Methods of Analysis of the Association of Official Analytical Chemists*, 15th ed., two volumes, Association of Official Analytical Chemists, Arlington, VA.

Lajunen, L.H.J. 1992. *Spectrochemical Analysis by Atomic Absorption and Emission,* The Royal Society of Chemistry, Cambridge, England.

Metcalfe, E. 1987. *Atomic Absorption and Emission Spectroscopy*, John Wiley & Sons, New York.

Montasar, A. and D.W. Golightly, Eds. 1987. *Inductively Coupled Plasma in Analytical Atomic Spectrometry,* VCH Publishers, New York.

Rump, H.H. and H. Krist. 1988. *Laboratory Manual for the Examination of Water, Waste Water, and Soil,* VCH Publishers, New York.

Ruzicka, J. and E.H. Hansen. 1988. *Flow Injection Analysis,* 2nd ed., John Wiley & Sons, New York.

Smoley, C.K. 1992. *Methods for the Determination of Metals in Environmental Samples,* CRC Press, Boca Raton, FL.

Thompson, M. and J.N. Walsh. 1983. *A Handbook of Inductively Coupled Plasma Spectrometry,* Blackie & Son, Glasgow, England.

van Loon, J.C. 1985. *Selected Methods of Trace Metal Analysis; Biological and Environmental Samples,* John Wiley & Sons, New York.

Varma, A. 1991. *Handbook of Inductively Coupled Plasma Atomic Emission Spectroscopy*, CRC Press, Boca Raton, FL.

Walsh, L.M., Ed. 1972. *Instrumental Methods for Analysis in Soils and Plant Tissue,* revised ed., Soil Science Society of America, Madison, WI.

Appendix G

Definitions

These definitions are oriented primarily to their relationship to techniques of soil and plant analysis, although some are universal and have wider applications.

AB–DTPA extraction reagent. An acronym for an extraction reagent of 1 *M* ammonium bicarbonate (NH_4HCO_3) in 0.005 *M* diethylenetriaminepenta-acetic acid (DTPA) that has a pH of 7.6 and is used for the extraction of P, K, Na, Fe, Mn, and Zn from alkaline soils.

Absorbance. The amount of light absorbed by a chemical substance, which can be either in solution (spectrophotometry, colorimetry) or a vapor of atoms (atomic absorption). Absorbance is expressed as the logarithum of the ratio P_o/P.

Acidity. Refers to the pH of a soil or solution in which the hydrogen (H^+) ion concentration exceeds that of the hydroxyl (OH^-) ion concentration, and, therefore, the pH value is less than 7.0 (*see* **Alkalinity**).

Adams–Evans buffer. A buffer solution used for measuring exchange-able hydrogen (H^+) and determining the lime requirement (LR) for acid, low cation exchange capacity (CEC) soils (*see* **SMP buffer**).

AgLime (agricultural limestone). Refers to cacitic or dolomitic lime-stone that has been crushed to a certain fineness making it capable of neutralizing soil acidity.

Alkalinity. Refers to the pH of a soil or solution in which the hydroxyl ion (OH^-) concentration exceeds that of the hydrogen ion (H^+) concentration, and, therefore, the pH value is greater than 7.0 (*see* **Acidity**).

Anion. An ion in solution having a negative charge. In chemical notation, the minus sign indicates the number of electrons the compound will give up (*see* **Cation**).

AOAC. An acronym that refers to the Association of Official Analytical Chemists, publishers of the *Journal of Association of Official Analytical Chemists* (JAOAC) and *Manual of Official Analytical Methods* located in Arlington, VA.

Arc spectrometry. An analytical technique based on the principle of light emission by excited atoms generated by placing the analyte into either an AC or DC electrical arc discharge.

Atomic absorption spectrophotometry — flame. An analytical technique based on the principle of light absorption by ground-state atoms in which the wavelength of light absorbed is of the same wavelength as that emitted by a hollow cathode lamp whose filament is of the determined element in which the assayed element is dispersed in a flame. Best suited for the determination of Ca, Cu, Fe, Mg, Mn, and Zn in soil extracts and plant tissue digests. Referred to by its acronym AAS.

Atomic absorption spectrophotometry — flameless. An analytical technique based on the principle of light absorption by ground-state atoms in which the wavelength of light absorbed is the same as the wavelength of that emitted by a hollow cathode lamp whose filament is of the same element as that being assayed, the assayed element by high temperature volatilization into the light path from the hollow cathode lamp. Referred to by its acronym AAS-FL.

Atomization. A procedure for producing a fine aerosol of a solution carrying elements into a flame or plasma for elemental concentration determination.

AutoAnalyzer. A trade name for an automated analytical system manufactured by the Technicon Corporation and used primarily for elemental assay of solutions.

Available. A term used to indicate that an element is in a form and position for plant root absorption.

Ball mill. A mechanical device for reducing a sample to small particles by means of crushing.

Base saturation percentage. Refers to the percentage of the cation exchange capacity (CEC) of the soil colloids occupied by the cations Ca, Mg, K, and Na.

Beer's law. Describes the linear relationship between absorbance (*A*) and concentration (*c*) of an absorptive compound in solution as shown by the equation: $A = abc$.

Blank. A solution containing all the necessary reagents and lacking only the substance to be determined.

Block digestor. An aluminum heated block with access ports for inserting digestion tubes for conducting Kjeldahl or wet oxidation digestions. Normally the block can be carefully temperature controlled.

Bouyoucus hydrometer. A hydrometer calibrated in grams per cubic centimeter (g/cc) used in the mechanical analysis technique (*see* **Mechanical analysis**).

Bray P1 extraction reagent. An extraction reagent of 0.03 *N* ammonium fluoride (NH_4F) in 0.025 *N* hydrochloric acid (HCl) for determining soil-extractable P in acid soils of moderate cation exchange capacity (CEC).

Bray P2 extraction reagent. An extraction reagent of 0.03 *N* ammonium fluoride (NH_4F) in 0.1 *N* hydrochloric acid (HCl) for determing soil-extractable P in acid soils of moderate cation exchange capacity (CEC) that have been either fertilized with rock phosphate or that have a sizable content of calcium phosphate.

Buffer pH (pH_B). pH of soil in a buffer solution that can be used to measure exchangeable hydrogen ions (H^+) and calculate the lime requirement (LR) (*see* **Adams–Evans buffer** and **SMP buffer**).

Calcareous soil. A soil having a pH above 7.0 that effervesces when a drop of 6 *M* HCl is placed on it.

Calcitic limestone. Mainly calcium carbonate ($CaCO_3$) finely ground and applied to soil to neutralize soil acidity.

Calcium carbonate equivalent (CCE). An expression of the neutralizing capacity of a liming material relative to pure calcium carbonate ($CaCO_3$), which is 100%.

Calomel reference electrode. Electrode used with either a glass electrode to measure pH [hydrogen (H^+) on concentration] or with a specific-ion electrode for measuring ion concentrations in solution.

Carbon rod atomization. Volatilization of an analyte by high-temperature heating used for the elemental determination by atomic absorption spectrometry (*see* **Graphite furnace** and **Atomic absorption spectrophotometry — flameless**).

Cation. An ion in solution having a positive charge, the plus sign(s) indicating the number of electrons the element will accept (*see* **Anion**).

Cation exchange. The interchange among cations in the soil solution with other cations taking place on the surface of any surface-active colloidal material, such as clay or humus.

Cation exchange capacity (CEC). The total negative charge of colloidal clay and humus in soil measured in terms of exchangeable cation concentration in meq/100 grams (meq/100 g) of soil.

Chelate. A type of chemical compound in which a metallic atom such as iron (Fe) is firmly combined with a molecule by means of multiple chemical bonds. The term refers to the "claw of a crab," which illustrates the way in which the atom is held.

Chromic acid digestion. Soil organic matter determination by the oxidation of potassium dichromate ($K_2Cr_2O_7$) in concentrated sulfuric acid (H_2SO_4) (*see* **Walkley–Black**).

Colorimetry. An analytical procedure for the determination of elements of compounds based on the absorbance or transmittance of prepared complexes (*see* **Spectrophotometry**).

Combustion techniques. Methods of analysis using high-temperature decomposition for the assay of C, N, and S in soil and plant tissue (*see* **Dumas** and **LECO analyzer**).

Composite soil sample. A soil sample consisting of several single cores taken to a specific depth that, mixed together, represents a given area to that specified depth.

Conductivity. A measure of the electrical resistance of a soil–water extract, or irrigation water, used to determine the level of ions in solution. Conductivity may be expressed as specific conductance as mhos/cm (micro- or milli-) or decisiemans/m (dS/m) (*see* **Specific conductance**).

Conductivity cell (*see* Standard cell).

Detection limit. The lowest concentration of an element or substance that can be measured by an analysis procedure.

Devarda's alloy. An alloy of Cu, Al, and Zn that, when in an alkaline solution with applied heat, converts NO_3 to NH_4 for the determination of N by NH_3 distillation.

Direct current plasma emission spectrometry. An analytical technique using a direct current plasma as the excitation source; the principle of detection is based on light emission for the elemental determination in water, soil extracts, and plant tissue digests. Referred to by its acronym ICP-DC.

Dolomitic limestone. Limestone that contains magnesium carbonate ($MgCO_3$), which may range from 4.4 to 22.6%. Pure dolomite is 54.3% calcium carbonate ($CaCO_3$) and 45.7% magnesium carbonate ($MgCO_3$).

Dry ashing (combustion). A procedure of organic matter destruction by high-temperature (450 to 550°C; 842 to 1022°F) oxidation (*see* **Muffle furnace**).

Dumas. An analytical procedure for the determination of total N in a substance by high-temperature reduction.

Effective calcium carbonate equivalent (ECCE). An expression of AgLime effectiveness based on the combined effect of chemical purity (CCE) and fineness. Other similar terms are effective neutralizing power (ENP), total neutralizing power (TNP), and effective neutralizing material (ENM).

Emission spectrophotometry. An analytical technique using the principle of light emission at specific wavelengths from excited atoms (ions) as they return to their neutral or ground-state condition (*see* **Light emission**).

Excitation potential. The amount of energy required to remove electrons from their orbital position in an atom.

Extractable elements. Those elements removed from the soil or dried plant tissue by means of an extraction reagent.

Extraction reagent. A reagent that may be pure water, or a mixture of acids, or buffered salts used to extract elements as ions from soil or plant tissue.

Filtrate. The liquid that has been passed through a filter.

Flame spectrophotometry. An analytical technique using a flame (normally acetylene or natural gas and oxygen) as the excitation source; the principle of detection is based on light emission. Best suited for the determination of K and Na in water, soil extracts, and plant tissue digests. With special sample preparation, Ca and Mg can be determined.

Flow injection. A method of automated analysis using precise rapid movement of small quantities of solutions through narrow-gauge tubing for determination of analytes in solution, using primarily colorimetric methods of determination.

Glass electrode. An electrode used in conjunction with a Calomel reference electrode for measuring pH (*see* **Calomel reference electrode**).

Graphite furnace. A programmed heated graphite rod used as an attachment to an atomic absorption spectrophotometer for determining elemental concentration in substances by high-temperature volatilization (*see* **Carbon rod atomization** and **Atomic absorption spectrophotometry — flameless**).

Grating. A ruled mirror that reflects by interference light in distinct wavelengths, providing monochromatic light for use in spectrophotometers (*see* **Interference filter** and **Prism**).

Heavy metals (elements). Those elements with atomic weights greater than 55, normally referring to the elements that are considered toxic to plants, animals, and human, i.e., arsenic (As), cadmium (Cd), chromium (Cr), and lead (Pb).

Hollow cathode lamp. A source of specific wavelength radiation corresponding to the emission characteristics of the element constituting the cathode. Lamps are used in atomic absorption spectrophotometers as the source of radiation for absorption.

Humus. Colloidal and chemically stable end product of the decomposition of organic material in soil.

Indicator. Refers to a dye that when in solution changes color with a change in pH and, therefore, is able to determine soil water pH or the end point in acid-base titrations.

Inductively coupled plasma emission spectrometry. An analytical technique using an inductively coupled plasma as the excitation source; the principle of detection is based on light emission for elemental determination in water, soil extracts, and plant tissue digests. Referred to by its acronym ICP, ICAP, or ICP-AES.

Interference filter. A filter that permits light of a certain narrow range of wavelength to pass through it. Filters are used in some types of spectrophotometers (*see* **Grating** and **Prism**).

Ion. An atom or group of atoms having either a positive or negative charge from one or more lost or gained electrons (*see* **Anion** and **Cation**).

Ion chromatography. An analytical procedure based on the separation of ions by means of an ion-exchange resin column.

Junction potential. Electrical conductivity between two electrodes placed into a solution with particular reference to the operation between a glass and calomel electrodes for measuring pH.

Kjeldahl digestion. An analytical procedure converting organic N to NH_4 by high-temperature digestion in concentrated sulfuric acid (H_2SO_4) in the presence of a catalyst.

Leaf analysis. A method of determining the total elemental content of a plant by means of assaying its leaves to evaluate its well-being (*see* **Plant analysis** and **Tissue testing**).

LECO analyzer. An analytical instrument for the determination of N, S, and C by high-temperature oxidation converting organic N into N_2 gas, S into SO_2, and C into CO_2 gases which are passed through a detector and its concentration measured (*see* **Combustion techniques**).

Light emission. The release of radiant energy of a specific wavelength as the result of electron movement during the establishment of a stable atom.

Lime requirement (LR). The amount of AgLime (agricultural limestone, either calcitic or dolomitic) required to neutralize soil acidity by raising an acid soil to a higher pH level.

Major elements. The nine essential elements, Ca, C, H, O, Mg, N, P, K, and S, required by plants found in relatively large (%) concentrations in plants. The elements C, H, and O constitute about 90 to 95%, the remaining elements 0.2 to 5% of the dry weight. In the past, Ca, Mg, and S had been identified as secondary elements.

Matrix. The combination of substances that compose a sample.

Mechanical analysis. Technique for the determination of the percent of sand, silt, and clay in a soil sample to identify its textural class (*see* **Bouyoucus hydrometer** and **Texture**).

Mehlich buffer. A buffer method for determining exchangeable acidity and the lime requirement (LR) for both mineral and organic soils.

Mehlich No. 1 extraction reagent. An extraction reagent of 0.05 *N* hydrochloric acid (HCl) in 0.025 *N* sulfuric acid (H_2SO_4) for extracting P, K, Ca, Mg, Na, and Zn from acid sandy soils of low cation exchange capacity (CEC) and low organic matter (OM) content.

Mehlich No. 3 extraction reagent. An extraction reagent of 0.2 *N* acetic acid (CH_3COOH) + 0.015 *N* ammonium fluoride (NH_4F) + 0.25 *N* ammonium nitrate (NH_4NO_3) + 0.013 *N* nitric acid (HNO_3) + 0.001 *M* EDTA for extracting P, K, Ca, Mg, Na, B, and Zn from acid to neutral pH soils.

Micronutrients. The elements B, Cl, Cu, Fe, Mn, Mo, and Zn essential for plants but only required in relatively small concentrations (<0.01%) of the dry weight in plants. These elements have been previously identified as trace elements (*see* **Trace elements**).

Mineral nutrition. The study of the essential elements as they relate to the growth and well-being of plants.

Monochromatic light. A light beam of a single wavelength.

Monochromator. A device that disperses light by means of a prism or grating into its component wavelengths and isolates them into the desired wavelength.

Morgan extraction reagent. An extraction reagent of 0.7 *N* sodium acetate ($NaC_2H_3O_2$) and 0.54 *N* acetic acid (CH_3COOH) buffered at pH 4.8 for determining soil-extractable P, K, Ca, and Mg in acid soils of moderate cation exchange capacity (CEC) (*see* **Wolf modification**).

Muffle furnace. An oven that can be heated to very high temperature (*see* **Dry ashing**).

Nebulizer. A device for producing a fine aerosol of an analyte solution that can be carried into a flame or plasma to conduct an assay of elements in the solution.

Near infrared reflectance (NIR). A nondestructive analytical technique for the determination of moisture, crude fiber, and protein-nitrogen in feeds and plant tissue. Frequently referred to by its acronym NIR.

Olsen extraction reagent. An extraction reagent of 0.5 *N* sodium bicarbonate ($NaHCO_3$) at pH 8.5 for determining extractable P in alkaline soils.

Organic matter. That portion of the soil or plant tissue that contains combined C, H, and O, and that can be easily naturally decomposed when in soils, and destroyed by high temperature or wet acid oxidation in both soils and plant tissue.

pH. The negative logarithm to the base 10 of the reciprocal of the hydrogen (H^+) ion concentration in solution on a scale of 1 to 14.

Plant analysis. A method of determining the total elemental content of a plant or one of its parts, and then relating the concentration found to the well-being of the plant in terms of its elemental requirement (*see* **Leaf analysis** and **Tissue testing**).

Plant nutrition. The study of the effect of the essential and nonessential elements on the growth and well-being of plants.

Plasma. A type of excitation source created by passing an excited gas (usually agron) through an electromagnetic field, which forms a very hot (8000 to 10,000 K) and quite stable plasma suitable for elemental excitation (*see* **Inductively coupled plasma emission spectrometry**).

Primary standard. A solution containing a single element of specifically known concentration.

Prism. An optical device triangular in shape which, when light passes through it, is dispersed into distinct wavelengths. Prisms are used in some types of spectrophotometers (*see* **Interference filter** and **Grating**).

Quality assurance/quality control (QA/QC). A management system to ensure reliable analytical performance in the laboratory.

Reference method. A method officially recognized and established as the basis for measurement.

Reference solution. A solution containing a known concentration of a pure substance or element that is used for the determination of the concentration of that substance or element in an unknown sample.

Reference standard. The pure substance used for comparison of unknown quantities of the same substance.

Ruggedness. A term that refers to the range of specific requirements of an analytical procedure, that is, what level of variance exists that will not significantly affect the final assay result (pH, time, temperature, as examples).

Saline soil. A soil containing sufficient soluble salts to impair plant growth, having an electric conductivity (EC) greater than 4 dS/m in a saturation extract.

Saline/sodic soil. A soil containing a sufficiently high combination of both salts and Na to impair plant growth.

Salt pH. Measurement of the soil water pH in a solution of either 0.01 *M* calcium chloride ($CaCl_2$) or 1 *N* potassium chloride (KCl).

Saturation extract. Solution drawn by vacuum from a saturated soil paste or soilless growth medium.

Sensitivity. The ability of a method to detect a substance at low concentration levels.

Separates. Refers to three sizes of particles in soil: sand, silt, and clay (*see* **Mechanical analysis** and **Soil texture**).

SMP buffer. A buffer solution used to measure exchangeable hydrogen ions (H^+) for determining the lime requirement (LR) of acid soils with cation exchange capacities (CEC) greater than 10 meq/100 g (*see* **Adams–Evans buffer**).

Sodic soil. A soil containing sufficient exchangeable Na to affect its physical properties and impair plant growth, usually having greater than 15% exchangeable cations.

Soil pH. A measure of the hydronium ion (H_3O^+) activity in the soil solution.

Soil solution. The liquid (water) portion of the soil that contains ions in soluble form.

Soil test (analysis). A series of analytical procedures used to determine the fertility status of a soil in terms of plant growth, and the need for lime and fertilizer additions.

Soil test (analysis) calibration. A two-stage procedure to determine first the agronomic meaning of a soil test value (index) in terms of a particular crop response, and then to establish the amount of nutrient element(s) required for specific crops within each category to achieve optimum yield.

Soil test (analysis) correlation. A determination of the relationship between plant nutrient element uptake or yield, and the amount of nutrient element extracted by a soil (analysis) procedure.

Soil test (analysis) deficiency critical level. That concentration of an extractable nutrient element below which deficiency occurs and above which sufficiency exists.

Soil test (analysis) toxic critical level. That concentration of an extractable nutrient element above which toxicity is likely to occur.

Soil test (analysis) value. A nutrient element level expressed in either concentration or as an index value.

Soil test interpretation category. An interval of soil test (analysis) values associated with corresponding probabilities of response by a specific crop to a nutrient element application.

Soil texture. A method of soil classification based on the percentage of sand, silt, and clay found in the soil (*see* **Mechanical analysis** and **Separates**).

Soluble salts. Total soluble ions (anions and/or cations) in a soil and measured as the conductivity (EC) of a soil–water suspension or extract (*see* **Conductivity**).

Spark emission spectrophotometry. An analytical technique based on the principle of light emission of excited atoms generated by placing a solution of the analyte in an AC spark discharge (*see* **Emission spectrophotometry**).

Specific conductance. The reciprocal of the electrical resistance of a solution measured using a standard cell, expressed as mhos/cm (dS/cm) at 25°C (77°F) (*see* **Conductivity**).

Spectrometry. An analytical method employing the emission of a specific wavelength of light (*see* **Emission spectrophotometry**).

Spectrophotometer. An analytical instrument consisting of a light source, a means of producing monochromatic light, a cell holder, and a detector.

Spectrophotometry. An analytical method employing the light absorbing or transmitting qualities of a solution for concentration determination of elements or compounds.

SRMs. An acronym referring to standard reference materials used for calibration and available from the U.S. National Institute for Standards and Technology (NIST) (formerly the National Bureau of Standards, or NBS, Gaithersburg, MD).

Standard cell. Refers to a conductivity cell containing two electrodes each with a total surface area of 1 cm^2 and spaced 1 cm apart, which is used in conjuction with a conductivity meter for measuring the resistance of a solution.

Standard curve. A line graph used to determine the concentration of a substance in solution from its resultant emission or absorbance measurement.

Steam distillation. Evolution of gaseous NH_3 by heating a solution or digest containing NH_4, which has been made alkaline, for the determination of the N content of the original solution or sample.

Tissue testing. A method for determining the concentration of the soluble forms of an element in the plant by assaying cellular sap that has been physically pressed from primarily stems and petioles (*see* **Plant analysis** and **Leaf analysis**).

Trace elements. Elements that are found at low concentration in soils and plant tissues. A term formerly used to identify those elements known today as micronutrients (*see* **Micronutrients**).

Transmittance. The ratio P/P_o of the amount of light transmitted or absorbed by a chemical substance that is usually in solution or a precipitate suspended in solution (*see* **Colorimetry** and **Spectrophotometry**).

Turbidity. An analytical procedure for determining elemental concentration in solution by means of producing a precipitate and measuring the transmittance of the solution containing the suspended precipitate. Example: determination of SO_4 by means of barium sulfate ($BaSO_4$) precipitate.

Valance. The combining capacity of atoms or groups of atoms. The change from one valance state to another involves the transfer of an electron.

Walkley–Black. A method of soil organic matter determination using the heat of reaction from mixing concentrated sulfuric acid (H_2SO_4) with a standard potassium dichromate ($K_2Cr_2O_7$) solution (*see* **Chromic acid oxidation**).

Water pH (pHw). A measure of the hydrogen ion (H^+) concentration in the soil solution on a log scale from 0 to 14.

Wavelength. A portion of a spectrum of light having a specific energy. Wavelength is usually expressed in length as nanometers (nm).

Wet acid oxidation. A procedure of organic matter destruction using high-temperature acid oxidation; the acids are various mixtures of nitric (HNO_3), sulfuric (H_2SO_4), and perchloric ($HClO_4$) with or without the addition of hydrogen peroxide (H_2O_2). Oxidation may be conducted in either an open or closed container.

Wheaton bottle. A straight-walled bottle used as a soil extraction vessel usually of 30-L capacity.

Wiley mill. A mechanical device for reducing by cutting action a bulk sample (normally plant tissue) to small particles by passing the sample between a set of fixed and moving rotating blades.

Wolf modification. The addition of diethylenetriaminepentaacetic acid (DTPA) to the Morgan extraction reagent to obtain the extractable micronutrients (*see* **Morgan extraction reagent**).

Working standards. A set of standard solutions ranging in concentration of one or more elements used to calibrate an analytical procedure. Working standards are usually made from primary standards and in the same matrix as unknowns.

Appendix H

Conversion Factors

Common Prefixes

Factor	Prefix	Symbol
1,000,000	mega	M
1,000	kilo	k
1/100	centi	c
1/1,000	milli	m
1/1,000,000	micro	μ

Metric Conversion Factors (approximate)

	When you know	Multiply by	To find	Symbol
Length	inches	2.54	centimeters	cm
	feet	30	centimeters	cm
	yards	0.9	meters	m
	miles	1.6	kilometers	km
Area	square inches	6.5	square centimeters	cm^2
	square feet	0.09	square meters	m^2
	square yards	0.8	square meters	m^2
	square miles	2.6	square kilometers	km^2
	acres	0.4	hectares	ha
Weight	ounces	28	grams	g
	pounds	0.45	kilograms	kg
	short tons (2,000 pounds)	0.9	metric tons	t

(continued)

	When you know	Multiply by	To find	Symbol
Volume	teaspoons	5	milliliters	mL
	tablespoons	15	milliliters	mL
	cubic inches	16	milliliters	mL
	fluid ounces	30	milliliters	mL
	cups	0.24	liters	L
	pints	0.47	liters	L
	quarts	0.95	liters	L
	gallons	3.8	liters	L
	cubic feet	0.03	cubic meters	m^3
	cubic yards	0.76	cubic meters	m^3
Pressure	inches of mercury	3.4	kilopascals	kPa
	pounds/square inch	6.9	kilopascals	kPa
Temperature (exact)	degrees Fahrenheit (after subtracting 32)	5.9	degrees Celsius	°C

Useful Information and Conversion Factors

Name	Symbol	Approximate size or equivalent
		Length
meter	m	39.5 inches
kilometer	km	0.6 mile
centimeter	cm	Width of a paper clip
millimeter	mm	Thickness of a paper clip
		Area
hectare	ha	2.5 acres
		Weight
gram	g	Weight of a paper clip
kilogram	kg	2.2 pounds
metric ton	t	Long ton (2,240 pounds)
		Volume
liter	L	1 quart and 2 ounces
milliliter	mL	⅕ teaspoon

Name	Symbol	Approximate size or equivalent
		Pressure
kilopascal	kPa	Atmospheric pressure is about 100 kPa
		Temperature
Celsius	C	5/9 after subtracting 32 from °F
freezing	0°C	32°F
boiling	100°C	212°F
body temp.	37°C	98.6°F
room temp.	20–25°C	68–77°F
		Electricity
kilowatt	kW	
kilowatt-hour	kWh	
megawatt	MW	
		Miscellaneous
hertz	Hz	One cycle per second

Yield or Rate

Ounces per acre (oz/acre) × 0.07	= kilograms per hectare (kg/ha)
Tons per acre (ton/acre) × 2240	= kilograms per hectare (kg/ha)
Tons per acre (ton/acre) × 2.24	= metric tons per hectare (kg/ha)
Pounds per acre (lb/acre) × 1.12	= kilograms per hectare (kg/ha)
Pounds per cubic foot (lb/ft^3) × 16.23	= kilograms per cubic meter (kg/m^3)
Pounds per gallon (lb/gal) × 0.12	= kilograms per liter (kg/L)
Pounds per ton (lb/ton) × 0.50	= kilograms per metric ton (kg/MT)
Gallons per acre (gal/acre) × 9.42	= liters per hectare (L/ha)
Gallons per ton (gal/ton) × 4.16	= liters per metric ton (L/MT)
Pounds per 100 square foot (lb/ft^2) × 2	= pounds/100 gallons water (assumes that 100 gallons will saturate 200 square feet of soil)
Pounds per acre (lb/acre)/43.56	= pounds per 1000 square foot (lb/ft^2)

Volumes and Liquids

1 teaspoon = ⅓ tablespoon	= 1/16 ounce
1 tablespoon = 3 teaspoons	= ½ ounce
1 fluid ounces (oz)	= 2 tablespoons = 6 teaspoons
1 pint/100 gallons	= 1 teaspoon per gallon
1 quart per 100 gallons	= 2 tablespoons per gallon
3 teaspoons = 1 tablespoon (tsp)	= 14.8 milliliters (mL)
2 tablespoons (tsp) = 1 fluid ounces	= 29.6 milliliters (mL)
8 fluid ounces (oz)	= 16 tablespoons (tsp) = 1 cup = 236.6 milliliters (mL)
2 cups = 32 tablespoons (tsp)	= 1 pint = 473.1 milliliters (mL)
2 pints = 64 tablespoons (tsp)	= 1 quart (qt) = 946.2 milliliters (mL)
1 liter (L) = 1000 milliliters (mL)	= 1000 cubic centimeters (cc) = 0.264 gallons (gal) = 33.81 ounces (oz)
4 quarts (qt) = 256 tablespoons (tsp)	= 1 gallon (gal) = 3785 milliliters (mL)
1 gallon (gal)	= 128 ounces (oz) = 3.785 (L)

Elemental Conversions

$P_2O_5 \times 0.437$	= Elemental P	Elemental P × 2.29	= P_2O_5
$K_2O \times 0.826$	= Elemental K	Elemental K × 1.21	= K_2O
CaO × 0.71	= Elemental Ca	Elemental Ca × 1.40	= CaO
MgO × 0.60	= Elemental Mg	Elemental Mg × 1.67	= MgO
$CaCO_3 \times 0.40$	= Elemental Ca		

Weight/Mass

1 ounce (oz)	= 28.35 grams (g)
16 ounces (oz)	= 1 pound (lb) = 453.6 grams (g)
1 kilogram (kg)	= 1000 grams (g) = 2.205 pounds (lb)
1 gallon water	= 8.34 pounds (lb) = 3.8 kilograms (kg)
1 cubic foot of water (ft^3)	= 62.4 pounds (lb) = 28.3 kilograms (kg)
1 kilogram of water (kg)	= 33.81 ounces (oz)
1 ton (t) = 2000 pounds (lb)	= 907 kilograms (kg)
1 metric ton (MT)	= 1000 kilograms (kg) = 2205 pounds (lb)

Volume Equivalents

1 gallon in 100 gallons	= 1¼ ounces (oz) in 1 gallon (gal)
1 quart in 100 gallons	= 5/16 ounce (oz) in 1 gallon (gal)
1 pint in 100 gallons	= 3/16 ounce (oz) in 1 gallon (gal)
8 ounces (oz) in 100 gallons	= ½ teaspoon in 1 gallon (gal)
4 ounces (oz) in 100 gallons	= ¼ teaspoon in 1 gallon (gal)

Temperature

°C	°F	°C	°F
5	40	120	248
10	50	125	257
19.4	67	180	356
20	68	200	392
21	70	330	626
23	73	350	662
25	77	370	698
27	80	400	752
32	90	450	842
38	100	500	932
40	105	550	1022
50	122	600	1122
80	176	900	1652
100	212	1350	2462
110	230		

Degrees F = (Degrees C + 17.78) × 1.8
Degrees C = (Degrees F – 32) × 0.556

Elemental Equivalent Calculations

Calculation of Milliequivalents and Microequivalents/100 Grams from Percentages and Parts per Million, Respectively[a]

Element	Converting from	Valence	Equivalent weight	Factor[b]
Nitrogen (N)	% to me.	3	4.6693	214.6
Phosphorus (P)	% to me.	5	6.1960	161.39
Potassium (K)	% to me.	1	39.096	25.578
Calcium (Ca)	% to me.	2	20.040	49.900
Magnesium (Mg)	% to me.	2	12.160	82.237
Boron (B)	ppm to μe.	3	3.6067	27.726
Copper (Cu)	ppm to μe.	2	31.770	3.1476
Iron (Fe)	ppm to μe.	3	18.617	5.3726
Manganese (Mn)	ppm to μe.	2	27.465	3.6410
Zinc (Zn)	ppm to μe.	2	32.690	3.0590
Sulfur (S)	% to me.	2	16.033	62.377
Sodium (Na)	% to me.	1	22.991	43.496
Chloride (Cl)	% to me.	1	35.457	28.175

[a] Milliequivalents can be converted to percentages by multiplying by (equivalent weight)/1000, and microequivalents can be converted to parts per million by multiplying by (equivalent weight)/100.

[b] Factor x = me./100 g and factor x ppm = μe./100 g.

To convert lb/acre to milliequivalents/100 g

Element	Multiply by
Calcium (Ca)	400
Magnesium (Mg)	780
Potassium (K)	240
Sodium (Na)	460

Index

A

AB-DTPA (see Ammonium bicarbonate–DTPA)
Acetic acid (2%) extraction reagent, 223, 228
Acidity
  Exchangeable, 53–56, 328
  Titratable, 53–54
Adams–Evans Buffer, 46–48
Alkalinity, 155
Aluminum
  Effect of pH on availability, 39
  Exchangeable, 54–56
Aluminum sulfate (0.025 *M*) extraction reagent, 222
Ammonium–nitrogen determination
  Plant tissue, 226
  Soil, 328–329
Ammonium–nitrogen extraction, 318–319, 328–329
Ammonium acetate (1*N*, pH 7.0) extraction reagent, 81–82, 311–312
Ammonium acetate (0.5 *M*)/acetic acid (0.25 *M*)
  extraction reagent, 129–130, 320–321
Ammonium bicarbonate–DTPA extraction reagent, 72–73, 80–81, 86–87, 101–103, 314
Ammonium sulfate (0.04 *M*) extraction reagent, 124, 319–320
ASI extraction reagent, 327–331
  Interpretation, 331
Atomic absorption spectrophotometry, 266–267
AutoAnalyzers, 263–264

B

Base saturation (soil), 93
Blanks, 308
Bray P1 extraction reagent, 67–68, 316
Bray P2 extraction reagent, 68–69, 316–317
Boron
  Conversion factors (plant), 215–216
  Determination (soil), 95–97
  Methods of expression
    Plant, 215–216
    Soil, 109
  Soil extraction reagents, 95
    ASI, 329–330
    $CaCl_2$ (0.01 *M*), 104–105, 321
    Hot water, 95–96
    Mehlich No. 3, 99–101, 313
    Morgan, 104–105, 313–314
Buffer soil pH, 42–46
Buffers
  Preparation, 298–300
  Reagents for preparation, 297

C

Calcium
  Conversion factors
    Plant, 215–216
    Soil, 89
  Interpretation (soil), 89–90
  Methods of determination (soil), 88–89
  Methods of expression
    Plant, 215–216
    Soil, 89

Soil extraction reagents, 80–81
Ammonium acetate (1*N*, pH 7.0), 81–82, 311–312
ASI, 328–329
$CaCl_2$ (0.01 *M*), 87–88, 321
Mehlich No. 1, 82–83, 312
Mehlich No. 3, 83–84, 313
Morgan, 84–85, 313–314
Water, 87, 315
Calcium chloride (0.01 *M*) extraction reagent, 75–76, 87–88, 104–105, 117, 123, 321
Calcium hydroxide (saturated) extraction reagent, 107–108
Calcium nitrate (0.01*M*) extraction reagent, 105–106
Calcium phosphate (500 mg P/L) extraction reagent, 320
Calcium sulfate (0.01 *M*) extraction reagent, 123, 319
Cation exchange capacity (CEC), 92–93
Chloride
Conversion factors (plant), 216
Soil extraction reagents
$Ca(NO_3)_2$ (0.01 *M*), 105–106
$Ca(OH)_2$ (saturated), 107–108
K_2SO_4 (0.5 *M*), 106–107
Cleaning glassware, 109–110
Copper
Conversion factors (plant), 216
Effect of pH on availability, 39
Methods of expression
Plant, 215–216
Soil, 109
Soil extraction reagents, 95
Ammonium bicarbonate–DTPA, 101–103, 314–315
ASI, 327–329
$CaCl_2$ (0.01 *M*), 104–105, 321
DTPA, 103, 318
Morgan, 104–105, 313–314

D

DTPA extraction reagent, 102–103, 116–117, 318

E

Electric conductivity units/conversions, 154
Emission spectrophotometry, 264–266, 268–274
Flame emission, 265–266
Inductively coupled plasma (ICP-AES), 268
Calibration techniques, 271–272
Common operating problems, 272–273
Important general points, 273–274
Operating characteristics, 269–270
Spectrometer designs, 268–269
Standard preparation, 270
Extraction procedures, 26, 311–321
Extraction reagents, 25–26, 311–321

F

Fertilizer
Acidification effect, 57
Availability affected by pH, 44
Flame emission spectrophotometry, 265–266
Flow injection analyzers, 263

H

Heavy metals
Description, 114
Interpretation, 118–120
Methods of determination, 117–118
Soil extraction reagents
Ammonium bicarbonate–DPTA, 114–116, 314–315
$CaCl_2$ (0.01 *M*), 117, 321
DTPA, 116–117, 318
Humic matter determination
NaOH (0.2 *N*) extraction, 144–147, 330

I

Inductively coupled plasma (ICP-AES), 268–273
Calibration techniques, 271–272
Common operating problems, 272–273
Important general points, 273

Operating characteristics, 269–270
Spectrometer designs, 268–269
Standard preparation, 270
Instrumental analysis, 259–261
Atomic absorption spectrophotometry, 266–267
Emission spectrophotometry, 264–265
Flame emission, 265–266
Inductively coupled plasma (ICP-AES), 268–273
Calibration techniques, 271–272
Common operating problems, 272–273
Important general points, 273–274
Operating characteristics, 269–270
Spectrometer designs, 268–269
Standard preparation, 270
Spectrophotometry terms, 274–275
Ion chromatography, 277
Methods of analysis by elements, 260–261, 274
Specific-ion electrodes, 275–277
UV-VIS spectrophotometry, 262–264
AutoAnalyzers, 263–264
Flow injection analysis, 264
Ion chromatography, 277
Iron
Conversion factors
Plant, 215–216
Soil, 89
Effect of pH on availability, 39
Extractable
Plant, 229
Soil, 313–315, 318, 321
Methods of expression
Plant, 215–216
Soil, 109
Soil extraction reagents, 95
Ammonium bicarbonate–DTPA, 101–102, 314–315
ASI, 327–328
$CaCl_2$ (0.01 *M*), 104–105, 321
DTPA, 102–103, 318
Morgan, 104–105, 313–314

K

Kjeldahl methods, 127, 211–212

L

Lime requirement determination
Adam–Evans buffer, 46–48
Mehlich buffer, 48–53
SMP, 42–46
Liming
Concept, 56–57
Rates, 44–45, 48, 51–52, 56–57
Liming materials
Definitions, 57–60
Depth effects, 61–62
Fineness, 62
Neutralizing value, 60–61

M

Magnesium
Conversion factors
Plant, 215–216
Soil, 89
Effect of pH on availability, 39
Interpretation (soil), 89–91
Methods of determination, 88–89
Methods of expression
Plant, 215–216
Soil, 89
Soil extraction reagents, 80–81
Ammonium acetate (1*N*, pH 7.0), 81–82, 311–312
ASI, 328–329
$CaCl_2$ (0.01 *M*), 87–88, 321
Mehlich No. 1, 82–83, 312
Mehlich No. 3, 84–85, 313
Morgan, 84–85, 313–314
Water, 87, 315
Manganese
Conversion factor (plant), 215–216
Effect of pH on availability, 39
Methods of expression
Plant, 215–216
Soil, 109
Soil extraction reagents, 95
Ammonium bicarbonate–DTPA, 101–102, 314–315
ASI, 327–328
$CaCl_2$ (0.01 *M*), 104–105, 321
DTPA, 102–103, 318
Morgan, 104–105, 313–314

Major cations (soil), 79–81
Matrix effects, 307
Matrix modifiers, 307–308
Mehlich Buffer, 48–53
Mehlich No. 1 extraction reagent, 70–71, 82–83, 97–98, 312
Mehlich No. 3 extraction reagent, 73–75, 83–84, 84–85, 99–101, 313
Micronutrients, 93–94
  Boron
    Determination, 96–97
    Soil extraction reagents
      ASI, 329–330
      $CaCl_2$ (0.01 *M*), 103–104, 321
      Hot water, 95–96
      Mehlich No. 3, 99–101, 313
      Morgan, 104–105, 313–314
  Chlorine
    Soil extraction reagents
      $Ca(OH)_2$ (saturated), 107–108
      $Ca(NO_3)_2$ (0.01 *M*), 105–106
      K_2SO_4 (0.5 *M*), 106–107
  Copper
    Soil extraction reagents, 95
      Ammonium bicarbonate–DTPA, 101–102, 314–315
      ASI, 327–328
      $CaCl_2$ (0.01 *M*), 103–104, 321
      DTPA, 102–103, 318
      Morgan, 104–105, 313–314
    Crop sensitivities, 112–113
  Interpretation, 109–111
    Soil/crop conditions, 111–112
  Iron
    Soil extraction reagents, 95
      Ammonium bicarbonate–DTPA, 101–102, 314–315
      ASI, 327–328
      $CaCl_2$ (0.01 *M*), 104–105, 321
      DTPA, 102–103, 318
      Morgan, 104–105, 313–314
  Manganese
    Soil extraction reagents, 95
      Ammonium bicarbonate–DTPA, 101–102, 314–315
      ASI, 327–328
      $CaCl_2$ (0.01 *M*), 103–104, 321
      DTPA, 102–103, 318
      Morgan, 104–105, 313–314
  Methods of determination, 109, 260–261
  Methods of expression
    Plant, 215–216
    Soil, 109
  Methods of soil extraction, 95
  Zinc
    Soil extraction reagents, 95
      Ammonium bicarbonate–DTPA, 101–102, 314–315
      ASI, 327–328
      $CaCl_2$ (0.01 *M*), 103–104, 321
      DTPA, 102–103, 318
      HCl (0.1*N*), 98–99
      Mehlich No. 1, 97–98, 312
      Mehlich No. 3, 99–101, 313
      Morgan, 104–105, 313–314
Morgan extraction reagent, 66, 84–85, 104–105, 313–314

N

NCR-13 scoops, 24
Nitrate–nitrogen (plant)
  Extraction procedures, 222, 228
  Interpretation, 224–225
Nitrate–nitrogen (soil), 121
  Interpretation, 129
  Ion strength adjusting solution, 127
  Methods of determination
    Kjeldahl distillation, 128
    Specific-ion electrode, 127
    UV-VIS spectrophotometry, 126–127
  Soil extraction reagents
    Ammonium bicarbonate–DTPA, 122–123, 314–315
    Ammonium sulfate (0.04 *M*), 124, 319–320
    $CaCl_2$ (0.01*M*), 124–125, 321
    $CaSO_4$ (0.01 *M*), 123–124, 319
    KCl (2*N*), 123, 318–319
    $(NH_4)_2SO_4$ (0.01 *M*), 124, 319–320
  Standards, 125–126
Nitrogen release from organic matter, 138
North American Proficiency Testing Program (NAPT), 323–326
Nutrient element availability
  pH effect, 39–40

O

Olsen P extraction reagent, 71–72, 317

Organic carbon
$CaCl_2$ (0.01 *M*) extraction, 148–150, 321
Organic matter determination, 139–140
Loss-on-ignition (LOI), 143–144
Soil extraction procedure (ASI), 330
Soluble organic carbon, 148–150
Wet digestion, 140–143

P

pH
Buffers, 30–31
Definition, 27–28
Determination in soil
$CaCl_2$ (0.01 *M*), 33–34
Indicators, 35–36
KCl (0.1 *N*), 34–35
Water, 31–33
Effect on soil nutrient element availability, 39–40
Interpretation, 36–41
pH meter, 28–30
Electrodes, 275
Phosphorus (plant)
Extraction procedures, 223, 228
Interpretation, 224–225
Plants, 215
Phosphorus (soil)
Conversion factors, 216
Methods of determination, 76–77
Methods of expression, 78
Soil extraction reagents, 65
AB–DTPA, 72–73, 311–312
ASI, 327–328
Bray P1, 67–68, 316
Bray P2, 68–69, 316–317
$CaCl_2$ (0.01 *M*), 75–76, 321
Mehlich No. 1, 70–71, 312
Mehlich No. 3, 73–75, 313
Morgan, 66–67, 313–314
Olsen, 71–72, 317
Soil forms, 62–64
Soil test interpretation, 78
Soil test levels (U.S.), 64
Plant Analysis
Comparative sampling, 199–200
Compound leaves, 199
Elemental analysis, 207–209
Elemental conversion factors, 215–216
Elemental distribution, 199
Extractable elements, 221–222
Ammonium–nitrogen, 226
Chloride, 227
Acetic acid (CH_3COOH) (2%), 228
Water, 228
Interpretation, 223–225
Iron, 229
Phosphate–Acetic acid (CH_3COOH) (2%) extractable, 223, 228
Potassium–Acetic acid (CH_3COOH) (2%)extractable, 228
Nitrate–nitrogen
$Al_2(SO_4)_3$ (0.025 *M*) extractable, 222
CH_3COOH (2%) extractable, 228
Sulfate–sulfur, 226–227
CH_3COOH (2%) extractable, 228
Inappropriate plant tissue, 200
Interpretation, 216–221
Lack of homogeneity, 198–199
Methods of expressing elemental content, 215–216
Number of plants to sample, 197–198
Organic matter destruction
Dry ashing, 203–205
Wet-acid digestion, 205–207
Particle size reduction (grinding), 202–203
Petioles, 197
Purpose, 151–152
Reference plant composition, 219–220
Sample preparation
Initial handling, 200–201
Decontamination (washing), 201–202
Moisture removal (oven-drying), 202
Particle size reduction (grinding), 202–203
Sampling procedures, 152–197
Total nitrogen (N) determination, 209–211
Ammonium determination, 212
Kjeldahl methods, 211–212
Non-Kjeldahl methods, 212–213
Total sulfur (S) determination, 213–215
Plant tissue standards, 309
Potassium
Conversion factors
Plant, 215–216
Soil, 89
Interpretation (soil), 89–91
Methods of conversion, 216

Methods of determination, 88–89
Methods of expression
Plants, 215–216
Soils, 89
Soil extraction reagents, 80–81
AB–DTPA, 86–87, 314–315
Ammonium acetate (1*N*, pH 7.0), 81–82, 311–312
ASI, 327–328
$CaCl_2$ (0.01 *M*), 87–88, 321
Mehlich No. 1, 82–83, 312
Mehlich No. 3, 83–84, 313
Morgan, 84–85, 313–314
Water, 87, 315
Potassium chloride (2 *M*) extraction reagent, 318–319
Primary standards, 301–305
Sources, 308
Proficiency testing, 290
NAPT, 323–326

Q

Quality assurance/Quality control (QA/QC), 285–289
Accuracy and precision, 287
Instrumentation, 289
Laboratory procedures, 289
Proficiency testing, 290
Standards, 287–289, 301–310

R

Reagents, 26–27, 293–300
Reference plant composition, 219–220

S

Salinity, 153–155
Scoops (soil), 23–24
Procedures of use, 24–25
SMP buffer, 42–46
Sodium
Conversion factors
Plants, 215–216
Soils, 89
Methods of expression
Plants, 216
Soils, 89
Soil extraction reagents, 80–81
Ammonium acetate (1*N*, pH 7.0), 81–82, 311–312
ASI, 328–329
$CaCl_2$ (0.01 *M*), 87–88, 321
Mehlich No. 1, 82–83, 312
Mehlich No. 3, 83–84, 313
Water, 87, 315
Sodium Adsorption Ratio (SAR), 151
Soil Analysis
Buffer pH, 41–42
Crushing/Grinding/Sieving, 21–22
Drying, 20–21
Extraction reagents, 25–26, 65, 80–81, 95, 311–321
History/Purpose, 11–15
Methods of expression
Cations, 89
Micronutrients, 109
pH, 27–36
Reference Methods, 2–3
Sampling, 16–17
Scooping, 24–25
Transport to laboratory, 19–20
Weighing vs. scooping, 22–24
Soil standards, 309
Soil storage, 27
Soil texture (mechanical analysis), 155–159
Method of determination — hydrometer procedure, 156–159
Separate characteristics, 155
Textural classifications, 156–158
Soilless media determination, 134–135
DTPA extraction (0.005 *M*), 136–137
Interpretation, 137–138
Water extraction, 135–136
Interpretation, 137
Soluble carbon, 148–150
Soluble salts, 150–151
Calibration solution, 152
Conductivity meter, 152
Determination
1:1 water: soil extraction, 152
2:1 water: soil extraction, 152
Saturation paste method, 153
Interpretation, 153–156
Sodium Adsorption Ratio (SAR), 151
Units and conversions, 154
Specific-ion electrodes, 255–256, 275–277
Spectrophotometry terms, 274–275

Standards
Plant, 309
Preparation, 270, 302–305
Primary, 301
Reagents, 296–297
Soil, 309
Working, 306–307
Sulfate–sulfur (soil)
Determination procedures, 130–134
Interpretation, 129–130, 134
Soil extraction reagents
Ammonium acetate (0.5 *M*)/acetic acid (0.25 *M*), 129–130, 320–321
ASI, 329–330
$CaCl_2$ (0.01 *M*), 130, 321
Monocalcium phosphate (500 mg P/L), 129, 320
Standards, 131

T

Tissue testing, 247–248
Methods of interpretation, 253–255
Preparation of reagents, 249–250
Sampling techniques, 250
Testing kits, 248–249
Source, 256
Testing procedures
Phosphorus, 252
Potassium, 252
Nitrate–nitrogen, 251–252
Specific-ion nitrate meter, 255–256

U

UV-VIS spectrophotometry, 262–264
AutoAnalyzers, 263
Flow injection analyzers, 263

W

Water extraction reagent, 87, 315
Water quality, 4
Working standards, 306–307
Sources, 308

Z

Zinc
Conversion factor (plant), 216
Effect of pH on availability, 39
Methods of expression
Plant, 215–216
Soil, 109
Soil extraction reagents
Ammonium bicarbonate–DTPA, 101–102, 314–315
ASI, 327–328
$CaCl_2$ (0.01 *M*), 104–105, 321
DTPA, 102–103, 318
HCl (0.1 *N*), 98–99
Mehlich No. 1, 97–98, 312
Mehlich No. 3, 99–101, 313
Morgan, 104–105, 313–314